A GUIDE TO CAREERS IN
COMMUNITY
DEVELOPMENT

Paul C. Brophy and Alice Shabecoff

About Island Press

Island Press is the only nonprofit organization in the United States whose principal purpose is the publication of books on environmental issues and natural resource management. We provide solutions-oriented information to professionals, public officials, business and community leaders, and concerned citizens who are shaping responses to environmental problems.

In 2000, Island Press celebrates its sixteenth anniversary as the leading provider of timely and practical books that take a multidisciplinary approach to critical environmental concerns. Our growing list of titles reflects our commitment to bringing the best of an expanding body of literature to the environmental community throughout North America and the world.

Support for Island Press is provided by The Jenifer Altman Foundation, The Bullitt Foundation, The Mary Flagler Cary Charitable Trust, The Nathan Cummings Foundation, The Geraldine R. Dodge Foundation, The Charles Engelhard Foundation, The Ford Foundation, The German Marshall Fund of the United States, The George Gund Foundation, The Vira I. Heinz Endowment, The William and Flora Hewlett Foundation, The W. Alton Jones Foundation, The John D. and Catherine T. MacArthur Foundation, The Andrew W. Mellon Foundation, The Charles Stewart Mott Foundation, The Curtis and Edith Munson Foundation, The National Fish and Wildlife Foundation, The New-Land Foundation, The Oak Foundation, The Overbrook Foundation, The David and Lucile Packard Foundation, The Pew Charitable Trusts, The Rockefeller Brothers Fund, Rockefeller Financial Services, The Winslow Foundation, and individual donors.

About the National Congress for Community Economic Development

The National Congress for Community Economic Development (NCCED), founded in 1970, is the trade association representing over 3,600 organizations committed to the development of economically distressed urban and rural communities across America. NCCED's principal constituency is the industry of nonprofit community-based development corporations (CDCs). NCCED is proud to have sponsored this publication.

NCCED's mission is to promote, support, and advocate for the community economic development industry. NCCED seeks to ensure the allocation of resources for affordable housing and to assist low- and moderate-income communities, families, and individuals in generating wealth and achieving lasting economic viability through employment opportunities and entrepreneurial enterprise. For additional information, call 877-446-2233 (toll-free) or 202-289-9020, or visit the NCCED Web site: http://www.ncced.org.

A GUIDE TO
CAREERS IN
COMMUNITY
DEVELOPMENT

Paul C. Brophy and **Alice Shabecoff**

ISLAND PRESS
Washington • Covelo • London

Book design and production by Stephen Kelly-Judd, Tempest Design Group LLC
ISLAND PRESS is a trademark of The Center for Resource Economics.

Library of Congress Cataloging-in-Publication Data

Brophy, Paul C.
A guide to careers in community development / Paul C. Brophy amd Alice Shabecoff.
p. cm.
Includes bibliographical references and index.
ISBN 1–55963–749–8 (cloth : acid-free paper)—ISBN 1–55963–750–1
(pbk. : acid-free paper)
1. Community development—Vocational guidance —United States.
2. Community development corporations—Vocational guidance—United States.
3. Social service—Vocational guidance—United States. I. Shabecoff, Alice. II. Title.
HN90.C6 B754 2001
307.1'4'02373—dc21   00–010478

Printed on recycled, acid-free paper
British Library Cataloging-in-Publication data available.
Printed in Canada
10 9 8 7 6 5 4 3 2 1

To Msgr. Geno Baroni—
and the thousands of committed
community development workers
carrying on his legacy

A GUIDE TO CAREERS IN COMMUNITY DEVELOPMENT

Contents

Foreword

Neal Peirce and Carol Steinbach

Community development supporters everywhere owe a debt of gratitude to Paul C. Brophy and Alice Shabecoff for writing this book. These experienced practitioners have produced a clear and objective map for navigating the uncharted waters of a career in community development. To our knowledge, nothing quite like this has ever been written before.

This guide is sure to be of great benefit to any young person contemplating a career in the field. It should also help community development specialists seeking to move around within the industry, as well as professionals from other fields who think community development might be right for them. Brophy and Shabecoff have created content that is highly informative, well presented, and chock full of darned good stories.

What's more, the book could hardly arrive on the scene at a more auspicious time. Demand for talented community development professionals has swelled over the past decade. Career opportunities are more diverse than ever before, and the field is gaining recognition for its contributions in the United States and abroad. The results produced by community development organizations are tangible and growing. Professionalism and performance standards within these organizations are on the rise. While salaries within traditional community-based nonprofits remain lower than in the private sector, a career in community development can be both rewarding and fun, too. It is still a place where one person can make a genuine difference in the lives of others.

The number of community development organizations has exploded in recent years. Thousands of community development corporations—CDCs—have sprung up in poor areas from Appalachia to Watts. While there is no precise count of the number of CDCs operating today, estimates range between four thousand and eight thousand groups. The bulk of this growth has taken place during the past ten to fifteen years. More groups means more demand for community development expertise from within the industry.

Community development is fully in tune with the times. The field is entrepreneurial; it is concerned with building and investing capital and with increasing the well-being of people living in distressed communities by creating work opportunities. Community development's approach spans the political spectrum and is popular because it combines doing good for people and communities with the ethic of self-help and self-improvement.

Demand for community development expertise is growing outside of traditional nonprofits, too. Bolstered by a strong national economy, many of America's distressed places are finally starting to rebound, especially in our central cities. For the first time in U.S. history, a majority of our urban residents are homeowners. Violent crime in our cities has fallen to its lowest point in thirty years. As city balance sheets turn from red to black, community development projects that formerly sat on drawing boards are finally underway—new parks and housing, schools and commercial projects.

Demand for community development expertise isn't coming just from a recharged public sector. In a growing number of corporate boardrooms, lagging city neighborhoods, rather than being viewed as places to avoid, are touted as targets of opportunity, as untapped markets for consumer spending, as alternatives to costly sprawl, and as promising locations for new job growth, new residents, and new commercial development. Corporations such as the Pathways supermarket chain and Rite Aid drugstores are leading the charge.

The range of career opportunities in community development is expanding. Working in community development no longer means a career spent almost exclusively in public agencies or nonprofit institutions. Today's community development professionals increasingly work in banks, corporations, utility companies, foundations, universities, and hospitals, too. There's also movement across the sectors as practitioners move from community development jobs in nonprofits to similar positions in the private or public sector and vice versa. These cross-sector movements, enlarging careers, can be expected to multiply in the next years.

Professionalism in the community development field has attained a high level. More CDCs are embracing today's performance-oriented culture. After years of experimentation, community developers have created strategies that work and can be adapted from one place to the next. Gradually, the old image of the well-intentioned but poorly managed nonprofit organization is giving way.

Community development has made important contributions to poverty fighting. A criticism of community development since its earliest days has been that CDCs have never achieved sufficient scale of activity to make much impact on poverty levels and distress. While CDCs may have succeeded in building some housing and commercial projects, critics argue, they've had little success at turning around inner-city economies.

It is true that community development has never been about replacing private market forces in poor neighborhoods. Rather, community development has had a different goal: To entice private investment back into lagging areas and spread the benefits of economic growth to people and places too often left behind. Through their housing and commercial projects, CDCs remove

community eyesores and try to demonstrate that profitable investment is possible. Through their capital and technical assistance, they support private businesses and micro-enterprises that create jobs and new opportunity. Through their initiatives to promote homeownership, CDCs help to stabilize communities. And virtually all complement their "hard" development activities with community-building and social service activities—ranging from organizing and advocacy to crime control and health care programs.

"You know CDCs are making a critical difference when you go to Roxbury in Boston or Hough in Cleveland, or any of these once utterly desolated communities, and see how the graffiti has dissolved, supermarkets are reappearing, and enterprises like Magic Johnson theaters are springing up," says Paul Grogan, former president of the Local Initiatives Support Corporation and now a vice president at Harvard University.

Grogan's prime example is the South Bronx. "In 1979 it was rubble, the dustbin of history," he told us. "Today, it's still poor. But housing is booming, crime is down, and you can see the start of economic revival. . . . Imagine a child who walks to school with broken glass and drug dealers as opposed to a still-poor child who walks to school where life is normal, a new supermarket is opening, there's a Little League team playing. Which would you rather have? That's been the achievement of CDCs."

Community development is becoming far more influential in public policy. For much of their history, community development organizations had little impact on policy at a national level. Today that's beginning to change. A growing number of domestic programs—in the areas of welfare, government-assisted housing, job training, and business development—are incorporating community development methods because they appear to have more lasting results.

Community development is no longer a theoretical, untested

model. It is now a field guided by experienced professionals. Across the nation, community development professionals can take credit for turning renters into homeowners, creating hundreds of thousands of jobs, investing in entrepreneurs, and sparking revitalization through their commercial developments. Community development workers are making an important contribution to a better quality of life for many low-income Americans.

What this means is ever expanding opportunities for men and women brought by idealism and hope into this fascinating field.

A joke made of the ancient Quakers was that they came to do good and ended up doing well. One can only wish the same for those who make community development a serious life pursuit.

Neal Peirce, the foremost U.S. journalist covering developments in grassroots America, writes a syndicated national newspaper column on state and local themes and is the author of many major newspaper series on selected U.S. cities as well as fifteen books. Carol Steinbach has been writing books and articles about community development for two decades. Together, Peirce and Steinbach authored the Ford Foundation's landmark 1987 study of community development, *Corrective Capitalism*.

—*July 2000*

Preface and Acknowledgments

Between the two of us, we have spent over half a century in the community development world. We certainly hope our love of this work animates this guide. We wrote this guide as one way of helping the next generation to find its place in this work. We thought readers would like to know how we got into community development and why.

Paul's Story:

The Hunting Park neighborhood in Philadelphia—now one of many neighborhoods struggling for survival—was a multi-European ethnic village in the 1950s and early 1960s when it hugged me through my youth. The Irish, Italian, Eastern European, and German immigrants who shared the row-house neighborhood shared similar values, including those of hard work and mutual help in an effort to get ahead.

This was a neighborhood of working-class people. Virtually every dad was a blue-collar worker, and many moms worked in nearby factories, at least part-time. When it came time to go to college, for economic reasons, I joined many other Philadelphia working-class kids in commuting to college and living at home.

My neighborhood helped me do it. Each semester I borrowed tuition money from the credit union operated by my church (in today's jargon, a faith-based community development financial institution) and earned enough working in a family-owned pizza shop to pay back the loan week by week. The neighborhood worked. It gave its youth opportunities to grow intellectually and financially.

At some point during my college days, I learned about a field called city planning. I immediately knew I was called to it. Maybe I could help strengthen and create neighborhoods that nurture people as much as Hunting Park had nurtured me. So off to city planning school I went.

As my career has evolved, I have been fortunate to work at neighborhood improvement—knowing now how much more challenging it is than I ever thought in my bright-eyed youth. And as I became more senior in community development work, more and more young people have approached me to help them understand what the field is all about. I can see in them the anticipation I felt about working on something as important as improving places where people live—and can also see the difficulty they are faced with in navigating the complex field that community development has become.

Hence, this career guide: an effort to provide a map and tool to those who would consider joining the work of improving communities, including my old Hunting Park, where there is now an active community development corporation working diligently to bring the community to a level of health—not a return to the village of the 1950s—but a community for the future, in whatever shape that should take.

Alice's Story:

I fell into the community development world by chance, but once I realized where I had landed, I held on to it for life. Maybe because I was the daughter of an honest stockbroker, maybe because I grew up in New York City in the era when it was a hotbed of idealism, the notion of reconciling capitalism with economic justice and democracy seemed doable.

I worked for economic equity through the consumer movement

in its heyday. I came from a position as executive director of the nation's oldest, longest-standing consumer group to HUD, to create a new consumer information function there. A journalist by trade, I believe that information is critical to people who wish to govern themselves. When Monsignor Geno Baroni arrived at HUD to start a new neighborhoods program, I joined him as director of his program's information function. Once inspired by Geno's vision, I have used the pen (word processor) in its service since then.

Community development, when it works, represents to me the fulfilled promise of what a good society and our country can be. People do things for themselves and their neighbors, and as the direct outcome of their determination and effort, they can live a good and useful life without doing harm to others (including the natural world around us). It's an ideal. Perhaps that's why so many people working in community development feel it as a spiritual dimension. Community development is not about "helping the poor," though it does; it is, rather, a way to organize our economy and our social interactions on a human scale, helping each other with a sense of a larger purpose. It recognizes the value of the individual, not as a frenetic consumer but as a productive citizen and neighbor.

I hold fast to this ideal because it is the way to build the kind of place I want to live in and I want my grandchildren to live in. I hope, as does Paul, that you, the reader, will find the motivation from this guide to find your place in community development work.

Acknowledgments

The National Congress for Community Economic Development offered its leadership in sponsoring this project. Its experienced and visionary president, Roy Priest, recognizes the vital importance of developing the people who will develop communities in years to come. Special thanks to Patricia Smith, the director of NCCED's Human Capital Development Initiative, for her expertise and support. Funding for the book came from the National Community Development Initiative, a remarkable coalition of foundations, corporations, and the federal government stimulating community development.

Throughout this book are the stories of people who do community development work in all its many aspects, and we interviewed hundreds of others whose wisdom and points of view gave us helpful facts and insights. We are grateful for the time all those people generously gave this project.

We thank the Community Information Exchange for permitting us to use its great store of information as we launched this project. In the considerable task of original research that went into this book, we had able assistance from Kathy Andrews, Anne Carrabino, Sandra S. Chung, Nikki Flionis, Laura Gates, and Donna Newton. Thanks to Susan Edelmann for hours of talented editing and to our fine editor, Heather Boyer, at Island Press.

A group of practitioners advised us as we researched and wrote this guide. Their jobs demonstrate some of community development's variety.

Sandra Sanchez Almanzan, Director of Public Policy, the Greater El Paso Chamber of Commerce

Rick Cohen, President, National Committee for Responsive Philanthropy, Washington, D.C.

Lisa Cunningham, Associate, ICF Consulting, San Francisco, California

Denise Fairchild, President, Community Development Technologies Center, Los Angeles Technical Trade College

Pete C. Garcia, President and CEO, Chicanos Por La Causa, Phoenix, Arizona

Lynette Lee, Executive Director, East Bay Asian Local Development Corporation, Oakland, California

Woullard Lett, Administrator, New Hampshire College's Community
Economic Development Program, Manchester, New Hampshire

Madeline McCullough, Vice President and Director, Washington, D.C.,
Office, The Enterprise Foundation

Joseph B. McNeely, President, Development Training Institute,
Baltimore, Maryland

Monroe Moseley, Vice President and Director, Isaacson, Miller,
Boston, Massachusetts

Ernest L. Osborne, Senior Consultant, SEEDCO, New York City

Patricia J. Rumer, Director, Community Programs, School of
Extended Studies, Portland State University, Portland, Oregon

Patricia L. Smith, Senior Director, Human Capital Development
Initiative, National Congress for Community Economic
Development, Washington, D.C.

Clarence Snuggs, Deputy Executive Director, Neighborhood
Reinvestment Corporation, Washington, D.C.

Karen Stokes, Executive Director, Coalition for Low-Income
Community Development, Baltimore, Maryland

Michael Swack, Founder and Director, Community Economic
Development Program, New Hampshire College, Manchester, New
Hampshire

Vickie B. Tassan, Senior Vice President, Bank of America Community
Development Banking Group, Washington, D.C.

Photo credit: James Warden, courtesy of National People's Action.

A GUIDE TO
CAREERS IN
COMMUNITY
DEVELOPMENT

Chapter 1

Just What is Community Development?

Community development is the economic, physical, and social revitalization of a community, led by the people who live in that community.

Community developers work in community-based organizations; banks; city, state, and federal government; foundations; real estate development companies; social service agencies; job training and placement organizations; investment firms; and think tanks. Community developers do things with, not to or for, the community.

They transform a brownfield into a neighborhood shopping center, creatively finance low-cost mortgages, shape public policy, develop community health centers, build housing to shelter battered women and their families, train residents for good-paying jobs, start new local businesses, organize tenants to convert their apartments to cooperative ownership, create a joint venture with developers to develop a local supermarket, counsel families to move off welfare, assist farm workers to build their own houses, create new enterprises with community youth.

Community development may receive relatively little public attention, but it is a rapidly growing, multitrillion dollar field with about 400,000 jobs today. This guide is intended to introduce you to this world and to open the eyes of and doors for people who are looking for a way of life that is much more than a career. Readers who already know something about the field will find new information and ideas here.

When you finish using this guide, you will understand the ideas, goals, and strategies that make up community development, the pros and cons of community development careers, and what it is like to work in the various places and positions this field offers. You will know how to find the organizations and agencies where the jobs are, how to prepare for and land a job, and which colleges and graduate schools offer programs in community development studies. You will see what entry- and higher-level jobs are like and how you can build a satisfying life within community development or use it as a springboard to other careers. You will find both the "big picture" and practical details about community development and its many practices.

The Way Community Development Works

Community development emerged as a field about

Career Story: Msgr. Geno Baroni

Neighborhood Development Pioneer

Msgr. Geno C. Baroni was born in 1930 in Acosta, Pennsylvania, the son of an immigrant Italian coal miner. Ordained a Roman Catholic priest in 1956, just about the first thing he did in his first parish job in a small Pennsylvania town was to instigate the setting up of a community credit union so that the struggling families of the town could get loans with lower interest rates. When in 1960 he was sent to St. Augustine's, a church serving poor black Catholics in de facto segregated Washington, D.C., he turned a vacant convent into a community center with a basketball court for kids, a shaded sitting area for the elderly, and a small theater for community productions.

Geno often said, "It is better to beg forgiveness than to ask permission," and that was the principle he followed when he joined the civil rights movement early on. He lobbied the halls of Congress hard for the Civil Rights Act, and then he joined the march in Selma, Alabama, one of the first nonviolent marches against segregation. With his prayer book inside his clerical hat so his skull would not be fractured by a billy club, Geno marched hand in hand with four hundred ministers, rabbis, other priests, nuns, laypeople, and Dr. Martin Luther King. He was the Catholic Church's coordinator for Dr. King's march on Washington in 1963.

Geno was a creator, a risk taker, always

pushing the edge. "There are only two lasting things we can leave our children. One is roots, the other is wings," he said. From the power of his ideas and his rambling yet convincing ways of presenting them, his ideas took off on others' wings. The first institution that sprang from his convictions was the Campaign for Human Development, set up by but operating independently of the U.S. Catholic Conference in 1969 after the urban crisis that followed Dr. King's assassi-

> "Let us pray that there will be new gifts to meet new needs. That there will be new voices of justice."

nation. The Campaign is a multimillion dollar fund, raised from Catholic parishes across the country for local nondenominational projects of social and economic jus-

tice. "I saw that my responsibilities and my position as a priest did not stop at the altar rail. I couldn't talk about heaven inside the church while all around on the outside, where people lived, it was hell."

Geno was also the force behind the founding of the National Neighborhood Coalition, the National Cooperative Bank, the Italian-American Foundation, the National Center for Urban Ethnic Affairs, and Network, an organization of nuns that lobbies on social policy questions.

One of the reasons for his success in lobbying was that he understood that all policy and all politics are personal. He had wit and the ability to explain policy in personal terms, almost like talking in parables. Geno believed in democracy and its institutions for self-government. He worked with Gale Cincotta, founder of National People's Action in Chicago, leading the fight for legislation to end bank discrimination in home mortgage lending. Geno loved politics and those who practiced it and worked to raise the art of politics to the highest level of service. He profoundly influenced local and national elections, including Jimmy Carter's run for the White House.

Geno was appointed, during the Carter administration, to the post of assistant secretary for neighborhoods, voluntary associations, and consumer affairs in HUD. That made him the high-

Career Story: Msgr. Geno Baroni, cont.

est-ranked Catholic priest ever to serve in the U.S. government's executive branch. While working in that position, he wrote to a colleague, "Because all of the economic, social or political issues raised in the community are ultimately ethical or moral issues of justice, equality and charity, there is an implicit necessity that clergy people become involved." As the advocate for neighborhoods, he eventually convinced the HUD secretary, the White House domestic staff, and then President Carter that federal policies should be reshaped to preserve and improve neighborhoods. His vision was not for another top-down federal program imposed on cities, but for a new initiative that would give neighborhood groups some funds with which they could leverage partnerships from the private for-profit and voluntary institutions in the private sector and from local government to improve their communities. He was a consummate lobbyist, and his vision became HUD's $25 million Neighborhood Self-Help Development Program, passed by Congress in 1978.

Geno's effect on individuals was as profound as his impact on ideas and institutions. One part of his genius was the gift of confirmation—he could spot talented people and confirm in them the talent and leadership they might not yet have shown. He inspired hundreds of people, many of them political and community leaders today, to do the work he thought was needed in public affairs. Barbara Mikulski, now Democratic Senator from Maryland, referring to herself and Congresswoman Marcy Kaptur, has said, "How remarkable it was that an Italian-American priest could convince two Polish-American women to run for Congress." He was a great connector, connecting people to people, connecting his experience to that of others, and connecting people to their own life experience. He had a unique empathy and could always understand and relate to what each person was, whether male or female, black or white, Christian, Muslim, or Jew.

Geno saw the plight of the working-class, nearly poor, white ethnic neighborhoods placed under siege by the decline of the cities in the 1960s and 1970s. He thought that the image of America as a melting pot was flawed. The son of an immigrant, he was deeply aware of his own roots. He wanted all of us from every background to respect and revel in our heritage. He started a movement for ethnic identity that swept across the nation.

But he also expected the differences to be extinguished when it came to jobs and housing and education and the basic ability of Americans to live in a multicultural society. Because of his links to the civil rights movement, he was able to build bridges between white ethnics and minority groups. Geno was named a Prelate of Honor by Pope Paul VI in 1970.

When he was dying of cancer, he organized a group of other cancer sufferers to bring the powers of will and spirit to bear on their affliction. His friends and colleagues came in droves to visit him in the hospital. They gained strength and a renewed determination to stick with it—the "it" being whatever Geno had first inspired or told or asked us to do. We all have. He died in 1983 at the age of fifty-three. [1]

"Lord, I pray: Help me to know that our limited charity is not enough. Lord, help me to know that our soup kitchens and second-hand clothes are not enough. Lord, help me to know that it is not enough for the Church to be the ambulance service that goes about picking up the broken pieces of humanity for American society. Lord, help us all to know that God's judgment demands justice from us as a rich and powerful nation. "

thirty years ago, as a grassroots movement to improve life, most often in low-income communities. The ideas that drive it are intensely American, rooted in our history and diversity. Picture the pioneers helping each other with their barn building; update the picture and add to it the need for multilayered financing, zoning approval, and other complexities of today's life. Mix into it the heritage of communal life and neighborliness from the cultures and races that make up this country, and you have today's community development field. At their core, community developers carry the same philosophy as the pioneers: there is strength in all of us, and no one else is going to build the kind of community that we want. We are all in this together.

Community development has an unusual and challenging three-pronged mission. One goal is *to change the economy* of a community for the better, increasing the income and wealth of the residents and stimulating investments in the community, while placing the assets and the economic fate of the community within the control of the residents as much as possible. A second goal is *to improve the physical nature* of the neighborhood, from its housing to its shopping areas, transportation, public spaces, and environment. The third is *to strengthen the social bonds* among the people in the neighborhood—their neighborliness, their readiness to talk to, help, work with, and socialize with one another; those are the human ties that hold a neighborhood together.

To accomplish these goals, community development works through grassroots groups, often called community development corporations or CDCs. CDCs are neighborhood-based organizations that usually originate from and are controlled by residents determined to turn their neighborhoods into healthy, thriving communities. Most CDCs are incorporated as nonprofit, tax-exempt organizations. All CDCs have a board of directors that includes local residents; other board members might be public officials, bankers, relevant professionals, and funders. CDCs engage in development activities, including creating affordable housing and economic development; many also provide community services, such as day care programs. CDC staff members are knowledgeable about the steps involved in development, and they hire consultants to fill any gaps in their expertise. CDCs are often successful where private developers and public agencies have failed because CDC staff often live where they work, and they draw on the indigenous leadership of the neighborhood in determining the types of economic developments and housing needed for successful revitalization. CDCs are also able to obtain low-interest loans and grants for many projects that would be too expensive for the neighborhood with market-rate financing.

These local groups are often called by other names—such as neighborhood housing services, or community-based development groups, and more—but we use the term *community development corporation* and mean them all.

Community development is usually focused on a place, which might be an inner-city neighborhood, a deteriorating suburb, a small town, or a rural community. Connection to place—the neighborhood or the rural community—is important in today's fragmented, impersonal, and electronically isolated world. The place, the community, is where neighbors know their neighbors, greet them at the grocery store, interact with each other through the Scout troop, crime watches, school benefits, and other activities that bring people together to pursue the common good. Sometimes, instead of focusing on place, community development groups form to serve a specific group of people, such as Hispanics, senior citizens, or the homeless, and these groups may work citywide rather than in a neighborhood.

To improve the community's economy and physical sur-

Career Story: Franklin Thomas

Neighborhood Champion and CEO, Ford Foundation

Frank Thomas went up the community development ladder in a big way—from running a community development corporation in the Bedford-Stuyvesant section of Brooklyn to head of one of the largest foundations in the nation, the Ford Foundation.

Actually, the starting point wasn't the neighborhood. It was the Air Force, followed by a law degree, that led Frank to the appointed position of deputy police chief in New York City, the department's top legal official. While he was working as the deputy police chief, on a host of issues including police-community relations, Senator Robert Kennedy and Senator Jacob Javits developed an antipoverty idea that at the time—1966—was novel: To tackle the problems of the inner city, why not create a partnership between the business community, local government, and the city government to improve distressed areas—rather than using the harsh and destructive tool of urban renewal, clearing slums, and disrupting the lives of people and whole communities. The idea of what we now call a CDC was to improve the neighborhood using the best business practices available.

Frank Thomas was the right person for the job because he was the candidate acceptable to all of the partners. He agreed to take the job of creating that partnership for a two-year period, set it up, and get it run by someone over the longer haul. Frank became the first CEO of the Bedford-Stuyvesant Restoration Corporation. The group tackled every aspect of community life: education, housing, business and job development, health, recreation, cultural affairs. As Frank and the board envisioned it, all of these elements needed to be addressed to add up to a

comprehensive approach.

"The process of doing the work was itself an industry," remembers Frank. "We worked hard to place people from the community in the jobs that were being created to accomplish the mission of turning the neighborhood around. If the people in the neighborhood could learn the processes, develop the sites, learn the financing approaches and devices—then the neighborhood itself would have the capacity and the work could continue. This philosophy resulted in a racially mixed staff. The quality of the staff and board made working at Bed-Sty highly desirable. We had people leave big law firms to join the corporation."

Then, after ten years, not two, Frank left the corporation and went into law practice. He was asked to join the board of the Ford Foundation, and when the presidency of the Foundation became vacant, he agreed to take on the CEO role, a position he held for sixteen years. While at Ford, Frank led the Foundation into supporting the emergence of community development corporations and community development generally.

"At the core, anyone wanting to work in community development has to ask themselves, 'What special talents do I bring to the field?' Don't abandon whatever your skill base is. Use it. Figure out what you don't know and get yourself schooled so that you can be a real asset to the community group or organization you're trying to work for. And there's no point in your personal career that's too late to make the move. I've seen many mid-career people make the move into community development successfully."

Frank isn't fully out of the community development field now. He spends much of his time as a corporate director for some of America's largest companies and heads the TFF Study Group, a nonprofit initiative focusing primarily on assisting the development process in South Africa.

roundings, community developers try to intervene in the forces of the marketplace through a variety of business, capital, and real estate strategies intended to connect the neighborhood to mainstream institutions and to produce tangible things—more affordable housing, more and better jobs, new local businesses, and new money invested and retained in the neighborhood. Accomplishing these goals often means addressing related problems. Local residents may need skills training to qualify for a

> **"The political left and the right meet in the neighborhoods."**
> — Geno Baroni

CDC Industry Profile[2]

- 52 percent of reporting CDCs serve urban areas; 26 percent serve rural areas; 22 percent serve mixed urban-rural areas.
- 550,000 units of affordable housing have been developed by CDCs.
- 71 million square feet of commercial and industrial space has been developed by CDCs.
- CDCs have $1.9 billion in loans outstanding to 59,000 businesses as of December 31, 1997.
- 247,000 private-sector jobs have been created by CDCs.

job, or transportation to get to jobs outside their neighborhood. Parents cannot work unless day care is available. Businesses cannot thrive unless the neighborhood is safe. Families with children will not buy a home in a neighborhood unless the quality of schools is at a critical minimum. The wealth of the residents needs to stay in the neighborhood in community-controlled financial institutions. Community development must deal with this entire set of issues and problems—so people working in the field must be versatile and

able to connect effectively with others.

Community development tries to increase the opportunities available to the people in the neighborhood and to make sure that people have a sense of opportunity. In addition to talking about a neighborhood's needs and problems, community developers focus on assets, market niches, and opportunities.

If the words *assets, market niches,* and *opportunities* sound like business terms, it is because community development is a business—the business of improving places and the lives of the people in them. Community developers are "social entrepreneurs," and many of the skills and traits needed to succeed in community development are similar to those needed to succeed in business.

Here are just a few of the thousands of examples of physical and economic redevelopment led by CDCs. The Development Corporation of Columbia Heights, in Washington, D.C., has developed a large retail center whose many tenants include a community development bank and a laundromat. The CDC also operates a business loan fund, teaches neighborhood youngsters business management and leadership skills, and is transforming vacant houses into homes for first-time homeowners. The Greater Germantown Housing Development Corporation, in Philadelphia, has renovated and built hundreds of apartments and houses, Recently, it took a 3.2-acre blighted and heavily polluted site in the heart of its neighborhood, cleaned it up, and redeveloped it into a 20,000-square-foot retail and commercial space with homes and apartments for the elderly.

If done well, physical and economic recovery efforts accomplish something else: They make it more likely for neighbor to cooperate with neighbor. To development strategies are joined other tools: community organizing, advocacy and public policy, and social services. These actions taken to improve a neighborhood also strengthen the social bonds among the people in the neighborhood.

In this way, community development represents a reworking of American enterprise—the tools, language, and mind set of capitalism—tied to communal values. That is why community development is sometimes described as "corrective capitalism" or "community capitalism."

The People and Institutions That Make Up the Field of Community Development

While the core of community development is people working at the community level for themselves and each other, they cannot do it alone. This is where the barn-building image of the last century fails. Today's society is too complex, the problems to be overcome too vast for community groups to succeed entirely on their own. A remarkable set of organizations has developed over the years that supports and partners with local community development initiatives. These organizations—the community development system's infra-

All the Places Where You Can Do Community Development

CDCs, neighborhood housing services, nonprofit housing development groups
Community action agencies
Community development financial institutions
Community-based organizing groups
Tenant organizations
Public housing resident management corporations
Community development intermediaries
Nonprofit job training and placement organizations
Nonprofit enterprise development organizations
YouthBuild groups
Community land trusts
Bank community development lending departments
City and county community development departments

State community development agencies
Foundations funding affordable housing, antipoverty efforts, community development, community building
Consulting firms directly dealing with affordable housing and community development
Think tanks directly dealing with affordable housing and community development
Private-sector companies dealing with affordable housing and community development
Housing counseling agencies
Neighborhood design centers
Regulatory agencies
Elected officials and staff offices
City and county housing and community development agencies
City and county planning departments
Empowerment zone agencies

State housing finance agencies
State housing departments
Land-use agencies and companies
Federal agencies dealing with housing, community development, workforce development
Faith-based community groups
Community health clinics
Youth-related organizations
Secondary market institutions
Low-income housing tax credit syndication companies
Mortgage banking companies
Pension funds that invest in affordable housing or community economic development
Community gardening groups
Community arts groups
Hospitals and other large institutions with community development jobs
Academic institutions

structure—represent the nonprofit private sector, the for-profit business world, and, of course, the government sector:

- At the grassroots level, local nonprofit groups other than CDCs are active in pursuing a specific agenda, such as helping people with mental or physical disabilities or organizing or running job training programs for at-risk youth.
- "Intermediaries" intermediate between neighborhood-based groups and national sources of grants and other capital, providing funding and technical assistance to local groups.
- Big business and for-profit investment banks, law firms, lenders, mortgage companies, and even venture capital pools invest the capital necessary to produce the physical and economic change. For some banks, community development finance has turned out to be an important new business line and a fine avenue for growth, while it also gives them great publicity.
- For-profit developers, property management companies, and business owners are now partnering with community development organizations in specific housing and economic development ventures.
- Community development financial institutions have been set up to make equity or debt available to ventures in situations where the chances of making a profit look slim and risks look high. Their technical expertise and marketplace know-how, which could power a *Fortune* 500 venture, work here on behalf of small-scale, often experimental ventures.
- The public sector works in the community development field, providing subsidies and funding services, regulating and monitoring, and even partnering in "deals" to make direct loans or to reduce risk.

- Research centers, policy institutes, think tanks, educational institutions, and consulting companies are involved, to look at the big picture and draw lessons for the future.
- Foundations, both national and community, play a role too, offering grants or loans or even investing their endowment money in development ventures.

Chapter 2 is a map of the field that fills in the details about what these institutions are and how they work together.

Community Development to Counter Social and Economic Problems

The impoverished community's problems—dilapidated housing, a low level of home ownership, lack of jobs, high crime and pregnancy rates, bad schools, and the absence of businesses as essential as grocery stores and banks—are interrelated. They are a result of broad forces in the American social, economic, and political system that have led to suburban sprawl, the abandonment and distress of many older city neighborhoods, and the depopulation and economic decline in many rural areas.

At its best, community development recognizes this systemic interrelationship and responds with holistic strategies to build the strengths of the place that is the community and the people living there.

Community development is in some ways different from other antipoverty efforts to improve the status of low-income people and distressed communities:

- It is different from community or labor organizing, which usually involves a struggle to influence outside powers around a particular issue, to get someone else to fix the problem. But it uses organizing to get the neighborhood involved and to compel the attention of outside powers.
- It is different from welfare's focus on "treating" the indi-

vidual and the individual family on a casework basis—but it addresses the needs of families through all the comprehensive economic, physical, and social development tools it uses.

- It is different from an income support program that would give people more money but not necessarily more power or skills. Community development is definitely an approach intended to empower people and communities.
- It stresses community control of assets in contrast to traditional economic development, which relies on the economic decisions of investors and developers, usually from outside the area. Community economic development seeks to improve the economic status of residents.
- It focuses on improving places as well as the people and institutions in those places, a unique and powerful combination.

The Story of Our Neighborhoods

Most of the great American cities developed during the 1800s as trading centers. Midway through the nineteenth century, emerging technology began to reshape the cities. Railroads and steam vessels, electrification, the telegraph, and later telephones all permitted large-scale industrialization. Massive new plants created the classic core industrial city around rail nodes and ports.

Cities grew rapidly again in the 1920s as centers of mass industrial production. Improved transportation made it easier to ship goods long distance, allowing corporations access to a national marketplace from ever larger centralized manufacturing plants. Around the plants, low-cost, high-density housing served a blue-collar workforce.

After 1945, cities began to experience a major loss of manufacturing jobs. Deindustrialization and a fundamental decentralization of economic and residential life into the suburbs and exurbs over the past decades have torn the fabric of once strong cities. In a series of mini-migrations, white families who could afford to left the old neighborhoods behind and moved to the suburbs.[3] New businesses were built on cheap former farmland (greenfields) rather than on older, reusable sites inside the city. Once thriving blue-collar neighborhoods suffered from a net loss of population and from disinvestment.

The result is what one observer of this transformation calls the "twin horsemen of the apocalypse: decay in the center and sprawl at the edge."[4] Some of this decentralization is the effect of public policy—the easy availability of federal housing insurance for suburban homeownership and the federal funding of highway systems (one example: between 1985 and 1995, the Philadelphia region spent over $1,041 per person for highway transportation in the suburbs, compared with $424 per city resident).[5] Racism lies just beneath the surface of some of these changes: as African-Americans moved up from the South to find jobs in the cities in a huge swell of migration in the 1950s and 1960s, white families left the neighborhoods.

Urban renewal in the 1950s meant economic initiatives to help the central business districts. But it also meant the bulldozing of many viable low-income neighborhoods, along with their churches, small businesses, and community fabric, while subsidized housing was isolated in places with little or no job or commercial base.

Now our nation has shifted from a manufacturing to a high-tech information economy and from a local or national to an international economy. In this new reality, the inner-city, suburb, and rural communities are tied together, their destinies interwoven. Central cities are the focal point of every metropolitan area, and the great majority of American people live in these areas.[6] Rural areas, already struggling with a poverty rate higher than other parts of the nation—16 percent compared with 11 percent for cities and suburbs[7]—are strongly affected by the economies and development patterns of nearby cities.

Community development is now trying to connect the neighborhood and rural communities to the realities of the metropolitan marketplace and metropolitan politics and to deal with the new challenges facing our economic system and civil society as a whole.

These fundamental changes are felt in every corner of America. Gale Cincotta, who years ago organized her Chicago neighborhood to combat its disintegration and after that founded a nationwide organization known as National People's Action, describes the deterioration she has seen firsthand over the past thirty years: "In my time, you had time. All the women were home with the kids. Now the mothers are too tired to go to a meeting. They have no extended family. Drugs and crime are different, and that's in part because of air conditioning and in part because of TV. No one's out on the streets, but there are more cars. My son used to tell me, 'That woman is always looking at me'; I told him, 'Good.' The neighborhood park was full at night because it was too hot to be in the apartment. Today everyone is isolated, watching the crap that's on TV." [8]

It is clear that community development has not solved the problems of our nation's most troubled places or people. Nor, given our history and economic system, could we expect it to. Yet it has improved and stabilized neighborhoods and does deal directly and successfully with major issues facing society today, from economic equity and the structure of welfare to environmentally sustainable development and the nur-

turing of social relationships. That is why community development work is so challenging and so satisfying to the people working in the field.

A Brief History of Community Development

When the community development movement began about thirty years ago, it caught the attention of high-level politicians and policy makers looking for solutions to the problems of poverty. This was the 1960s, a time of great ferment. The riots that rocked cities across America were fueled by frustration and rage at a century of segregation and racism, at private-sector practices that milked the poor and cut off the lifeblood of capital to their homes and businesses, and at flawed public policies that devastated neighborhoods and relegated their citizens to handouts. The anger was further flamed by the assassination of the Reverend Dr. Martin Luther King, Jr., in 1968. Communities of all colors, urban and rural, sought a new economic life, not just federal aid. Robert Kennedy, for example, while a U.S. senator (D-NY), not only sponsored community development legislation (cosponsored by Senator Jacob Javitz, R-NY) but also took a direct hand in helping a local neighborhood organization in Brooklyn's Bedford-Stuyvesant attract new business investment. As this movement gained skills, attention, and funding—and as the needs of poor inner-city and rural communities became ever clearer—the self-help approach began to flourish and spread.

The community development movement's roots are as diverse as our country. Alexis de Tocqueville, a perceptive young Frenchman visiting in the 1830s, wrote about the American talent for innovative civic associations. Booker T. Washington and Marcus Garvey organized collective business

Resources for Finding Your Core Beliefs

There are many different points of view about society and the causes of poverty and many philosophies about how to make the world a better place. In reading this career guide, you may decide that you'd like to know more about these different belief systems. Some of the books that will give you that broad view of the world are listed below. (Appendix L lists books specifically about community development.)

Black Wealth/White Wealth: A New Perspective on Racial Inequality, Melvin L. Oliver and Thomas M. Shapiro (Routledge, New York. 1996). By analyzing private wealth—total assets and debts rather than income alone—the authors, document the huge racial disparity in accumulated resources, caused not by a poor work ethic or an inability to defer gratification but by the deep and persistent racial inequality in America. They show how public policies fail to redress the problem.

Community and the Politics of Place, Daniel Kemmis (University of Oklahoma Press, Norman, OK. 1990). This beautifully written book holds that people can govern themselves only when they feel their connectedness to real, identifiable places. Kemmis reaffirms the "republican tradition" that rests on the ability of large numbers of ordinary citizens to rise above a narrow self-centeredness and identify and pursue the common good. He says that our politics—our public life—and our struggle to live together must occur in the context of very specific places.

Democracy's Discontent: America in Search of a Public Philosophy, Michael J. Sandel (The Belknap Press/Harvard University Press, Cambridge, MA. 1996). In search

> "The civic society occupies the middle ground between government and the private sector. It is not where we vote and it is not where we buy and sell; it is where we talk with neighbors about a crossing guard, plan a benefit for our community school, discuss how our church and synagogue can shelter the homeless, or organize a summer softball league for our children. Civil society is thus public without being coercive, voluntary without being privatized."
> —Benjamin Barber, Jihad vs McWorld.

of a public philosophy to replace today's impoverished vision of citizenship and community, shared by Democrats and Republicans alike, Sandel ranges across the American political experience, recalling the arguments of Jefferson and Hamilton, Lincoln and Douglas, Holmes and Brandeis, FDR and Reagan, and offers a new interpretation of the American political and constitutional tradition to rejuvenate our civic life.

Double Exposure: Poverty and Race in America, Chester Hartman, editor (M.E. Sharpe, Armonk, NY. 1997). This collection of essays by leading activists and scholars, drawn from the publications of the Poverty & Race Research Action Council, addresses major issues that affect community development: the permanence of racism, immigration, the "underclass" debate, multiculturalism, affirmative action, and economic inequality in a democratic society.

Jihad vs. McWorld: How the Planet Is Both Falling Apart and Coming Together, Benjamin R. Barber (Ballantine Books, New York. 1996). Barber describes the two great opposing forces at work in the world today—global consumerist capitalism and religious and tribal fundamentalism—that are the two biggest threats to democracy. These diametrically opposed but strangely intertwined forces are tearing apart the world as we know it, undermining democracy and the nation-state on which it depends. This is a brilliant look at the context confronting community development (and all other efforts to bring about citizenship and economic equity).

Losing Ground, Charles Murray (Basic Books, New York. 1984). One of the intellectual mainstays of the conservative movement, this book contends that welfare programs have had a perverse effect, worsening instead of improving conditions for the poor. The author contends that by blaming poverty on society's structural weaknesses, Great Society programs removed responsibility from the individual and made poverty less immoral than it had been. Fur-

Resources for Finding Your Core Beliefs, cont.

ther, the programs provided negative incentives, making it rational to remain on welfare.

Metropolitics: A Regional Agenda for Community and Stability, Myron Orfield and David Rush (Brookings Institution Press, Washington, DC. 1998). This book describes the destructive land-use practices and policies that have brought about the decline of inner cities and older, inner-ring suburbs, sprawl, and the destruction of rural areas. It introduces new concepts and practical tools to bring together cities, suburbs, and rural communities to pursue a productive metropolitan or regional agenda.

Regulating the Poor: The Functions of Public Welfare, Richard A. Cloward and Frances F. Piven, editors (Vintage Books, New York. 1993, 2nd edition). The premise of this book is that public welfare's main function is as a safety valve that supports the prevailing capitalist system. Welfare programs mollify protestors, restore social order, support the supply of low-paid labor, and deflect the need for systemic changes.

The Other America: Poverty in the United States, Michael Harrington (Collier Books, New York. Reprint 1997). Originally published in 1962, this classic reveals the depth of poverty in America and analyzes why such "invisible" citizens as minorities, the elderly, and children are not given adequate opportunities.

When Work Disappears: The World of the New Urban Poor, William Julius Wilson (Knopf, New York. 1996). For the first time in the twentieth century, most adults in many inner-city ghettos are not working in a typical week. This book describes how technological advances and global economic shifts have resulted in the disappearance of blue-collar jobs from the inner city, the suburbanization of employment, and the devastating effects on the urban poor.

Worlds Apart: Why Poverty Persists in Rural America, Cynthia M. Duncan (Yale University Press, New Haven. 1999). Here is an explanation of how persistent patterns of power and inequality, rooted in class and race, erode the fabric of community and undermine the civic culture that is the prerequisite for building the community institutions that make it possible for the poor and powerless to change the conditions of their lives. Though focusing on rural communities, this thesis holds for urban neighborhoods as well.

efforts to provide economic opportunities for former slaves after the Civil War. Settlement houses, such as Hull House founded by Jane Addams in Chicago, sprang up around the turn of the twentiethth century; they were a kind of community development institution, where trained workers settled in a disadvantaged community in order to be an integral part of its revitalization. They tried to help improve neighborhood life as a whole, through recreational, educational, and health care programs, as well as activities to promote neighborhood cooperation. Part of the agenda of Progressive reform of that era, settlement houses did not primarily draw on the indigenous leadership of the community they served.

In the late 1930s, a time of great unemployment and horrific workplace conditions, Saul Alinsky began organizing the poor to fight for their rights in Chicago's Back of the Yards slum neighborhood behind the city's slaughter and meat-packing houses, using strategies pioneered by the labor movement's Congress of Industrial Organizations to organize packinghouse workers. Alinsky drew on the CIO's use of clear, specific goals and systematic approach to building a local organization that would understand and deal with power.

Most historians of the community development field think it was the convergence of the civil rights movement and community organizing strategies, such as those championed by

Saul Alinsky, that gave final shape to today's community development field. "That moment of truth, when residents realized that they could actually organize and change what was destroying their community, is really the seed of the community development movement," explains Hipolito Roldan, a longtime Chicago community development leader.[9]

A remarkable number of new community development groups have grown out of church-based initiatives for social and economic justice. Religious institutions in this country have always organized to help the community and the needy. But a turning point occurred when a group of four hundred black ministers in Philadelphia, organized by the Reverend Leon H. Sullivan, pastor of the Zion Baptist Church, mounted one of the most successful civil rights boycott campaigns of the era. From 1958 to 1961, using their pulpits as a common base, the ministers persuaded the city's black community to use its purchasing power in "selective patronage" against businesses and industries that discriminated against black Americans in hiring. To train workers for the skills they needed to get jobs, Reverend Sullivan founded the Opportunities Industrialization Centers, still active in many cities. Money to support the program came in part from the "10-36" initiative, which continues to this day, in which congregants contribute $10 a month over thirty-six months.

A new community development approach started in the 1960s with the Ford Foundation's experimental Gray Areas program, which gave substantial funding to five cities to form new nonprofit organizations that would try to create a comprehensive vision and plan for community self-help renewal.[10] The Johnson administration's War on Poverty, launched in 1964 with the Economic Opportunity Act, also focused on the community as the best means for tackling poverty. That legislation supported the founding of community

action agencies in poor rural and urban communities to deliver direct human services, from Head Start to community health centers.

With Robert Kennedy's sponsorship in 1966, the community-based approach to fighting poverty was expanded. A new Special Impact Program, targeted to community development corporations, was created in the Office of Economic Opportunity (OEO); some of the CDCs that received funding had been part of the Ford Foundation's Gray Areas program. The Special Impact Program grew into a full-scale program in the 1970s (and became known as the Title VII CDC program), funding about forty-five urban and rural organizations.

Thousands of new community organizations sprang up in the 1970s, to counter disinvestment, to oppose freeways that would cut through their neighborhoods, to represent the interests of new Hispanic and Asian immigrants. These grassroots efforts could count on the help of a broad array of federal programs. One was the Community Services Administration, created in 1974 as OEO's successor. Another source of grants and technical assistance was the new Neighborhood Self-Help Development program created in the U.S. Department of Housing and Urban Development (HUD) and headed by Monsignor Geno C. Baroni, an inspiring civil rights and community activist.

The Neighborhood Self-Help program proved a turning point for grassroots community development. It created a significant enough mass of groups to raise the capacity and accomplishments of the movement, enabled those groups to serve as models and teachers of others, and proved that neighborhood-centered public-private partnerships could work. On that base, the Local Initiatives Support Corporation, The Enterprise Foundation, the Neighborhood Reinvestment

Corporation, and other intermediaries could take the community development industry to subsequently higher levels of funding and achievement.

By the end of the Carter administration, one study estimated that $2.6 billion in federal funds was flowing each year to community development projects. Then, in 1981, the Reagan administration closed down the Community Service Administration and other neighborhood-oriented programs. By 1985, the level of federal funding for community development had dropped to $1.6 billion.[11] But even in the face of the massive dismantling of federal government support in the 1980s, the community development field continued to grow in size and sophistication, led by self-help efforts and funded by national foundations.

In the 1970s, banks began to try different ways to reinvest in low-income communities. First they did so under pressure from citizen groups fighting with high-visibility organizing tactics against the disinvestment of their neighborhoods. Citizen action, in fact, led to new national legislation in 1977, the Community Reinvestment Act, that prevents many types of financial institutions from "redlining" (banks literally drew red lines on maps around neighborhoods to whose residents and businesses they would deny credit). Gradually, the banks realized that loans to low-income communities for housing and businesses could be a profitable business line.

Financial institutions were drawn deeper into community development deals by carrots and a stick. The stick was the increasing use by grassroots groups of the provisions of the Community Reinvestment Act, such as filing objections to bank mergers. The carrots were the community programs that all the federal financial regulatory agencies set up, starting with the Federal Reserve Community Affairs program in 1981, to educate and encourage community lending. By the mid-1990s, so many banks were engaged in lending for affordable housing that they formed a trade association, while the American Bankers Association opened a new Center for Community Development.

At the same time, cities and states created new tools to attract private investments, such as giving a credit against state income tax for corporate funding of community development projects. The key national intermediaries were created in the early 1980s to support the grassroots efforts, pulling in private corporate contributions and more foundation support. Various national financial mechanisms were devised to attract major private investment, such as a Low-Income Housing Tax Credit program (authorized by Congress in 1986 to give tax credits to corporations making equity investments in low-income rental developments). Though direct federal funding is minimal, the legitimacy of a community development approach has been recognized at the national level: a number of major federal programs, such as HUD's Community Development Block Grant program, now set aside money to be used by community development groups.

Today we estimate that well over 400,000 people are involved in community development work, from the grassroots organizer to the affordable housing director at city hall and the community development lender at the bank.[12]

Community Development's Track Record

Community development has built up a remarkable record of accomplishment, especially when you consider that it usually focuses on the most disadvantaged urban and rural places in America.

Community development corporations produce more housing for low-income people than the federal government does each year—over 30,000 units annually. They run health clinics and nursing homes. They have built, financed, and now operate supermarkets and shopping malls. They pull together the money and offer the training that enables women providing child care in their homes to get licensed and to make improvements to their facilities. At the end of 1997, they had loans totaling $1.9 billion invested in 59,000 small and micro businesses. They find out from local industry what jobs are going to be in demand, then train unemployed people to fill such jobs. They have created 247,000 private-sector jobs. They guard their neighborhood's environment and take on the cleanup of industrial waste sites and air-polluting incinerators. They partner with the local schools to improve the level of teaching.

Matching the achievements of local community development groups are the remarkable strides made by other development partners. One example: Over the past thirty years, financial institutions have committed to increase their community development investments in low- and moderate-income neighborhoods by $1.03 trillion dollars. Most of those commitments have been made since 1992, an indication of how rapidly the field is growing.[13]

The Future of Community Development

When community development first emerged some thirty years ago, it was "a movement." It was characterized by charismatic leaders, grassroots efforts isolated and unconnected, a great deal of righteous passion, little in the way of remuneration, and everyone struggling to figure out the very basic stratagems. Thirty years ago, no one could have written a career guide.

Today community development is called an industry by many if not most of its adherents. It has form and structure, a history, and a body of knowledge (evolving as that may be). It has discernible careers and experienced practitioners who can serve as role models. Yet one notable and appealing factor remains pretty much the same: it is still a calling, not just a job.

Where Is the Field as a Whole Heading?

When the Reagan administration zeroed out most of the neighborhood-oriented programs of the federal government in the 1980s, everyone projected the demise of the community development field. Instead, it leapt forward. Nor have the boom times of the 1990s made community development obsolete. Concentrated poverty still exists in America, many urban and rural communities could still use the corrective capitalism of community development, and areas such as inner-ring suburbs are now facing their own economic and social disturbances.

Because community development has weathered storms from many different directions and grown despite inclement conditions, because it has proved its usefulness, the field is likely to expand no matter what the outcome of national elections or federal policies.

There will be shifts and changes, however, responding to new knowledge and new needs—or opportunities, to think like a community developer. While housing construction or renovation has been the most frequent community development activity in the past, experience has shown that housing alone cannot alleviate poverty or turn around neighborhoods. In the coming years, housing development will continue, but it will be part of a comprehensive "community-building" agenda. Under that agenda, economic development, commercial development (getting supermarkets and other stores back into the neighborhood), and workforce development (preparing and training people for jobs and supporting them once they have jobs) will increasingly take center stage. That has already begun: in the past four years, CDC commercial and industrial development has doubled. The community-building agenda will also pay attention to a broad range of human services and to the social fabric of the community (organizing, vigorous community participation in planning, measures to reduce crime and drugs, and support for neighbor-to-neighbor relationship building).

The environment, from brownfields redevelopment to new enterprises based on environmental technology and environmental awareness, has just begun to engage the community development world. Growth management is an issue that so much affects communities that it is bound to become central to community development policy and initiatives. CDCs are beginning to take a place at the table when the public and private sectors discuss metropolitan economics and politics, and to serve the essential role of connecting the discussion to the neighborhood level.

Another definite trend is the growth of partnerships, bringing in for-profit businesses, nonprofit institutions, and public-sector agencies to participate in community development initiatives.

These trends affect all the organizations and institutions that do community development, and the jobs you are likely to find in the coming years.

How Will Various Institutions Grow?

Community development corporations stand at the center of almost all the initiatives in the field, so it makes sense to start with them in looking at the future.

In the past four years, the number of community-based development groups grew by a stunning 64 percent, according to the latest census of that part of the field. Continuing rapid growth in the number of CDCs is likely. Those that al-ready exist are probably going to add more staff. What CDCs do will change. They will continue to create housing but on a larger scale; larger groups will probably produce more than one hundred units a year.

They will use their real estate development know-how to develop and manage or provide financing to facilities such as child care centers and community health clinics. Some may venture into the high-tech world. In 1999, for example, New Economics for Women, a CDC in Los Angeles founded to provide housing and job opportunities for women who are single heads of households, renovated a building to serve as an information technology incubator for women-owned and minority-owned start-up companies geared to developing and commercializing NASA technologies. NEW's partner is the University of Southern California's School of Engineering Technology Transfer Center. NEW will earn income by taking an equity position in each business that the incubator helps to develop.

Because of new opportunities generated by a strong national economy and the consequences of welfare reform, combined with the central importance of jobs in low-income neighborhoods, CDCs will increase initiatives in economic development as well as in job training and placement.

Over the years, CDCs have experimented with starting and running businesses that generate revenue. Though that has proved harder than it looks, once again fee-generating ventures are on the agenda. Practitioners are looking at ways to make at least part of their core work pay for itself (as Pioneer Human Services in Seattle, profiled in chapter 5, has done). The experience and know-how gained the hard way as well as the respect that the private sector now holds for community development approaches indicate a higher level of success for future CDC-sponsored enterprises.

International community development

Internationally, community development has a strong presence in industrialized nations as well as emerging democracies and underdeveloped countries. Over the years, there has been a steady stream of technology transfer in both directions between countries overseas and the United States. The settlement house movement sprang up in England in the late 1800s, then spread to the United States. From Bangladesh's Grameen Bank, the United States learned about one powerful way—through peer lending circles—to make small loans to mom-and-pop micro-enterprises; and before that, our country drew heavily from micro-enterprise lending history in Central and South America. The U.S. cooperative movement has taken many principles and approaches from the decades-old Mondragon model pioneered in the Basque region of Spain. The American land trust movement is built on Indian land reform policies led by a colleague of Ghandi. Our country has learned from overseas—from Italy, for example—about the critical importance of a strong social fabric for building wealth.

In turn, when England and Scotland faced deindustrialization and lay-offs in the 1980s, they called on U.S. community development practitioners to help figure out alternatives to spur local economic development. Today, many of America's community developers have piled up frequent-flier miles in trips to Russia, Poland, and other countries trying to privatize industries and struggling to make their political and economic institutions work democratically.

Other nonprofits, especially those that once concentrated on social service delivery (for example, United Way agencies and groups helping new immigrants), have begun and will accelerate their expansion into housing development and job creation.

Faith-based groups, which have organized in remarkable numbers over the past decade for the purpose of community development, are likely to accelerate in number. Their agenda of development and community building will probably follow the kind of path that CDCs are taking.

Intermediaries will also experience growth at the local and regional levels to keep pace with the growth in grassroots organizations.

Community development financial institutions, which offer financing and technical assistance to community businesses, now manage more than $3.25 billion in assets.[14] This is another form of organization with likely prospects for growth in numbers and size.

Universities as helpful community development agents were the exception rather than the rule at the beginning. And it was the rare planning school that offered a course relevant to community development theory or practice. Now, however, more and more universities and community colleges are creating true partnerships with grassroots developers to revitalize communities. And more and more universities and community colleges—over a hundred and rapidly growing—have developed certificate and degree programs that help to prepare people for community development work.

For-profit businesses were at first pushed by government incentive and regulation to enter the field. But several business sectors, such as banks and supermarket chains, have found this work a profitable business line. Now it seems that the city neighborhoods and rural areas where community developers have been active are being viewed generally as "rediscovered markets." And grassroots development organizations are now seen as savvy partners that bring a lot of advantages to the table.

According to President Clinton, "The largest pool of untapped investment opportunity and new customers is not overseas; it's in our backyard—in Harlem or Watts or Appalachia or our Native American reservations. Undeveloped communities in America still control more than $85 billion in purchasing power. That's more than the entire retail market of Mexico, our second largest trading partner."[15] General Motors Acceptance Corporation, the auto company's finance arm, has an "emerging markets" section in its home mortgage lending department targeted to low-income neighborhoods.

State and city agencies have an increasingly important role, as major programs such as welfare reform and tax incentives are implemented on the state level (for example, it is the states that allocate welfare funds under the new Temporary Assistance for Needy Families program; similarly, states allocate tax credits for low-income housing). Cities and city agencies are central to new welfare-to-work programs, new ways to deal with subsidized housing, new incentives for corporate community involvement, and leadership in regional land-use and smart-growth strategies.

How Will Jobs Grow?

The jobs will multiply in the institutions just described. The exact nature of the new jobs is impossible to predict and is less important than the nature of the skills and mindset they will call for. As has always been true, community development will call for resourcefulness, flexibility, entrepreneurship, and commitment. Business skills and knowledge can only be a plus, but there will be opportunities for newcomers of all disciplines and leanings.

A Mirror for America

When community development organizations first emerged in the 1960s, they were designed as remediating institutions for disadvantaged neighborhoods. The rest of the nation was presumed to be intact. What's different today is that most Americans feel there is something wrong throughout our society.[16] Like poor neighborhoods, middle-income neighborhoods and the suburbs are often isolating, socially fragmented places; many Americans feel that we have lost a sense of purpose and that the moral fabric of community—from neighborhood to nation—is unraveling.

The way community-based development works and the values it illuminates are models for a more equitable nation with a stronger civic fabric. It is a way to understand the interwoven destinies of low-income and more affluent communities, of poor and better-off people.

> " All life is interrelated. Whatever affects one directly, affects all indirectly."
> — Rev. Dr. Martin Luther King Jr.

Community Story: Bethel New Life, Chicago

Comprehensive Community Building

In the whirlwind of black migration from the South that started in the 1940s and continued through the 1960s, Chicago was one of the prime destinations in the North. But instead of finding a welcoming environment, the newcomers faced racism and antagonism. Blockbusting and racial steering by sleazy realtors changed many Chicago neighborhoods from all white to all black. This was the fate of the West Garfield Park neighborhood, among others. Then the city and the banks cut off their lending, investment, and support to the communities. In the decline that followed, the West Garfield Park neighborhood lost two hundred housing units a year, uninhabitable victims of tenant abuse and landlord neglect. By 1970, the neighborhood had become a picture of decay and disorder, a mirror of many other urban neighborhoods.

The local Lutheran Church had shrunk to a congregation of forty when Mary Nelson moved to West Garfield Park with her brother, the new pastor. She joined a hand-

ful of other people who would not give up, and they began to organize community meetings. They set up an after-school tutoring program for high school dropouts as their first step in improving the neighborhood. By 1979 the congregation, growing

once again, embarked on its first housing venture. With $5,000 raised in collections from congregants, they bought an abandoned three-unit building. Volunteers pitched in to renovate it. When the project ran out of money for materials and banks refused to make loans to the church, church members took out loans in their own names. Mary Nelson bought the last building materials needed by charging them to her credit card. The house was sold, the money recaptured was put into the next housing venture, and Bethel New Life was born as a community development organization.

Twenty years later, West Garfield Park is noticeably though incompletely revitalized. First, the social changes: People now have a sense of opportunity, of hope for a better future. Many of the sons and daughters have become doctors, lawyers, and accountants. Residents no longer think of themselves as clients or victims but as citizens and participants. People there say they feel West Garfield Park is a community.

There have been enormous economic and physical improvements to the neighborhood. Many of its major community buildings, once sliding into rot and ruin, have been restored. Foremost among these are the local library (a treasure that is listed on the National Register of Historic Places) and the elegant Garfield Park Conservatory that faces onto a pastoral lagoon complete with ducks. Bethel New Life bought and renovated the huge old, abandoned St. Anne hospital complex, and the formerly garbage-strewn empty buildings are now humming with life. Through adaptive reuse, the buildings now house a small-business incubator center with services to nurture start-up and expanding local enterprises, 125 apartments with supportive services for seniors, and a small business center. The complex also includes the Mother Hen Day Care Center, where eighty children are cared for and twenty-five former welfare recipients are trained to be licensed day care providers. Gardens flourish where once there were vacant lots.

A Few Statistics

- Over 1,000 housing units have been built or renovated. This remarkable number includes 150 new homes, condos, and co-ops, constructed for first-time homeowners and clustered relatively close together to make the strongest impact. The housing changes throughout West Garfield Park represent dramatic before and after pictures. The median value of owner-occupied homes has increased 20 percent over the past seven years.
- Over $87 million has been brought into this once disinvested neighborhood since 1980, in public- and private-sector investments in its housing, commercial, and industrial development.

Unemployment has dropped from a high of 24.4 percent (in 1993) to 15.5 percent (in 1997).
- The high school dropout rate has gone down 10 percent.
- Pre-employment training prepared 375 welfare recipients and 103 ex-offenders for jobs in one year alone.
- Over the last ten years, Bethel New Life has placed about 400 people *a year* in full-time employment.
- Bethel New Life itself now employs 305 people, as lawyers, accountants, program directors; as property managers, construction workers, security guards; as nurses, home health care aides for seniors, and physician's assistants.
- Among the neighborhood's elderly, about 650 every week receive adult day care and in-home services.
- The Wholistic Health Center, located in one of Bethel New Life's new apartment buildings, provides family health services to over 1,000 patients a month. The Healthy Moms, Healthy Kids preventive health care initiative helps over 2,000 families a year. The Center's Women, Infants & Children (WIC) program serves 3,500 families a month.
- Voter turnout increased 67 percent on average in the five precincts that BNL focuses on, from the 1994 to the 1998 general election.
- Community organizing events drew in 1,361 neighborhood residents last year alone. There are 25 block clubs.

Just about every conceivable kind of institution—public, private, and nonprofit—has taken part in West Garfield Park's revitalization. In these institutions, people in community development jobs from all sectors interact with each other and with the grassroots Bethel New Life in a forward-looking holistic approach to rebuilding a neighborhood.

Private Sector
- Private corporations, in return for credits against their corporate income taxes, have invested in Bethel New Life's renovation of apartment buildings (in fact, Bethel New Life was part of the first pool in the nation to use Low Income Housing Tax Credits as a housing finance tool).
- Other for-profit partners include the LakeShore Development & Construction Company, which built Parkside Estates, a cluster of brick townhomes developed by Bethel New Life as part of its scheme to encourage homeownership.
- City colleges have collaborated with Bethel New Life to develop a curriculum and train residents as certified nurses aides; some of the graduates are hired to work in the organization's elderly care program, or in nursing homes.
- The Chicago Public Art Group helped teach painting to high school students who then painted a mural covering one wall of the day care center and designed and made a mural for an outside wall.
- The local music school staffs a music pro-

gram at the Beth-Anne Cultural & Performing Arts Center, including a summer chorus and cello lessons for neighborhood youth.

- Helene Curtis, the cosmetics company (now part of Unilever), worked with Bethel New Life on innovative ways to recycle its manufacturing by-products to create new enterprises for the community.
- Argonne National Laboratory worked with Bethel New Life to develop a new methodology for identifying the vacant industrial sites in the neighborhood with the best potential for reuse.

Public Sector

- The Chicago Transit Authority, in partnership with a developer and Bethel New Life, is rebuilding the T stop in the neighborhood as a "superstation," three stories high, with space for new businesses.
- The Chicago Department of Public Health contracts with Bethel New Life to train local companies and residents who are then licensed to remove or abate lead from buildings in the neighborhood, which was one of the most lead toxic in the city. This initiative has translated into new business for local companies and well-paying ($17–19 an hour) jobs for residents.
- The Chicago Department of Housing pays for a parallel asbestos abatement training course and recertifies the lead abatement workers.

- The city's New Homes for Chicago program makes subsidies available to moderate-income families to lower the purchase price for a development of newly built single-family homes in West Garfield Park.
- The city Department of Planning has included West Garfield Park in its Empowerment Zone and Enterprise Community, translating into valuable tax credits and other inducements for private-sector investment. It has also given the neighborhood special tax financing to rebuild the commercial-industrial areas.
- The state's housing finance agency has awarded Bethel New Life an allocation of Low Income Housing Tax Credits to attract equity investments in the development of multifamily housing in the neighborhood.
- The state's Medicaid office has given the organization approval for a pilot program that will pay for health care services for seniors living in the supportive housing apartments that Bethel New Life converted in a part of the Beth-Anne complex.
- HUD has funded the construction of three different apartment buildings with three hundred apartments for seniors in the neighborhood.
- The U.S. Department of Labor has designated Bethel New Life to run a welfare-to-work initiative intended to help 650 long-term welfare recipients become employed. The initiative includes pre-em-

ployment training; training for careers in health, the environment, and child care; and direct job placements.

- Bethel New Life helped with elections of the local public school councils for eight schools.
- The organization created youth enterprise networks in three schools (and is now expanding that to others); this means combining academic and vocational content in the curriculum and helping students open and run school-based businesses (a school supply store in one, a credit union in another). When one high school was slated to be closed, Bethel New Life joined forces with staff, parents, and students in a successful fight to keep it open. Bethel has brought in resources to the school and lobbies for their needs.

Other Partners

- Banks—including First Chicago National Bank (now Bank One), American National Bank, LaSalle National Bank, and FirstStar Bank—have made loans and grants to Bethel New Life and its projects. First Bank of Oak Park is the major lender to the rebirth of the abandoned hospital complex as the Beth-Anne Center.
- The Federal Home Loan Bank has repeatedly awarded Bethel loans and grants from its Affordable Housing Program for various housing developments.
- The John D. and Catherine T. MacArthur

Foundation, based in Chicago, and the Chicago Community Trust are the longest-standing funders of Bethel New Life's innovative projects and organizational costs. The Ford Foundation, Pew Charitable Trusts, and Annie Casey Foundation are recent Bethel grantmakers.

- The Local Initiatives Support Corporation and The Enterprise Foundation have worked with Bethel New Life in putting together the financing for a number of the neighborhood's housing ventures, as well as in the organization's Jobs Network that prepares residents for work and places them in jobs.
- Bethel New Life has been a long-standing member of the National Congress for Community Economic Development, the trade association for CDCs.
- Loyola University's Center for Urban Research and Learning has created a longitudinal study of specific key indicators of neighborhood change for the West Garfield Park neighborhood. From tracking these indicators over the years, Loyola and Bethel New Life can quantify the tangible outcome of the many neighborhood revitalization efforts (these are some of the figures used in this story).
- DePaul University is working with Bethel New Life on employment and training initiatives.
- Volunteers clean up garbage-strewn lots and transform them into community gardens. Block clubs adopt blocks and hold block parties. Neighbors clean up around the schools on the first day of school. Groups of neighbors, along with church congregants, have held prayer vigils at corners where drug dealing has been most rampant.
- Throughout the ups and downs, the people of the Bethel Lutheran Church have stood by the community. They have mortgaged the church building five times over the last eighteen years.
- Twenty churches in the West Side neighborhood are collaborating on the Westside Isaiah Plan to build 250 new homes in the area. Over 100 have been completed.
- Congregants from many different churches donate money to neighborhood projects and give hours of their time as volunteers—for example, as tutors. They mentor and sponsor families, to help them get on their feet. They put money into special bank accounts where the interest generated goes toward BNL projects. Whole congregations sponsor poor families from the neighborhood at Christmas time, giving them the makings of a Christmas dinner and a present for each child.

Is West Garfield Park the perfect place to live now? By no means. Someone stole the sod that had been freshly planted in the yards of the new condo townhomes. Forty percent of the residents still have incomes below the poverty level. Crack cocaine came to the neighborhood about seven years ago and has eaten at families in a terrible way. The commercial district needs much more attention. But West Garfield Park is now a neighborhood with tangible, measurable economic, physical, and social changes, driven by the residents and by an indomitable spirit that rescued it from that bleak picture twenty-five years ago of decay and despair.

Chapter 1 endnotes:

1. Lawrence M. O'Rourke, *Geno: The Life and Mission of Geno Baroni* (Mahway, NY: Paulist Press, 1991); and Philip Shabecoff, "Msgr. Geno Baroni, a Leader in Neighborhood Organizing," *New York Times,* August 29, 1984.

2. All data throughout this chapter describing the growth and current activities of community development organizations come from the latest survey of the field, *Coming of Age, 1999,* by Carol Steinbach (Washington, DC: National Congress for Community Economic Development, 1999).

3. This history is described in *The Old Neighborhood,* by Ray Suarez (New York: The Free Press, 1999).

4. Bruce Katz, director of the Brookings Institution's Center on Urban and Metropolitan Studies in Washington, D.C., makes this point repeatedly in his speeches and presentations.

5. *A New Partnership with Cities,* Paul C. Brophy and Mary Reilly (unpublished. Columbia, MD, 1999).

6. The connection between the health of cities and the health of rural areas is the theme of *Interwoven Destinies: Cities and the Nation,* Henry C. Cisneros, editor (New York: W.W. Norton, 1993).

7. Interview, August 1, 1999, with Professor Cynthia Duncan, rural sociologist, University of New Hampshire, Durham.

8. Interview, August 8, 1998, with Gale Cincotta, president, National People's Action, Chicago.

9. Alice Shabecoff, *Voices from Chicago* (Washington, DC: Community Information Exchange Strategy Alert, Summer 1997), p. 2.

10. The Gray Areas cities were Oakland, New Haven, Boston, Philadelphia, and Washington.

11. Neal Peirce and Carol F. Steinbach, *Corrective Capitalism* (New York: Ford Foundation, 1987), p. 57.

12. Our estimate is derived from the interviews we held as research for this guide with people working in the range of institutions that make up the community development field; the estimate is reinforced by extrapolation from recent surveys, as well as from the number of HUD-entitlement cities (metropolitan cities with populations over 50,000 and urban counties with populations over 200,000) and the average number of jobs related to community development in cities of that size.

13. *CRA Commitments* (Washington, DC: National Community Reinvestment Coalition, 1998).

14. Paul C. Brophy and Mary Reilly, "An Environmental Scan of the Community Development Field" (San Francisco: Low Income Housing Fund, unpublished report, 1999), p. 7.

15. President Clinton's State of the Union address, 1999.

16. The theme of social disarray and the loss of community and one's personal sense of meaning are discussed in *Habits of the Heart,* Robert N. Bellah et al. (Berkeley: University of California Press, 1996).

Chapter 2

A Map of the Community Development Field

Community development, perhaps more than any other field, is rich and complex because it deals with almost every aspect of community life—physical, economic, and social. The description of Bethel New Life, the CDC in Chicago profiled in chapter 1, illustrates the broad agenda of a community-based organization and the many partners who work together to improve opportunities for people and the neighborhood.

This chapter describes the components of the community development field and how the pieces fit together. Figure 2.1 shows the different parts of the field and how they relate to each other. The network of organizations described in this chapter is what makes the progress in communities possible; it has evolved over the past thirty years and is still evolving. It adds up to a system that has become large enough to qualify as a "field" or an "industry." This chapter describes the functions of the components of the field, and chapter 5 describes the kind of jobs that exist in the various parts of the field.

Community-Based Organizations

At the center of the field—and the figure shown below—are neighborhood-based groups. They come in many forms.

They can be as simple as an all-volunteer community association, a residents' association in an apartment complex, or a homeowners' association, working to solve particular problems in the immediate area and advocating for improvements with the responsible authorities. These groups are not usually incorporated and may come and go as the need arises. Some of them evolve into organizations that take on improvement programs for their communities—and in so doing become more sophisticated and often hire paid staff to carry out technical work.

If the neighborhood-based group is involved in real estate development in the neighborhood—building housing or commercial space that creates new local jobs and businesses—it is usually called a community development corporation. Its staff can number from one or two people to a staff in the hundreds, depending on the scale and longevity of the organization. The National Congress of Community Economic Development estimates that there are now 3,500 community-based groups actively engaged in development in their communities; others in the field estimate the number to be as high as 10,000.

Other neighborhood-based groups include:

• Neighborhood Housing Services (NHS) programs, which combine the efforts of bankers, city government, and neighborhood leaders to improve the housing stock in an area.

• Community action agencies, begun during the War on Poverty in the late 1960s, have a general strategy of improvement of housing, social services, and health of the low-income residents in their communities.

• YouthBuild, a newer network of community-based groups, uses federal and other funds to employ young people in communities to rebuild derelict homes in the area, achieving the purpose of training youth in rehabilitation and making affordable homes available to those who need them.

• Faith-based groups are developing housing, starting busi-

Figure 2.1

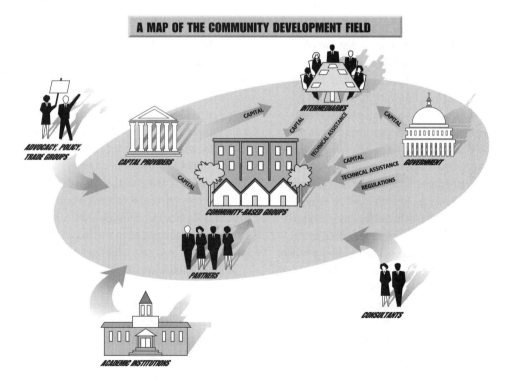

A MAP OF THE COMMUNITY DEVELOPMENT FIELD

nesses, fighting crime, and finding jobs for residents—as well as carrying out the more traditional roles of feeding the hungry and providing shelter to the homeless—in virtually every city. The movement of faith-based groups continues to grow with support from HUD, the Catholic Campaign for Human Development, the Congress of National Black Churches, and other large, mainstream religious denominations.

- Community-based arts programs aim at neighborhood improvement and the engagement of artistically oriented young people through arts and music activities. While far less prevalent than some of the other forms of nonprofits, groups like the Jazz District Corporation in Kansas City, Missouri, are working successfully at neighborhood improvement through an emphasis on arts and culture.

- Urban gardening groups have sprouted up, working with neighborhood residents to become more self-sufficient through the planting of vegetable crops in central city neighborhood vacant land, as a way of teaching skills, having fun, and saving money on food bills. Other aspects of this movement are aimed at neighborhood beautification and at open space preservation.

- Police-community crime watch groups are an emerging form of community group, working with police on crime watch programs, which combine trained community volunteers with police to reduce crime in neighborhoods.

- Land trusts are community-based organizations with the goal of preserving land for open space and/or for low-income housing. These groups work with local government to acquire land and hold it in trust for uses compatible with neighborhood plans and needs.

- Retail (merchant) associations exist in many neighborhood shopping areas and have the goal of maintaining or improving a shopping area so as to make it attractive for customers. Often these groups team up with community leaders working on other improvement programs to make more comprehensive improvements and to develop more space for retail use.

- Job training and placement organizations identify the sectors in their local economies that are likely to grow, then work with local businesses to train and place neighborhood residents in those jobs.

All of these diverse efforts have community improvement as their goal. These are the groups through which community people speak out and act for change.

Community-based groups get started in many ways. A common starting point is that someone or some group of people in a neighborhood responds to a threat to the neighborhood's well-being, or sees an opportunity for improvement—a cleaner park, an additional basketball court, cleaner streets. One concerned neighbor begins to talk to another, then another, and they hold a meeting in someone's kitchen. They may decide to meet with their elected representative to get something done. If that doesn't work, they might just organize a group to attend a City Council meeting or a meeting of the City Planning Commission—whatever is the right step to take to get action.

A group of neighbors might next decide to create a legal structure—some form of governance—so they can make decisions and carry on activities in a sustained way. To have a stronger base to work from, the group might raise some money to hire a community organizer who will get people involved in agreeing on what needs to be done and get them engaged in pushing for changes, creating a sense of movement in the neighborhood, and protecting the neighborhood from threats to its well-being. The larger the base of involvement,

Low Income Housing Tax Credits (LIHTC)

The Low Income Housing Tax Credit (LIHTC) is an important way to attract private-sector capital into affordable housing. First passed by Congress in 1986 as the result of work by The Enterprise Foundation and LISC, the program provides a credit on federal taxes over a ten-year period to companies that invest in qualified low-income housing. To limit the amount of loss in federal taxes collected, the program allocates a maximum amount of credit to each state, based on the state's population. Each state's housing finance agency is responsible for producing an allocation plan for the use of its state's share of the tax credit. Potential users of the available credit—like ABC—apply to the state and compete for credits with other applicants. Once a developer wins an allocation of tax credits, it then tries to find the most attractive source of equity for its deal—using the Enterprise Social Investment Corporation, a subsidiary of the Enterprise Foundation, or the National Equity Fund, a subsidiary of LISC, or through direct investors like Fannie Mae, Bank of America, or other large companies. These companies provide cash to the developer of the qualified housing in exchange for the tax credits they receive. The LIHTC program develops more low-income housing units a year than any other federal program—a total of almost one million since its inception.

the stronger the political clout of the group. Organizers know that numbers equal power and often work to build a power base for a community-based organization.

Many groups expand their agenda for action and find they need to hire a paid staff to carry out their programs and operations. Once an organization moves in that direction, it creates some formal organization and governance structure, usually that of a nonprofit corporation. That organization serves as the vehicle for community members—neighbors—to agree on what goals and strategies are needed to improve the neighborhood or improve conditions for a segment of people living in the community.

Many community groups have moved beyond advocating for their community's improvement and become directly responsible for that improvement. They've taken matters into their own hands. These community groups are variously called community development corporations and community-based development organizations (CBDOs). The terms are interchangeable. These groups build and rehabilitate housing, build space for businesses, help community residents find jobs, and carry out the many other activities described by individual community development professionals throughout this book.

Let's use an example to see how a CDC accomplishes something that may seem relatively simple—the development of a small rental building for low- and moderate-income families in need of affordable housing.

Let's give our CDC a name—A Better Community CDC—usually referred to as ABC. First, ABC's leadership—the resident and business leaders of the neighborhood—must decide that the building of rental apartments is something that will benefit the neighborhood, that it fits into a plan or strategy for the neighborhood. This means that the ABC leadership must be able to reach a consensus on that point. This sometimes takes time, since not everyone always agrees with whether a particular idea is right for the neighborhood. The planning process probably uses data from the U.S. census and from the local planning department and garners information about neighborhood needs and conditions from area residents. The better CDCs engage community leaders, including the heads of local businesses and institutions, to garner broad-based support for their strategies so that a wide group of "stakeholders" develop a sense of ownership of the CDC's development plans. There is also likely to be discussion with the city's planning department to see whether the idea of rental apartments fits within the city's plans for the area.

Second, ABC must determine whether it has the ability to produce the units—that is, does it have the skills (or can it access them), and can it find the necessary financing to produce the apartments? Here it might engage in fundraising to get the dollars needed to hire a staff, or turn to a local or national intermediary, or engage a technically capable consultant for help. Intermediaries (described below) provide technical assistance and some funding to help CDCs produce quality housing and other neighborhood improvements. If ABC does not want to hire a staff, it might engage a consultant to work on the project, thereby getting the job done without building a large staff. One way or the other, ABC has to have people who have technical skills—knowledge of the development process, the ability to talk with architects and contractors about prices, knowledge about local approval processes for land sales, zoning, and building permits. In addition, ABC must know how to access the different kinds of funds needed to finance the housing, so they cultivate a relationship with a commercial bank, or a community development financial institution, or other source of investment dollars.

Third, ABC must decide whether it is trying only to build the housing or also intends to help the people who will live in the apartments with the community aspects of living in the building and, perhaps, with other social issues. For example, ABC might decide that the apartments it is building are intended for short-term renters, people who want to be homeowners but do not have the financial resources or good credit history to do so. ABC might provide credit counseling and other homeownership counseling services to the residents in addition to managing the project. Or, to use other examples, ABC might be housing people with special needs—the physically or mentally disabled, single mothers with children, or the elderly. In those cases, ABC would provide special services to the people being housed.

Developing housing from start to finish—including the planning, acquisition of site, design, financing, construction, occupancy—usually takes a year or more, with a variety of skills called on at different times during the project. But this simple example doesn't capture the dynamism of all that's going on at ABC. The group usually has numerous projects in play simultaneously—some in planning, some in construction, others in occupancy—and that's not all. During the entire process, there is the day-to-day work of raising funds for the operations of ABC; continuing to build and hold a consensus in the neighborhood by getting people out to community meetings; handling emergency conditions that might arise in the community; working with partners such as local government, lenders, funders, and others. ABC is a very dynamic place.

The funding for a CDC's operations is seldom certain, nor does it ever seem to be enough. The challenges of small business are prevalent here—finding the people, negotiating with suppliers of goods and services, staying on track, being creative, maintaining a sustained energy level, and much more.

CDCs like ABC are one kind of group at the core of the community development field. These groups and others—NHS, YouthBuild, community action programs, and others—are accomplishing a great deal in their communities, but they need technical and financial help to accomplish their missions.

National and Local Intermediaries

The technical and financial help needed by CDCs often comes from major national institutions called intermediaries. The three biggest groups in the field are the Neighborhood Reinvestment Corporation, the Local Initiatives Support Corporation (LISC), and The Enterprise Foundation. These organizations are called intermediaries because they are bridges between the national level and the grassroots. They locate sources of funds (capital) at the national level, and provide those funds to grassroots groups that need the capital and the technical help in accomplishing their goals. There are other intermediaries at the local and state levels that also play this role; they also offer technical assistance to help neighborhood-based groups expand their capacity. The national intermediaries grew rapidly during the 1990s. At the end of the decade, LISC had 500 employees in forty-five locations, Enterprise had 375 employees in fourteen locations, and Neighborhood Reinvestment had 260 people in nine offices. All three groups work with networks of community-based nonprofits that are producing large volumes of housing and some commercial space and are providing services to people in communities undergoing improvement.

The national intermediaries play a crucial role in building the field of community development, provide leadership to see to it that the field thrives, and, on a day-to-day basis, help individual groups succeed at their goals and bring partners into the work of community development.

First, the intermediaries raise funds from philanthropies, corporations, and governments for community development. The funds raised cover the costs of technical assistance and are used to make loans to CDC projects and for grants to CDCs for their operations. LISC, the largest of these intermediaries, had provided $665 million in loans and grants from its founding in 1980 through 1998. LISC and Enterprise have also succeeded in raising billions of dollars in the form of equity for housing investments from corporations in the United States.

Second, they are advocates for community development, continuously promoting the successes of the field and reminding key outsiders that the problems are deep and need the commitment of resources. They lobby on behalf of the field, trying to get legislation and appropriations for the field from government at all levels. The national groups have public policy staffs that work on shaping up policies, programs, laws, and regulations that benefit the community development field.

Third, they provide technical help and training to community-based groups. To help ABC produce its housing project, an intermediary might assign a program person with experience in the development of rental housing to coach the group through the development process. This program officer might help produce the first levels of financial analysis for the project and might introduce ABC to sources of capital that it needs. The intermediary might lend working capital or construction funds to the group for the project. At a broader level, the program officer might help ABC with its internal systems of operations, with strategic planning, with personnel recruitment—whatever it takes to help ABC accomplish its mission. The large national intermediaries have staff located in target cities to offer technical help and training more easily.

Fourth, the intermediaries are continuously trying to advance the overall technical and programmatic expertise in the field, seeing to it that successes and innovations are disseminated as rapidly as possible, via newsletters, the Internet, and reports. The national groups sponsor conferences and training programs that build the capacity of ABC and hundreds of other CDCs to carry out their missions. This means that there is a research and development dimension to the intermediaries—they are inventing, promulgating, and extending the field, moving beyond housing, often the first starting point in neighborhood improvement, into commercial development, job placement, day care, crime prevention, and much more.

In addition to these community development intermediaries, there are national groups that play the role of "organizing intermediaries." These groups include ACORN (the Association of Community Organizations for Reform Now), IAF (the Industrial Areas Foundation), and COI (the Consensus Organizing Institute), all of which provide training, technical assistance, staffing, and funds to local affiliates who are involved in organizing community groups for action. They are typically less concerned with real estate development—although IAF's affiliates have been successful partners in housing development—and more with issues of power and economic justice. These groups push for living wages, fair housing, and environmental policies that are of benefit to all classes of people.

Still other national groups provide critical help to the field. Among the many (see appendix D):

- The National Congress for Community Economic Development is the trade association for CDCs in the field. Almost every state has an association of CDCs providing support in some way to CDC members.
- The Development Training Institute specializes in providing in-depth training courses to build the skills of community development professionals and provides technical assistance to community-based groups and other partners in the field.
- The National Council of La Raza specializes in providing help to a large network of Hispanic groups throughout the nation, groups that are building housing and are also ac-

tively engaged in economic development programs.

- Habitat for Humanity is a large, faith-based international organization that uses networks of volunteers organized in local chapters to build and rehabilitate homes for low-income people. Churches, corporations and businesses, schools, social organizations, and others organized in almost 1,500 affiliates have built 71,000 homes in the United States.

Other large nonprofit, nationally focused groups play the role of partners with local CDCs and housing providers. Mercy Housing, a large nonprofit based in Denver and operated by the Sisters of Mercy, has partnered with local organizations to develop 5,000 housing units. The Community Builders, a national nonprofit headquartered in Boston, has codeveloped 14,000 units and is involved with commercial development in some situations.

In addition to these national intermediaries and other groups, there are numerous local and regional intermediaries that play a similar role to the national intermediaries, albeit on a more limited basis. Groups such as Neighborhood Progress, Inc., in Cleveland, the Pittsburgh Partnership for Neighborhood Development, and the Indianapolis Neighborhood Housing Partnership provide funding and technical help, the funding coming from local foundations and corporations. Greater Miami Neighborhoods and the Greater Minneapolis Metropolitan Housing Corporation are both examples of regional nonprofit housing developers that partner with smaller groups.

Federal Government

Government is another of the key participants in the community development field. Without government support for community development activities, far less could be accom-

Career Story: Rebecca Adamson

Founder and Executive Director, First Nations Development Institute, Fredericksburg, Virginia

"So much of Native American life is community life, but it's been forced into a system that emphasizes individuals at the expense of community, so that our community fabric was unraveling. I had been working in the movement in the 1970s to help gain control of the Indian educational system. I was arrested three times during protests and demonstrations until finally the Education Reform Act was passed. It was a real turning point in Indian affairs. This movement sought to implement a new law, the Indian Education Self-Determination Act passed in 1975, granting Native American tribes control over their children's schools. Before that, native children had been sent away from their families and placed in government-run boarding schools where they were forced to abandon their language and customs.

"I thought at that time education was the key to how we as a people could move forward. But I realized that to achieve sovereignty and exert control over our lives, we as native people had to understand how to embody our culture in our economic development. I began to understand that, in all of the ways you can get engaged in community work, from schools to health care, economics plays the pivotal role.

"I came to realize that it was the force of the economy that was destroying the native way of life. We had to move into the twenty-first century controlling our economics and retaining the values of our native culture. It is virtually impossible to achieve economic success according to standard conventions and yet to remain stewards of what we care about. In conventional economics, a tree has no value unless it is cut down. To native peoples, a tree has intrinsic value, for today and for future

generations. I needed to understand how to translate and be that bridge between those two economic systems.

"I was born in Akron, Ohio. My mother was a Cherokee, and my father was of Swedish descent. I spent summers as a child in the Smoky Mountains of North Carolina with my maternal grandparents, who introduced me to Indian ways. After I dropped out of the University of Akron, I moved to the Lakota Indian reservation of Pine Ridge, where the massacre of Wounded Knee took place, and began organizing for change.

"But, as we came to realize, decades of stifling regulation by the Bureau of Indian Affairs and other federal agencies had denied tribes the right to control their own resources. We ended up only doing what the federal government said we could do. Whenever large-scale economic development efforts were tried, such as motels and industrial parks, they proved inappropriate and failed.

"So I founded First Nations in 1980. I cashed in my last unemployment check in 1980 to pay for a fundraising trip to New York City. It was rough going at first. People had rather dim expectations of Native American business acumen. But the Ford Foundation gave us a small grant, and that's how we started.

"In 1992, I earned a master's degree in economic development from New Hampshire College, and now I teach a course there on indigenous economics.

"First Nations' mission is to help tribes build sound, sustainable reservation communities by providing money and technical expertise to local enterprises. Each year since then, First Nations has worked with about 1,500 tribal groups and individual Native American entrepreneurs. First Nations also created the first micro-loan fund in the United States, as well as a national movement for reservation land reform.

"First Nations Development Institute has been the vehicle through which we have been able to have communities demonstrate this theory and this better practice."

plished, because often some level of subsidy is needed to make a project financially feasible. The roles that government plays vary considerably at the federal, state, and local levels.

The key federal agency involved in community development is the U.S. Department of Housing and Urban Development. HUD has a broad agenda. It is:

- a regulator of programs and activities in the real estate and housing area, including fair housing;
- a provider of federal insurance on single- and multi-family mortgages;
- a source for rent subsidies to low-income people who rent housing in the private market;
- a funder to cities and nonprofit groups for housing and community development activities;
- a source of funds to the thousands of public housing authorities in the nation that provide low-cost housing to very low-income people.

In the community development realm of its activities, HUD allocates funds to local governments through its Community Development Block Grant Program (CDBG) and its HOME program; provides technical assistance funds to nonprofit groups; and works to improve neighborhoods through programs to local governments that are awarded competitively, such as its Empowerment Zone and Enterprise Communities programs, which seek a more comprehensive approach to community improvement. A newer program, the HOPE VI public housing improvement program, provides funding to public housing authorities for substantial improvement of distressed public housing projects.

HUD helps CDCs and others in the community development field by providing:

- direct funding to CDCs and other groups building housing for special needs groups like the homeless and HIV/AIDS victims (If ABC decides that its ten-unit apartment building should be for special needs housing, it would apply directly to HUD for funds.);
- funds to local and state governments that are used for housing and community development programs and projects (ABC's local government might award it some of the funds it gets from HUD to finance ABC's apartment project.);
- funds to local and national intermediaries to provide technical assistance to CDCs and other groups;
- mortgage insurance to encourage lending institutions to provide mortgages to some rental projects, and, for individual home buyers, the mortgage insurance needed for many families to buy their first home;
- rental subsidies to thousands of low-income renters to help them afford to rent housing on the private market— perhaps some of the apartments built by ABC.

To a lesser extent, other federal agencies also have community development roles:

- The Treasury Department's Community Development Financial Institutions Fund provides capital and technical assistance to community development financial institutions—specialized capital providers.
- The Office of Community Service, a part of the Department of Heath and Human Services, administers the Community Service Block Grant Program, which provides direct support to community development groups. In addition, Health and Human Services operates a number of programs to test new approaches to reducing welfare dependency.
- The Federal Home Loan Bank System provides capital to thousands of banks that provide financing to community-

based groups and, of special importance, operates an Affordable Housing Program that makes below-market interest rate loans to groups such as ABC through participating banks.

- The Comptroller of the Currency regulates many banks and enforces Community Reinvestment Act regulations, which require banks to actively seek to make loans to low- and moderate-income households and in communities that are predominately low income.
- The Economic Development Administration (EDA), a division of the Department of Commerce, provides grants for infrastructure development, local capacity building, and business development to help communities alleviate conditions of substantial and persistent unemployment and underemployment in economically distressed areas and regions.
- Other federal agencies, listed in appendix G, also provide help to the community development field.

State and Local Government

State Government

At the state level, there are four primary kinds of agencies involved in community development:

- Most states have a department of community affairs that provides assistance to local governments and, sometimes, to community-based groups.
- All states have a housing finance agency that provides financing assistance to private for-profit and nonprofit organizations to produce multifamily and single-family housing.
- Most states have an economic development agency that is concerned with stimulating economic development in communities and the development of the labor force.

- All states have a welfare or social service department that provides welfare payments and, in recent years, programs to move welfare recipients from welfare to work.

The departments of community affairs generally provide funding and technical assistance to local governments and are concerned with local government problems, including neighborhood improvement. Departments vary widely in their size and interest in community development and, depending on the state, may be helpful to groups like ABC. A number of states offer special tax credits to companies that make charitable contributions to community-based groups.

The housing finance agencies (HFAs) at the state level are a major source of capital for the community development corporations that are working to build housing. HFAs provide three kinds of funds: below-market-rate debt for single- and multi-family housing, tax credits that are used to attract equity to multi-family housing, and grant funds for all kinds of housing for low-income people. ABC might apply to the state housing finance agency to receive an allocation of tax credits for its apartments. If it were awarded tax credits, then it would go to a national source for an equity investment in its project. In some states, ABC might also be eligible for a grant from the housing finance agency to help subsidize the rents in its apartments to a level affordable to low-income people.

State economic development agencies fund projects that create jobs in the community and offer loans and grants as incentives to businesses to expand or locate in communities within the state. These agencies may also work with other departments in the state to expand employment opportunities for people on welfare or other low-income workers. They can often be of assistance to community development groups that are working to bring companies into their neighborhoods by providing incentives to those companies.

Some states also use tax credits as an approach to community development. The Pennsylvania Neighborhood Assistance Program, for example, makes state tax credits available to community groups for qualified grants from private business, thereby increasing the financial benefits to companies that make investments. Missouri has a number of credits that make it attractive for business to invest in community economic development programs there.

Social service agencies have been an important part of state government since the passage of the welfare laws in the New Deal of the 1930s. These agencies provide cash support and family and children's services to families in need. During the 1990s, these agencies went through an important transformation as they focused on moving people from welfare to work. In many instances, these agencies can be partners with community-based groups that are working to improve the lives of people in need in their communities.

Local Government

In addition to their customary role of providing services such as police protection and street cleaning and garbage removal to neighborhoods, local governments act as regulators of land uses and zoning, provide funds to community groups for housing and economic development activities, provide technical assistance, and set an overall improvement strategy for the communities within a town or county.

City and County Planning Departments

These departments are generally responsible for the physical planning of a local jurisdiction, including zoning and land-use planning. Many local departments of city planning also undertake a level of neighborhood planning that involves considerable interaction with and support for neighborhood-based groups. Jobs at planning departments are most often filled with professionals with degrees in city planning or some related field, and the work involves everything from data analysis to planning meetings at the community level. It's very likely that before ABC decided on its apartment building, ABC staff would have met with its local planning department to see that ABC's plan was consistent with local zoning and other requirements.

Housing, Community Development, and Economic Development Departments

These organizations carry out local housing and community development programs, often funded with state and federal funds as mentioned above, that seek to achieve neighborhood improvement by producing better housing, more jobs, and better services in communities. They are sometimes a part of city government but can also be part of a redevelopment authority set up by a locality to carry out community improvement programs. ABC is likely to apply to the community development department in its town to get funds for the apartments it wants to build. When it does so, the staff at the community development department will compare its proposal to other requests it has for funds and will underwrite the project; that is, they will determine whether ABC has the financial aspects of the apartments well figured out. Will the people who rent the apartments in the neighborhood be able to pay enough rent to cover ABC's cost to maintain the building properly over a long time period? Will it be a stable building for the community? The staff at this department will work closely with others who may be asked to fund the project–the state's housing finance agency, a bank, and a foundation—to make sure that everyone who is being asked to participate in funding ABC's project is, in fact,

Career Story: Teresa Cordova

Associate Professor, Community and Regional Planning Program, University of New Mexico, Albuquerque

"I was born in southern Colorado. My forefathers and -mothers moved from northern New Mexico. My father, uncles, and brother were coal miners. I went to undergraduate school on a scholarship that paid for tuition, and I worked to earn money for my personal finances. I've never not worked. I got my Ph.D. in sociology from the University of California at Berkeley. Then I took a job as assistant professor in Latin American Studies at the University of Illinois at Chicago. I moved to the University of New Mexico because it was a way to get closer to home.

"I always liked school and learning, and I understood that part of why I was able to pursue my academic opportunities was because of the civil rights movement in general, and the Chicano movement in particular. I understood that a lot of people took risks and struggled in order that people like me could attend college and could work in the academic world. So I take very seriously my position in the university as a position from which I can serve my communities. I have a sense of social responsibility. My education is not my own—it belongs to my community.

When I earned tenure (in 1996), it was a community celebration. My tenure showed that I could be successful within the aca-

demic world without having to give away my connection to the community.

"I see the university as a base of operation. I think about how I can apply the skills and knowledge I've developed to deal with concrete issues in the community. I've always been driven, as far back as I can remember, by a desire to understand why there is social injustice, why there is poverty.

"It's a challenge to find an appropriate role as an academic in a community development effort, and at the same time it's a very exciting place from which to operate. One of the problems is that I have to overcome the negative stereotypes that people have of academics. In a very important sense, I've never left the community.

"One of the nice things about the field of planning is that your practice is part of your academic work. I bring my community work back to the classroom, including in a course I teach called the Foundations of Community Development. I can't measure how much time I give to academic work and how much to the community because my community and my academic work are so integrated. Also a lot of the students I work with are from those neighborhoods.

"I play different community roles. Sometimes I testify on behalf of a community issue. And I help communities with their planning initiatives. I also serve on the board of directors of two CDCs, including the Rio Grande CDC. I'm also now building a new organization, the Resource Center for Raza Planning, which applies planning processes and techniques for traditional communities here in New Mexico."

ready to do so.

Housing and community development agencies do more than fund projects. They work with their planning departments and others to plan for the future of communities, based on research and data analysis, so that the strategies for improving neighborhoods and meeting the needs of people who cannot afford to buy or rent housing on their own are based on the specific needs identified. These plans may take the form of broad community discussions (sometimes undertaken by housing and community development agencies, sometimes by the local planning agency), or they may be generated by the local government and put out for public comment.

These departments are often involved in the broad community development picture and the details of getting deals done. The focus of activity is not only housing, but might also include economic development, business relocation, commercial (retail) area improvement, and more.

Local Public Housing Authorities (PHAs)

PHAs are the public agencies responsible for developing, owning, and managing public housing—the housing aimed at the lowest-income populations. The PHAs manage housing that they own, and they are responsible for taking funds from HUD and providing vouchers to low-income people who are seeking to rent apartments from private landlords, including groups like ABC. The work undertaken by PHAs can have community development as its purpose, and those that view their role that way actively work with low-income residents to improve living conditions in public housing communities. Many housing authorities work closely with residents who live in the properties they own and manage and are involved in some of the same kinds of efforts that CDCs pursue to create opportunity for upward mobility.

Capital Providers

For community development to succeed, it needs capital to finance the housing, economic development efforts, day care centers, and other real estate and needed programs. Capital comes in the form of debt (loaned money), equity (an investment that is less secure than debt and expects to earn a return), and charitable grants (gifts). Governments and intermediaries provide some forms of capital, but there are other institutions whose primary function is community development finance.

Commercial Banks

Most banks in the United States are required to invest in distressed areas and make loans to low- and moderate-income people through a federal law called the Community Reinvestment Act (CRA). This has led many banks to set up community development departments or subsidiaries. Many

Bank of America

The largest bank in the country, Bank of America, also has the largest community development group and program. When formed by the merger of NationsBank and Bank of America in 1998, Bank of America made an astonishing commitment of $350 billion of community development lending over ten years. The bank provides debt, equity, grants, and technical help to a wide variety of community development organizations. This commitment is not charity. Bank of America—as do some other banks—sees an opportunity to bring capital to the community development field and make a profit in doing so.

Community Story:

Albuquerque: Neighborhood-Based Planning and Participation

Teresa Cordova, professor of planning for the University of New Mexico, worked on a project that is a fine example of how academics can assist communities without acting like the know-it-alls. In this project, the planner-professors and the community groups "were teaching and learning interactively," as Professor Cordova explained.

Professor Cordova and two other professors from the Planning Department of the University of New Mexico were invited to help in this project by fifteen neighborhood organizations. The organizations, all of them active in the older, less affluent neighborhoods of downtown Albuquerque, had formed an alliance to ensure that the city would allocate (as it was supposed to) a special stream of HUD funds for use in their neighborhoods.

In the past the neighborhoods had fought with each other over narrow self-interests, as shrinking funding pitted one against the other. But this time they worked together over the special problem and opportunity that these HUD funds presented them.

The three academics helped the city and the neighborhood groups hold a well-attended neighborhood summit to discuss the use of the HUD funds. After that, the planner-professors, serving as a technical assistance team with the alliance as a client, went through a "facilitated planning process" with the alliance members. "We worked together with mutuality and respect." The team designed the planning process, structured a series of workshops, helped the alliance members frame the agenda for each,

and facilitated the meetings. Through these steps, the process made it possible for the neighborhood groups to form a common community identity. Then they identified a general vision and organized a strategic plan based on their common interests and specific steps to fulfill those interests.

The outcome was successful: The alliance's strategic plan influenced the way the city structured the allocation of the HUD funds. So far, the city has made several grant awards to various neighborhood groups to carry out housing and economic development projects—from home buyer acquisition and rehabilitation to small business lending—and further allocations are expected.

banks in the nation, once unlikely to make loans to CDCs, now make them routinely, provided the risk is reasonable. Some banks, like the largest in the nation, Bank of America, have found that community development lending can be profitable.

If the apartments that ABC is building are likely to generate enough rent that after operating costs are paid there will be enough left over to repay a loan, then ABC might succeed at getting a mortgage from a commercial bank. The bank underwrites the loan; that is, the lending officers examine the risk of making the loan, looking at the probabilities that the

loan will be repaid, and the quality of the collateral behind the loan.

Some commercial banks provide capital to intermediaries, which, in turn, make it available to CDCs for project development. Some of the larger commercial banks also provide equity and charitable contributions to intermediaries and CDCs.

Community Development Financial Institutions

These institutions are specialized financial institutions, sometimes organized as banks and sometimes as nonprofit entities, with the mission of providing specific kinds of capital for community development. Some CDFIs receive capital from the Treasury Department's Community Development Financial Institutions Fund. Their role is to provide capital that might not meet the investment criteria of more general purpose commercial banks. Generally speaking, the kinds of CDFIs include:

- *Community development banks* are regulated financial institutions that commit themselves to lending in a particular neighborhood or to a particular group of customers. The longest-standing community development bank is South Shore Bank in Chicago, the first bank deliberately organized to use the tools of a private, for-profit commercial bank to stimulate market forces and improve a distressed neighborhood. It has become the model for other community development banks across the country and the inspiration for the creation of the federal Community Development Financial Institutions Fund. As a bank, it is owned by private investors and is government chartered and regulated. The bank has always shown a small profit, and its holding company now has twenty-three nonbank affiliates and commercial banks in Chicago, Cleveland, Detroit, and Washington State.

- *Community loan funds* are unregulated nonprofit organizations (or programs) that provide loans to small businesses and nonprofit community development groups with neighborhood improvement and job development in mind. The loans from these groups are typically made at an interest rate lower than what is available from a commercial bank or a community development bank—and the loans are usually smaller. These groups raise capital from philanthropies and from individuals who are willing to earn a less-than-market interest rate on their deposits.

- *Community development credit unions* are special credit unions that provide small loans to community members. Credit unions are a form of federally regulated mutual savings institutions, in which funds deposited by members are used to make loans to other members. Federal regulations say that credit union members must have some form of common affiliation, and, in the case of community development credit unions, that is living in a neighborhood together.

- *Micro-enterprise loan funds* provide very small loans to small (micro) businesses, sometimes as start-up loans. These funds have been particularly successful as part of community development strategies to help those not in the

Ford Foundation: A Community Development Leader

The Ford Foundation has supported community development for more than thirty years. In 1983, the Foundation began the Community Development Partnership Strategy, an initiative that pools resources from the private sector, foundations, government agencies, and institutions to revitalize neighborhoods. The Foundation has invested over $30 million in twenty public-private partnerships. These partnerships have helped more than 150 CDCs to construct or rehabilitate more than 17,000 affordable housing units and nearly 4 million square feet of commercial, industrial, and commercial space, leveraging more than $400 million in total investment.

The "Big Three" Intermediaries

Of the many organizations that have a mission to help community development groups, three have emerged as the largest and most successful at building a network of community development groups.

Founded in 1978, the Neighborhood Reinvestment Corporation is the oldest of the three. It began in 1974 under the name Urban Reinvestment Task Force, sponsored by the Federal Home Loan Bank System, to replicate throughout the country the first Neighborhood Housing Services program, organized in Pittsburgh in 1960. The NHS is a neighborhood-initiated program that combines commitments from the neighborhood group with those from lending institutions and city government to stimulate reinvestment in neighborhoods that are experiencing decline. The Neighborhood Reinvestment Corporation is a congressionally chartered corporation that has developed readily transferable training approaches that led to the creation of more than one hundred NHS programs around the county, each of which finances rehabilitation of homes and businesses. The corporation has brought this network of neighborhood groups together under the label NeighborWorks©. In 1999, the NeighborWorks organizations and their lending partners extended $1 billion in affordable credit. Today, under the directorship of Ellen Lazar, the Neighborhood Reinvestment Corporation is helping over 825 groups in two hundred cities.

The Local Initiatives Support Corporation was the brainchild of Mike Sviridoff, one of the key figures in the creation of the community development field. Mike had been at the Ford Foundation and, while there, saw the need for a national nonprofit organization with the single-minded mission of helping community development corporations thrive. Founded in 1979, LISC set out to build local alliances of business leaders, bankers, foundation officials, local governments, and others to build neighborhood groups and, through grants, loans, and technical help, assist these groups in accomplishing their missions of improving their communities. Now headed by Michael Rubinger, a community development professional who has worked at virtually every level in the field, LISC is involved in helping thousands of groups in forty-one states and metropolitan areas and fifty-nine rural communities. LISC has raised $3 billion in grants, equity, and loans for CDCs. The CDCs have used those funds to leverage over $4 billion for their community development purposes.

The Enterprise Foundation, founded in 1982 by the visionary real estate developer Jim Rouse and his wife, Patty, has as its mission, "To see to it that all low-income people in the United States have the opportunity for fit and affordable housing and to move up and out of poverty into the mainstream of American life." Enterprise has a network of more than 1,100 nonprofit community organizations in four hundred locations—and has helped create 100,000 new and renovated homes by raising and committing more than $2.7 billion in loans, grants, and equity investments. Enterprise's active network of job-placement nonprofits has placed almost thirty thousand people in permanent full-time employment. Bart Harvey, Enterprise's CEO, came into the community development field after a successful career in Wall Street.

These three organizations are led by executives with very different backgrounds. Their philosophies are similar, however—they build local capacity and partnerships and work to create innovation in the field, as well as pushing governmental bodies to adopt public policies and legislation that will help the community development field thrive.

workforce find a particular niche businesses that can support the owner and the owner's family. Often these businesses are home based: for example, day care, catering on a small scale, and household services. Micro loan funds usually get their capital from foundations, state and local governmental agencies, and, in limited instances, the federal government.

- *Equity investors* are nonprofit and for-profit companies that provide equity for housing and commercial development to the community development field. In the low-income housing field these companies use the availability of tax credits to attract investors into low-income housing. (See sidebar on page 28.) Equity investors will provide capital for community commercial development without tax credits if the equity investor judges that the investment promises to be profitable.

- *Philanthropies* are contributors to worthy causes. America is a nation of givers and one of America's trademarks—in part spurred on by the tax code—allows deduction for qualified charitable giving from federal tax payments. Charitable giving is very important for the community development field because—almost by definition—the work going on in community development cannot support itself. CDCs are almost always in need of income from charitable sources to pay for their staffs and other operating costs and sometimes to make the projects they are undertaking feasible.

National and locally based foundations continue to support community development activities. In fact, foundations that are part of the Council on Foundations have formed a separate group, the Neighborhood Funders Group, that meets periodically to compare notes and to learn about the community development field. This group's membership almost doubled in the three-year period 1996–99, and has 180 member foundations.

In virtually every city where community development is active, corporations and business have seen the benefit of the activities and have made contributions to community development groups like ABC. Individuals also provide support to community-based groups through fundraising drives and other related activities.

- *Fannie Mae and Freddie Mac* are two specialized institutions that play important capital roles in the community development field. Fannie Mae and Freddie Mac are publicly traded corporations established by the federal government to buy mortgages from banks and other mortgage issuers, thus replenishing the supply of money available to lenders. Because they are regulated by the federal government, and because they are held accountable for providing capital to low- and moderate-income home buyers and to central cities, both institutions are aggressive partners in the community development field.

Fannie Mae, the larger and more aggressive of the two, fulfilled a commitment to invest $1 trillion in community development over the period 1993–2000 and, in early 2000, committed $2 trillion over the period 2000–2010. Fannie Mae is prominent in virtually every part of the country in striving to attain that goal and has partnership offices in thirty-five cities. Fannie Mae also created the Fannie Mae Foundation, which is a major donor to community development activities.

Freddie Mac's commitment to community development has taken the form of national partnerships with the NAACP, the Enterprise Foundation, the Neighborhood Reinvestment Corporation, and others. The organization also works directly with CDCs in some locations.

Community Development Partners

The popularity of community development and the challenge of its goals have led to the establishment of many forms of partnerships with other sectors and entities, all operating within the community development system. Two of these are worth particular mention: business and service providers.

Business

Private business works with community developers in at least two ways. The first and most direct is that it is quite common for CDCs to partner with private real estate developers to bring about physical improvements in communities. When it wants to work on a larger project, ABC might find it advantageous to join forces with a local or national development company that brings skills to bear that ABC doesn't have. The development partners bring expertise, their own sources of capital, and the ability to attract capital because of their know-how and reputation. McCormack-Baron, for example, a St. Louis–based company, has developed approximately 8,500 housing units in twenty-two cities since its formation in 1979. In many instances, such developers work in partnership with community-based groups. In others, they develop consistent with a plan formulated in conjunction with neighborhood-based groups.

Some retail businesses are active partners with neighborhood groups. The Pathmark Supermarket company, for example, has entered into joint venture agreements with CDCs to develop and operate supermarkets in neighborhoods lacking supermarkets. Pathmark and the Abyssinian Development Corporation collaborated to produce a new supermarket in Harlem, creating over two hundred jobs for neighborhood residents and improving the quality of life for Harlem residents generally.

In these partnership situations, the private partner helps make a complex deal happen and happen faster than it would otherwise; it thus achieves a mission of serving the kinds of people it is trying to serve and usually earns a fee that supports additional developments.

But business can be involved in quite another way. Businesses also work with CDCs and others by donating the time of their employees. The form of that donated help can be anything from helping a CDC set up a new computer system, to having employees participate in tutoring programs or house-painting weekends. Habitat for Humanity is particularly skilled in this arena. Habitat now estimates that it has thirty-eight national business partners providing construction materials, tools, and appliances and thousands of businesses

James J. Johnson Fellows Program

Each year the Fannie Mae Foundation selects up to six seasoned professionals in the affordable housing and community field as James A. Johnson Community Fellows, honoring the leadership and distinguished service of the Foundation's past chairman. The program gives the Fellows a grant and education and travel expenses to pursue personal and professional development goals that will encourage them to contribute further to the field. Fellows design and pursue a personal and professional development plan that can include research, travel, study, and other activities to achieve their goals.

working with its local affiliates.

Major corporations like TastyKake in Philadelphia and Brooklyn Union Gas have, in effect, adopted the community surrounding their facilities, leading the way in improving the area. Companies such as these see a genuine benefit to improving the area from which many of their employees are drawn.

Service Providers

Most community development entities are concerned with the physical, economic, and social improvement of their communities, but it is impossible for any one organization to house all of the skills and resources to do it alone. It is common for groups like ABC to reach out to other groups whose mission complements theirs. ABC might form an alliance with a service provider to senior citizens if its apartments house elderly people—or with a local mental health provider if the building houses people with needs for services of that kind, or with a job placement and training program if the apartment houses welfare recipients looking for jobs.

Service providers are involved in a wide variety of specialties from programs like Meals on Wheels that provide meals to shut-in elderly people, to day care services, to a wide variety of self-improvement and support groups for recovering alcoholics and drug addicts, to job placement and training programs.

Academic Institutions

Large institutions—colleges, universities, and medical centers—are playing a major partnering role in many locations. Spurred in part by the need to keep their immediate environments healthy, many of these large institutions have become very active in community development. They invest in

real estate, provide funds for housing, make expertise available to community groups, and participate in planning as an active partner. For example, the University of Pennsylvania is involved in five initiatives in its West Philadelphia area: safe, clean, and attractive streets and neighborhoods; excellent school options; high-quality and diverse housing choices; reinvigorated retail options; and increased job opportunities through economic development. Other institutions, like Yale in New Haven, the University of Chicago, and Johns Hopkins Medicine in Baltimore, are engaged in similar broad activities.

Some of these activities are supported by a specialized national intermediary based in New York City, SEEDCO, which provides technical assistance and funding to partnerships between large institutions and neighborhood groups for community development.

As is clear from many of the profiles of community developers included in this book, many community development professionals have come into the field without specific training in community development and have learned on the job. While that is still possible, the learning-as-you-go phenomenon was more prevalent in the earlier stages of the community development field than it is today. More typically, community development organizations are looking for well-trained people, as the job descriptions in appendix A indicate.

Today, many institutions of higher learning have become training grounds for community development workers. The range of disciplines through which you can study community development underscores the extraordinary breadth of the field: planning, social work, urban studies, public policy, sociology, economics, business, nonprofit management, and law.

A growing number of universities complement their academic curricula with hands-on training and professional development programs in community development. In many cases, they offer a certificate at the end of the program. (See appendix B for a list of academic programs.)

Academic institutions are also the places where much of the research on the field is done. Academic researchers inform the field about what is being learned, trends, and their assessment of particular program approaches. These institutions are a source of people and knowledge for the field.

Policy and Advocacy and Trade Organizations

Community development grew out of the actions and vision of people who decided to change the difficult conditions of all too many communities. But grassroots action alone cannot possibly bring about enough change in a society and marketplace as complex as ours to solve all of the problems. Just as policies such as segregation, the placement of highways, and redlining by banks brought about some of the problems faced by communities, so too can changes in policies and laws remedy the root causes of some of those problems.

Policy and advocacy organizations push for changes that can help the entire field overcome some of the systemic handicaps faced by grassroots groups. For example, the National Community Reinvestment Coalition fights bank redlining, a practice in which banks choose not to lend to anyone in an entire area because they believe that area is too risky. NCRC has succeeded in getting the CRA law passed, requiring banks to invest in all areas.

Trade organizations, national, statewide, and local, are policy and advocacy groups for community development organizations. They include groups like the New Jersey Nonprofit Affordable Housing Network, the Massachusetts Association of Community Development Corporations, the California Association

of Community Development Corporations, and others—all of which are working for policies that support community development from a grassroots perspective.

Consultants

The technical nature of much of the work that must be accomplished by those in the community development field requires very specialized skills. This need has given rise to the presence of a large number of private consultants who work with the range of players both to help CDCs and others get their goals accomplished, and to provide advice and counsel to many others in the field—the intermediaries, suppliers of capital, and government. Mainstream national consulting organizations such as ICF, McKinsey and Co., and Abt Associates have staff specializing in community development. Other companies are singularly involved in community development. These include Shorebank Advisory Services in Chicago and Mt. Auburn Associates, based in the Boston area, both of which specialize in community economic development—helping foundations develop initiatives to improve job opportunities for low-income people, helping local governments analyze their regional economy, and creating local capacity to improve neighborhoods. The consulting world in community development is also full of capable individuals who provide specialized consulting.

The community development field is complex and multidimensional. It has evolved in such a way that it now has clear categories of players, all using their specialized roles to advance the goals of community development. As figure 2.1 indicates, the community development field is highly interactive, with much movement among the various sectors. From a career development perspective, each of these sectors has a set of jobs, and movement among the sectors is quite common. Chapter 5 describes the jobs, and chapter 7 the movement within the field.

Chapter 3

Choosing Community Development as a Career

"This job, this field, is the agony and the ecstasy," says Mary Nelson of her twenty or so years of running Bethel New Life, one of Chicago's largest and most successful CDCs.

This chapter looks at community development as a career—and whether the field is the place for you. Here we discuss the pros and cons of jobs in the field—and provide a self-assessment quiz, to see how you score as a candidate for a career in community development.

Most successful practitioners in the field would agree with Mary's description. There *is* ecstasy in a community development career—it comes in seeing the positive changes in lives and neighborhoods that result from the hard, challenging work of overcoming the deep forces in the American economy that cause decline and dissolution of neighborhoods. This positive side of community development work is similar to that often cited by other professionals who are successful in empowering people to live fuller lives for themselves and society, such as teachers who come back year after year because they see the growth in their students and medical professionals who experience the wonder of curing a

patient's illness and extending a life. These returns—the gratification that comes from making the world a better place—are the chief benefits of being in the community development field, as reported by many in the field. Another benefit is the satisfaction that comes from uniting people behind a vision, working against difficult odds to achieve shared goals, and working toward the implementation of a community improvement program.

Take one example: combining community development and environmental protection. In 1984, a grassroots coalition called the Calumet Project was formed to participate in the reclamation of northwestern Indiana's hundreds of brownfields—abandoned and often contaminated industrial sites. The role of the Calumet Project's community organizers is to provide information to the community and give the people a strong public voice through town meetings, surveys, and testimony before public bodies. The goal is to ensure that brownfields redevelopment results in local business creation and jobs.

But what about the agony part? There are many impedi-

Career Story: John Andrew Gallery

Community Development Consultant, Philadelphia

John Gallery describes discovering community development and its pull on him as a vivid, life-changing experience. Trained as an architect, and working for the Philadelphia City Planning Commission on the urban design of downtown projects, John was assigned the job of working on some neighborhood planning with North Philadelphia residents.

John moved on from the work at the planning commission to follow other areas of interest in planning and urban design, but he never lost his passion for community development. This led him to accept the offer from the mayor of Philadelphia to reorganize the city's housing programs. He subsequently became the city's first director of housing and community development, during a time when racial tensions were high. John's challenge was to plan a set of programs and initiatives in neighborhoods in partnership with neighborhood groups and sell them to a mayor and City Council that were often at odds. All the while, he was confronted with building a capable staff and a set of operations, so that the mix of institutions in Philadelphia government could work together effectively.

After leaving city government, John joined up with other community develop-

> "The poverty I encountered was a rude awakening. Working with people in the heart of North Philadelphia who were coping with tough neighborhood problems touched something very deep in me. Community development and working to overcome the problems of poverty in neighborhood's became my life's work. I've stayed with it ever since."

ment professionals to form a consulting and development firm called Urban Partners, which continues to provide planning and development services to governments and nonprofit groups working on community development projects.

John has spent considerable time as a professor in the City Planning Department at the University of Pennsylvania and at the University of Texas at Austin's School of Architecture and Planning. He continues to mentor young people entering the community development field. "I continue to be amazed by the enthusiasm and dedication to community development and work on overcoming poverty that I see from young people—in part from time I've spent with

the young people at YouthBuild [a national organization training young people in construction techniques]."

John's advice to people entering the field is to experience a local public agency—a city planning department or redevelopment authority, or community development agency—early in your community development career so that the connection between community development and the public policy choices that must be made can be clearly experienced.

As for skills, John encourages community development professionals to have (1) a basic knowledge of real estate finance; (2) group facilitation skills—the ability to work with boards, make good presentations, orchestrate complex meetings; and (3) the ability to manage people, particularly in the nonprofit world. "These are the skills that position someone in the field for advancement," John says, although he is quick to add that "like most fields community development has fewer jobs the higher you go—it's a typical job pyramid."

"I stay in community development because I feel I was led spiritually into the field, and there is the presence of a mission in it for me," he says.

ments to success. At its best, community development is trying to change the way things are now in communities, and there are massive forces that resist change. Community development's agony can be about the frustrations—day in and day out—of making positive change occur. The lives of many people in low-income communities are often caught up in a cycle of poverty, a cycle that can feed on itself. The challenge of overcoming this cycle of poverty can be the agonizing.

Unforeseen events can wipe out a great deal of progress at the community level. A sudden plant closing, eliminating many jobs and taking people with stable incomes into a more marginal income condition; the slow decline of a once successful commercial area; the sudden eruption of a crime wave, causing some long-term homeowners to put their homes up for sale, thereby causing home prices to decline—all present the community development professional working at the grassroots level with a set of challenges that can lead to frustration and even burnout.

People who succeed at community development as a career usually do so because they are suited for the work—and this book can help you determine whether your personality and community development are a good match. They are usually also motivated by the mission of improving communities and lives more than by career advancement and financial compensation. At the end of this chapter is a self-assessment quiz—Is Community Development for You? Take a few minutes to see how you score as a potential worker in the community development field.

Of course, as chapter 4 describes, there are many career possibilities in the community development field and not all of them are at the community level or contend with those challenges directly. Part of the challenge of moving into the community development field has to do with matching one's individual talents, knowledge, and temperament with a position that uses that particular mix to its maximum.

To many in the field, the work of community development is value driven. Although there is a need for technical ability, managerial talent, and a range of skills, the element held in common by most people in the community development field is a belief that the work is about improvement of communities, and, therefore, American society, in some important way. Different people use different language to express the value-driven nature of their work. Often, the words "social justice" or "economic justice" are used. To others, the value is expressed in words like "empowerment" and "self-help." And the value is not a political one held more by Republicans or Democrats. Both parties support community development because it embraces fundamental American values that include self-help (language found coming from conservatives) and economic justice (words more often spoken by liberals).

As Paul Niebanck, a leading community development professor in Seattle, puts it, "Some groups in America need to stand at an ethical center, where part of the reality is values that have to do with equality and fairness and distribution of opportunity. The community development field is part of that ethical center."

The Pros and Cons of a Career in Community Development

Pro: *Entrepreneurship.* Talk to the director of a community development corporation about what he or she does, and you will quickly understand that you are talking with an entrepreneur. CDC directors have much in common with small business owners. They are trying to succeed at community improvement using any technique they can uncover, always

"My job is full of excitement. The phrase 'Doing well by doing good' may be trite, but it's true. Community development lending—seeing homes being built and sold to low-income families—certainly differs from making a loan that will make some incredibly rich guy even more wealthy."

—Dan Nissenbaum, Chase CDC

being entrepreneurial: raising capital (fundraising), figuring out their products (what does the neighborhood need, what do community leaders want for their community, and how does it get delivered?), making their annual budgets balance, and more.

Elsewhere in the field—say, at a bank that provides loans in the community development arena—the same entrepreneurial spirit is present. The long-standing motto of one leading community development lender—Bank of America—has been "Find a way or make one" when it comes to community development investing.

Even in some parts of government—often reputed to be full of nay-saying bureaucrats— the spirit of community development entrepreneurship is alive and thriving. In 1998, Secretary Andrew Cuomo of HUD created an entire category of employees called community builders, with the mandate to expedite HUD's programs and to strengthen communities through an aggressive form of partnerships with others in the community development field. Some local and state government community development departments have been successful entrepreneurial partners for years.

Being entrepreneurial means that successful community development professionals develop skills in knowing how and when to take reasonable risks with their funds and programs, often pushing the envelope on what constitutes "reasonable" in defining risk in the community development world, overcoming stereotypes about what is safe and what is too risky. The field is marked by a spirit of entrepreneurship that can make it an attractive choice for people willing to take some risks.

Con: *Chance of failure*. In the business world, a breathtaking 62 percent of start-up businesses fail. While the track record is somewhat better in the community development field, failures are part of the work. And, while there is great opportunity in the field for an active, aggressive, entrepreneurial type, there is considerable uncertainty about whether one will always succeed in improving communities. It is hard work, and the forces of decline can win over the best efforts of community development professionals and volunteers, even when great talent and energy are placed in an effort. Many community development organizations are short lived because they do not succeed at continuing to raise the funds needed to operate their programs, so those entering the community development field should be the kinds of people that are not easily discouraged by setbacks.

Pro: The wide variety of people. Working in the community development field carries with it the benefit of connecting with many different kinds of people. While the specifics vary depending on where one is in the field, community development workers come into contact with people at all ranges of the income spectrum, of all races and classes, playing different roles in the community development field. Unlike many fields, where most people have the same background, your colleagues in the community development field are likely to be a very diverse group—from many fields and personal backgrounds: race, ethnicity, income class, gender, age, sexual orientation, and political views.

Take the case of the work underway in many of America's large cities to demolish the worst of the failed public housing and to replace it with sound, mixed-income residential neighborhoods. The community development professional engaged in this work is likely to be involved with very low-income public housing residents, architects and planners, lawyers, loan officers who are a source of financing for the replacement housing, city officials, officials from the housing authorities, residents in the area around the public housing (who may be of a different race and income group than the public housing residents), people who control charitable funds, social service agency professionals who are helping residents develop plans to move out of poverty, local police officials who are involved in crime reduction strategies, church leaders, business owners, and more. Quite an array of folks!

A community development professional needs to be the kind of person who is comfortable—or willing to learn to be comfortable—with such diversity. It is not easy for someone in the field to succeed if he or she harbors attitudes of racial prejudice, treats people in a condescending way, or is easily intimidated by rank or seniority.

Pro: On-the-job flexibility. While it is difficult to generalize about particular positions in the field, it is fair to say that those who succeed in community development do not view the work as a nine to five, Monday to Friday, desk job. Com-

Jim Rouse, the late real estate developer and cofounder of the Enterprise Foundation, tells a poignant story of how Enterprise came to be. Jim and his wife, Patty, were approached by a group of women living in a decrepit building in a low-income neighborhood in Washington, D.C. They wanted Jim's opinion on rehabbing the building they were living in. Jim's reaction upon seeing the building was that it was too far gone; they couldn't possibly improve the building using the volunteer help they said they had lined up. But Jim and Patty gave them some financial help anyway—and, much to the Rouses' surprise, some months later they were called back to walk through a successfully rehabbed building! The success of this group led Jim and Patty to decide that a new foundation was needed to help other groups—and Enterprise was born.

munity development professionals are usually expected to be available to get the job done at whatever times they are needed. At the community level this usually means attending evening meetings with the community leaders, or participating in a weekend cleanup program, or leading a bus tour of the neighborhood. In other parts of the field, it may mean working to meet a deadline to close on a real estate deal, or seeing to it that a contractor is on the job as promised on a Sunday to get the job done on time.

In exchange for this flexibility on the part of the commu-

Career Story: Stephanie Smith

Community Development Lending Manager, Bank of America Mortgage, San Francisco

Stephanie Smith had no idea the field of community development existed during her undergrad years at Stanford University. When she graduated, with a degree in history, she tried to find a job in alternative publishing but was turned down for lack of experience. By happenstance, she got a job as editorial assistant at the National Economic Development and Law Center, a national organization that supports legal service lawyers and CDCs working on economic development issues in low-income neighborhoods across the country. Within a year, the Reagan administration drastically cut funding for legal services across the country, which resulted in a drop in funding for the Law Center. "At twenty-five, I suddenly ended up running the publications department because there was no one else left," she explains.

"During my three years at the Law Center, I moved from wanting to write about community to wanting instead to do community development. That meant, I needed more skills, and that's why I went to UCLA, to get a master's in urban planning and an MBA.

"There, in the first quarter, I met the head of the LA Design Center (which helps nonprofit groups develop affordable housing), and she invited me to work at the center as an intern, which I did during school and for

a full year after I graduated. My original job at the Design Center was managing affordable housing projects. Over time, I became a senior project manager with my own projects, as well as supervising several other staff. In Los Angeles back then, you could see ten years in advance of other parts of the country many of the issues that would eventually impact other cities, from in-migration to disinvestment as the aerospace and other manufacturing industries left the area.

"Then, in 1987, the Design Center began working with the National Equity Fund, which had just been set up by LISC to finance affordable housing using the newly legislated Low Income Housing Tax Credits. After a year, I returned to San Francisco and went to work for NEF, underwriting their tax credit deals statewide. But I decided I wasn't cut out to be a project underwriter. It was like writing about community development rather than doing it. I switched jobs and took a position as a program officer for LISC, making recommendations to LISC for grants and loans to CDC projects in the area. I also ran the Bay Area Housing Support Collaborative, which provides multiyear operating support and technical assistance to select CDCs.

"In 1993, I got a phone call from Nic Retsinas, who had been appointed head of HUD's Federal Housing Administration under the new Clinton administration, asking me to serve as a special assistant to then HUD secretary Cisneros. My assignment was to improve the regulatory framework that would give greater creativity to two secondary market agencies, Fannie Mae and Freddie Mac, that HUD supervises. I had concerns about taking the job—one was that I had never worked in an organization with more than two hundred people or so, and I had never worked in such a political environment before. Both of those turned out to be equally fun and daunting challenges. But part of why I like this field is that I like the fact that the learning curve is often incredibly steep, that the field is never simple. You never stop facing challenges because it's always changing.

"After two years, when we finished putting in place the improved regulatory structure governing Fannie Mae and Freddie Mac, Nic asked if I would be the general deputy assistant secretary for housing, essentially his chief deputy and a totally different job. I stayed in that position until the fall of 1997. After four years in Washington, my partner and I decided it was time to come home to San Francisco.

"I was recruited by the Bank of America to manage community developing lending for the Mortgage Division. After fifteen years in the nonprofit and public sectors, I wanted to work in the private sector, to learn more about the kinds of tools they have available. In this field, the more you know, the more tools you have, the more effective you can be.

"The issues have been the same in all my jobs. You could say that, in a way, my whole career has been about access to capital for people and communities who have historically been denied that access. Then you could ask, 'Who's got the biggest treasury for this?' and it would be the Bank of America. This bank has made a $350 billion commitment to community development lending.

"I'm thoroughly enjoying this new role. Though this is a huge bureaucracy, in terms of how you get things done, it's much the same as HUD.

"We're having an impact. We're promoting economic justice, though not in the same way as a local nonprofit group. My job here is to increase the Bank's lending to low-income families and neighborhoods and to families of color. We are the third largest mortgage originator in the country, and about 40 percent of the mortgages we fund are to low-income and minority individuals. Day in and day out, we are institutionalizing community development."

nity development worker, most employers take an attitude of flexibility as well, often accommodating schedules around child care and other personal needs. Although it depends on the particular circumstances and the philosophy of the organization with whom one is working, generally community development jobs are more flexible than most.

Con: Long hours. In exchange for flexibility in most places in the field, community development means lots of hours—evenings to attend meetings, weekends to get special projects done. In general, the community development field is filled with dedicated hard workers, and employers have come to expect hard work and long hours from community development workers. The problems seem so urgent that moving fast is often the norm.

Pro: Growth in responsibility. At the ground level—in the community—community development feels like a small business. When lots of things need to get done, it often happens that the inexperienced community developer is asked to take on responsibilities that, in another field, would only be handed to people with considerably more experience. Like flexibility, this varies with the organization's culture and size.

Many experienced practitioners in the community development field look back on their early days with amazement at the level of responsibility they had at a young age—and their need to learn as they encountered problems. George Knight, for example, the former executive director of one of the largest community development intermediaries in the nation, the Neighborhood Reinvestment Corporation, recalls his early entry into housing development. "I was a community organizer who one day was given the assignment from my board to get homes built in a neighborhood where no homes had been built for twenty-five years—and I hadn't done anything like it before. But with some advice from others, and some hard work, we got six homes built in the first year. All of that at age twenty-three!"

Pro: Mobility. The community development field has two kinds of professionals: the long-termers and the movers-on. The people who stay in the field for a long time largely do so because, they often report, the work feeds their soul in some fundamentally important way. The day-to-day frustrations are more than offset by the satisfaction of seeing progress and movement toward a vision of a society that is more in keeping with their view of how America should be. Many report that their continued involvement has to do with economic justice, or opportunity to build capacity in individuals and groups so that the pace of improvement can be sustained and people can do more on their own..

For many others, work at the community level can be rewarding but can also lead to an invitation to leave it—often into something else in the community development field, like working in a bank or a mortgage company, as a home builder or real estate developer, or with a social service agency. Often, a skill developed in the community development field is transferable to other fields altogether. Learning how to sell in the community development field can lead to work in a public relations firm or a government agency. The community development field has been the root for at least one United States senator, numerous mayors, and a dozen or so members of Congress.

Con: Unclear career ladders. As chapter 7 describes, there are patterns of career moves in the community development field, but they are not as clear as those in better-established fields. It is the exceptional case in the community development field for someone to know how he or she expects to advance in the field, from junior to senior level.

Pro: A chance to be inventive. Successful small business entrepreneurs create innovative solutions to overcoming the obstacles to reaching their goals. The community development field has been marked by invention—whether that invention has been new financing techniques or new approaches to getting information. Bob Giloth, now a vice president with the Annie E. Casey Foundation in Baltimore, recalls that when he began his work for a CDC in rehabbing housing in Chicago, he didn't have good information on contractors, costs of materials, or the other basic elements needed for a property owner to handle housing rehabilitation successfully. Bob knew there were other community development specialists with the same needs, and out of that need came the creation of the Chicago Housing Rehab Network, an organization that provides technical help to groups in Chicago that are in the business of rehabilitating homes for low- and moderate-income people. That is a process invention—one that in this case led to the creation of an enduring association serving its members.

Pro: An environment that leads to learning and growth almost daily. Community developers comment on how much they learn on the job about many things—people, techniques, skills, how the world works, politics, and much more. Although virtually all jobs in community development have some routine, most call for daily learning and invention to get the job done. India Pierce Lee learned as she went to adapt herself from her background as an air traffic controller to a community development professional. So did Nancy O'Brien, a one-time volunteer for the Great Falls, Montana Neighborhood Housing Services Program who was asked to take the job as its executive director: "I learned on the job, and from my board, especially the bankers on the board who trained me."

Con: *Not as much money as in other careers*. If making lots of money is your primary goal, the community development field is probably not for you. As chapter 5 describes, salaries in community development are about even with jobs in other nonprofits but are lower than those for comparable work in the private for-profit sector. The returns in community development are as much in the satisfaction of improving people's lives and communities as they are in financial compensation. And, while many jobs in the field have fringe benefits, generally the field has lagged behind others in the quality of fringe benefits such as pension programs and health care coverage.

Fitting Your Interests and Talents into the Community Development Field

Although the community development field is broad in scope, there are basic skills that virtually everyone in the field needs. Perhaps the most common trait cited by those in the field is the ability to solve problems. This is followed closely by the ability to communicate in writing and speaking, relate well to a diverse group of people, think analytically, engender trust, work with people, be organized, and, at upper job levels, manage people and organizations well.

Problem solving is that trait that enables one to overcome obstacles to achieving a goal. In the community development field this can mean solving both big and small problems. The women that Jim and Patty Rouse encountered had to solve the big problem of making a building livable with little money or housing rehabilitation skills. They figured it out through creativity, hard work, some successful fundraising, talent, sheer determination, and lots of prayer. Other kinds of problems require no less skill: How does one get the five votes from City Council needed to get the zoning approved for the new building that you and your group know is just what's needed

to improve the key block in the community? Since many jobs in community development are essentially in small businesses, the problems to be solved can be as mundane as making sure that the Meals on Wheels van is going to be in good driving condition on Monday morning, with the insurance current and a properly licensed driver.

Those who succeed in community development—and who have the kind of talent being sought by employers—have a high degree of comfort with their problem-solving skills. Everyone gets discouraged from time to time in having to cope with problems, but if, when confronted with a problem, your typical reaction is to panic, move into denial, or fret about it to the point of inaction, you are likely to have a difficult time in most positions in community development.

Communication skills—the ability to write well, make a decent presentation, participate actively in a meeting—are all essential skills for virtually every job in the community development field. Some positions require lengthy proposal or report writing, but most do not. In most cases, the writing takes the form of letters and memoranda, brief reports, applications for funds, or newsletters. The writing skill needed is the ability to write clearly and reasonably quickly, so that not too much time is taken up with the writing part of the job. Laboring over a short memo for hours simply won't do—there is not enough time in the day to get everything done and spend that much time writing.

Similarly, communication through speaking is important. Community development people make lots of presentations, attend many meetings, deal with diverse groups, and must be able to communicate well in those settings. There is plenty of room for both extroverts and introverts in community development, but most positions do require that whether you tend to be shy or outgoing, you have the ability to handle yourself in those situations.

Career Stories: Community Development "Geniuses"

The John D. and Catherine T. MacArthur Foundation's prestigious Fellows designation—sometimes called the MacArthur "genius awards"—has gone to community development leaders almost annually. The award is based on criteria that include exceptional creativity and the promise of making a "significant and benign difference in human thought and action." As a group, the community development MacArthur Fellows show the remarkable diversity of approach in the community development field: potters who have built community development businesses, organizers who have reconfigured power relationships, business-oriented leaders who have brought needed capital to areas and groups, and grassroots leaders and clergy who are succeeding at rebuilding neighborhoods. A remarkable group of people...

Lourna Bourg (1992 recipient) is the cofounder and executive director of the Southern Mutual Help Association, an advocacy and self-help organization working with the rural poor in Louisiana. Using a comprehensive self-development approach, Bourg's program concentrates on rebuilding whole communities.

Elouise Corbell (1997) is the director of the Blackfeet Reservation Development Fund and the Blackfeet National Bank, both institutions providing much needed capital to Native American businesses and nonprofit groups.

Ernesto Cortes (1984) is a community organizer working with the Industrial Areas Foundation, founded by Saul Alinsky. Cortes is the founder of the San Francisco Antonio Communities Organized for Public Service (COPS), the United Neighborhood Organization in Los Angeles, and the Metropolitan Organization in Houston—all working to empower low-income people in their communities.

Martin Eakes (1996), currently the president of the Self-Help Credit Union in North Carolina, is a community development worker who combines hard-core business principles with his self-help strategy. Eakes's work pioneered the development of Community Development Financial Institutions.

Wesley Jacobs (1987), an Oglala Sioux, is a rural planner leading efforts to improve the living conditions and economies of tribal areas with high levels of poverty and unemployment.

Sakoni Karanja (1993) is the founder of the Centers for New Horizons, a multipurpose organization operating within Chicago's deeply impoverished public housing developments. New Horizons centers provide early education, mental health services, and youth training and employment programs.

Bill Linder (1991)—Monsignor Linder to his congregation at St. Rose of Lima Roman Catholic Church in Newark—is the founder of New Communities Corporation, a large-scale community development effort that includes housing and commercial development and has provided jobs to well over one thousand low-income people.

Carol McKecuen (1994) is a community development entrepreneur who has used the development of crafts as a vehicle to employ over seven hundred low-income artisans, promoting their connections to mainstream markets.

Otis Pitts (1990) is a community organizer who has helped to develop and implement an economic agenda for the Liberty City area of Miami. He cofounded the Tacolcy Economic Development Corporation, a successful developer of low-income housing and shopping centers in low-income areas.

Hipolito (Paul) Roldan (1988) is a community developer active in the revitalization of Latino low-income neighborhoods in Chicago, through the Hispanic Housing Development Corporation.

Dorothy Stoneman (1996) founded YouthBuild USA in her living room and turned it into a national organization training low-income youth in construction skills and building and rehabing housing for needy people in the process. YouthBuild is now in over one hundred cities nationwide.

William Strickland (1996) used his interest and skills in pottery as his way into training low-income youth from his Manchester neighborhood in Pittsburgh. An informal start led to the creation of the Manchester Craftsman Guild, then a fuller job training program, a performing arts space, and more—all combining jobs and community development.

Maria Varela (1990) is a community organizer, photographer, teacher, writer, and president of Rural Resources, an organization helping rural communities in the Southwest develop sustainable economies and environments.

Robert Woodson (1990) founded and is president of the National Center for Neighborhood Enterprise, an organization working to empower the poor to devise and implement solutions to community problems.

Awardees are from all parts of the country, are racially and ethnically diverse, are urban and rural, and are from all walks of life—a microcosm of the community development field.

Communication is the bedrock of William Bostic's job as executive director of the Pennsylvania Housing Finance Agency. It's Bill's way of influencing policy externally and ensuring that his staff understands the context of community and economic development within which his agency operates: "I spend my day communicating with agency staff; with other state agencies, such as community development, welfare, and aging; with our partners—the investment bankers and bond counsels; and with national organizations to check in on the national level. I communicate with twenty-one members of Congress and two senators on hot issues. I spend time educating them and others such as the state Homebuilders Association and the state Mortgage Bankers Association on what we need to do to accomplish our goals."

Analytical skills are needed for many, though not all, community development jobs. The field has become increasingly technically sophisticated in that a good bit of it has to do with improving places, and that usually means the development of real estate—residential and commercial. The development of real estate requires analytical skills in determining project feasibility, or in deciding whether to lend money to a project, or in understanding the value of an investment in a building or project. Some specializations in community development require well-established numbers skills—but even many of those that do not require extensive skills do require a familiarity with the world of finance. Even in parts of the field that don't deal with calculations of numbers, analytical skills may be needed. Gale Cincotta, the founder of a leading community development advocacy group, National People's Action, points out that an organizer has to do a lot of research of a practical nature to find out what the laws are and who's in charge of what. If you're an organizer, you have to know what to do if property insurance is canceled, for example—or how much the city is spending in each neighborhood. The analysis of such facts equips the organizer with what he or she needs to inform people about actions they should take.

One last generic skill often cited by employers as desirable is something community development professionals call *process skills*—the ability to manage complex tasks and processes to move an idea, a project, a program along its

Community Story: Harbor Point, Boston

Resident Leaders Transform Public Housing

The Columbia Point public housing project in the Dorchester neighborhood in Boston had become so awful as a place to live that by 1979, only 350 of the original 1,500 units were occupied. The project was drug infested, unsafe, and so unattractive that very poor people on the public housing waiting list in Boston refused to live there.

The Residents Council demanded that the Boston Housing Authority do something—and the residents wanted to convert the place into an attractive place to live, not only for poor people, but for people of all incomes. And in order to see to it that the vision actually happened, the residents insisted that they be part of the real estate partnership that would redevelop the project into a community.

After years of persistence, insistence, and negotiation, in 1986 the Harbor Point Apartment Company was formed, a partnership between the Residents Council and a local, private, for-profit real estate development company, Corcoran-Jennison. The developer worked with city government to arrange the complex financing needed to bring about the physical quality envisioned and to make the project succeed at the income mixing it was seeking. In short, the project had to be attractive enough to get people who could live elsewhere to choose Harbor Point, the new name for the community.

The state provided $154 million in loans and grants, and HUD provided $21 million through two different programs. The remaining $75 million came in the form of private equity. Ongoing operating subsidies are provided from the state and federal governments to keep a portion of the units affordable.

The deal made between the Residents Council and the private developer was to build a mix of units—70 percent at the market rate and 30 percent for the public housing residents, with both groups agreeing that this mix would produce a community that would be attractive for the foreseeable future. An important design consideration was that the mix created prohibited more than 50 percent of any building from being occupied by low-income tenants, to assure a mix throughout the development.

Emphasis was placed on making the units attractive, the community safe, and the amenity package sufficient to attract market-rate tenants. The project has a swimming pool, tennis courts, a fitness center, and free parking—all at a rent competitive with other projects in Boston.

The community that has resulted is very diverse. Many of the market-rate tenants are students, professionals, and foreign nationals living in Boston while studying. At times, forty different nations have been represented by the tenants in Harbor Point. The public housing residents continue to work on other aspects of community development—education improvements and job placement programs.

Harbor Point has worked. It isn't perfect. There continues to be a challenge to monitor teenage behavior to reduce vandalism, which occurs from time to time. In hindsight, it may have been helpful to the community to have a three-level income tier—low, moderate, and middle—creating an even better mix. But that is a refinement of a remarkable example of community leaders taking charge of their troubled situation and working hard enough, smart enough, and long enough to bring about a transformation of their neighborhood.

---

intended track and timetable. This skill includes both the ability to manage time and complex tasks and the ability to manage the process of involving the people needed in the proper way to move from a problem to a solution. These process skills often come through on-the-job experience, but they also are contained in an attitude and personal style that can be modified to suit different situations. Another way to say this is that the skill needed is the ability to know when to ask and when to tell, when to be conciliatory and when to be confrontational, when to involve someone in a process and when to leave them out of the decision.

Lots of these process skills have to do with building partnerships with others to accomplish goals. One of the process skills successful community developers have is to bring together the key elements of the community to work in harmony—helping people see the mutual self-interest in taking on projects or programs that can improve community conditions. This skill usually is based on another very basic one—building relationships of trust between the community developer and others. That means keeping your word, following through on commitments, respecting your partners even when you're not seeing everything eye to eye. The trust element is part of important people skills, needed by almost everyone who works in the community development field.

These general skills are needed in some combination in virtually all positions in the community development field. They are the building blocks that are necessary to handle the complexities of community development. The mix of these skills varies by one's place in the field, as do the more detailed skills one needs to handle specific jobs in the field.

A few examples might help.

A *community organizer* helps people see where their mutual interest lies in solving a problem or improving their community in some way. Organizers believe that there is power in the collective and that an organized community can demand more from government—like better schools, or health care, or homes for homeless people—and from the private sector—home mortgages and pollution reduction, for example. Organizers often work on the principle that an organized community is a better place to live because when neighbors help each other, the quality of life improves overall. The process of working together to solve problems produces a sense of empowerment that builds confidence to take on more difficult and complex challenges.

Usually employed by a community group or by an organizing intermediary like the Industrial Areas Foundation or the Consensus Organizing Institute, community organizers are often less concerned with what specific problem gets solved than with whether the problem being worked on is important to the community and will benefit people living in the community.

The skills that a good organizer needs are very high-level process and people skills. It is critical to be able to win the trust of those being organized. The organizer must successfully build consensus among people who have not previously been able to achieve it. The organizer must be able to motivate people to put time and energy into solving problems that they may have given up on. Barbara Schliff, an organizer for the Los Sures organization in Brooklyn, recalls her first assignment: "The landlord in this building had stopped even basic maintenance services. Because the parents spoke only Spanish, I looked to their children to help me communicate, and that helped the parents warm up to me. I also studied Spanish at night. I helped the tenants start a rent strike, then take the case to housing court—and the landlord was ordered to make repairs, which he finally did."

An organizer's skills are not so much quantitative skills; they are analytical, however—in the area of people skills, talent identification, "sales," and political judgments. Gary Delgado, a long-time organizer, describes what community organizers have accomplished as a process of finding and developing grassroots leaders rather than assuming that the only people that can solve problems are anointed experts. People have been helped to understand that their opinions count, that there is power in numbers, and that, even though there may be conflict within an organization, democratic decisions are possible. Organizers have contested for and won power—to reverse discriminatory loan policies, force the development of low-income housing, influence school curricula, stop illegal dumping, and enforce first-source hiring.

A *CDC housing developer* uses a combination of real estate development skills, financial analysis, and consensus building to get housing built or rehabilitated. The part of the community development field that builds or rehabilitates housing has grown rapidly over the past twenty years.

CDC directors are essentially in the real estate development business, and the work requires the skills, energy, and persistence that any real estate developer needs to be successful. The skills needed include knowledge of real estate finance, the real estate development process (How does one get a zoning variance? What is an option on land?), at least a passing knowledge of design and construction and bidding procedures, how to finance real estate, and how to sell or lease units.

It's a complex job, one filled with challenges and the ultimate reward of seeing families moving into new or rehabilitated affordable housing—people who would not otherwise have a decent roof over their heads.

A *CDC housing professional* needs to be a mix of technician and entrepreneur. Business skills are very important, but successful CDC housing professionals are not just techies—they are able to relate well to their "bosses," the leadership in the neighborhood that is setting the overall goals of housing production for the area. Sometimes this job also requires supervising others. It might require relating well to the many actors involved in getting real estate produced—architects, planners, lenders, investors, brokers, buyers, and property managers, for example.

A *community development lender* has strong financial analytical skills and understands how to measure risk. This kind of job calls for more skills in the quantitative analysis area than the jobs of organizer or housing developer do.

Usually, a person in this position—which could be with a commercial bank, a community development financial institution, or an equity investor—has training in finance. This person usually does not have contact on a daily basis with neighborhood leadership or those who might buy or rent housing. This person works with housing developers, developers of commercial real estate, or small business owners in community development areas.

A *policy analyst* in community development usually works on the formulation of appropriate public policies that will have positive effects on communities. This professional is usually trained in public policy or government and has strong research and analytical skills. This kind of job involves work with ideas and policies rather than the nuts and bolts of getting specific community improvements underway. The successful outcomes of this kind of job are usually new legislation or foundation initiatives—changes that are then used by practitioners who are working at the community level. Michael Barr, a deputy assistant secretary for community development policy in the U.S. Treasury Department, says, "I love the chance to make a difference. I am presented with an enormous opportunity to change policies in ways that affect people's lives."

A *service deliverer* is a hands-on people person working to help individuals and families overcome particular problems in order to participate in the American economy and the American dream—"the pursuit of happiness." Usually trained in the social sciences or as a social worker—but by no means always—community development–oriented service deliverers have a high level of day-to-day contact with people with needs. The skills needed here are both helping skills and a knowledge of how to get things done, like helping a family get food stamps or funds to relieve a particular family emergency. Service providers are often employed by community-based groups or by larger nonprofit agencies. In one emerging part of the community development field, service deliverers work to place unemployed people in jobs and then see to it that they have some level of support to stay in their jobs. Other service deliverers specialize in helping particular groups: youth, the elderly, AIDS victims, and others. All of this service is provided in the context of a community development approach, which means it is usually part of a larger set of activities being undertaken by a community-based group.

Community development is a very diverse field, with lots of pluses and minuses. Some skills are needed by almost everyone in the field: problem solving, the ability to communicate well, and analytical skills. More specific skills are needed for particular kinds of jobs. But how does one go about getting into the field in the first place? That's the subject of chapter 4.

Is Community Development for You? A Self-Assessment Quiz

Those who succeed in the field seem to have a set of personality traits and skills that equip them for the combination of social purpose and business acumen needed to succeed. Complete this quiz to see how you score.

Answer each question with the number closest to your personal feelings. There are no right or wrong answers. Try to be as honest as you can be. "1" means the statement is not at all true"; "5" means it is very true. Score yourself between those two points depending on how true the statement is for you. Circling a "3" indicates you're just about in the middle.

Not true		Neither true nor false		Very true

1. Working with people is a real passion for me.

    1   **2**   **3**   **4**   **5**

2. Making lots of money is one of my key life ambitions.

    1   **2**   **3**   **4**   **5**

3. If I could choose to succeed at only one thing in life, it would be to make the world a better place.

    1   **2**   **3**   **4**   **5**

4. I want my life to be safe, secure, and predictable.

    1   **2**   **3**   **4**   **5**

5. Most people who know me would say I'm easy to get along with.

    1   **2**   **3**   **4**   **5**

6. I like to get my work done on my own without a lot of involvement from others.

    1   **2**   **3**   **4**   **5**

7. Most people who know me would say I'm a high-energy person.

    1   **2**   **3**   **4**   **5**

8. My attitude toward work is that I'll give it eight hours a day, five days a week, but not much more.

    1   **2**   **3**   **4**   **5**

9. When I put my mind to get something done, I don't stop until I succeed.

    1   **2**   **3**   **4**   **5**

10. Once my mind is made up about something, it's difficult to change it.

    1   **2**   **3**   **4**   **5**

11. I trust my skills to get me through almost any situation.

    1   **2**   **3**   **4**   **5**

Not true		**Neither true nor false**		**Very true**

12. I have to admit it: I hate making decisions.

1	2	3	4	5

13. I'm really good about learning from my mistakes.

1	2	3	4	5

14. One of the things I really hate is making group presentations.

1	2	3	4	5

15. Generally, I'm willing to take risks that others aren't willing to take.

1	2	3	4	5

16. I simply can't stand dealing with government bureaucracies.

1	2	3	4	5

17. Compared to other people, I can keep lots of activities underway at the same time.

1	2	3	4	5

18. I'm not a very organized person.

1	2	3	4	5

19. Writing is something I like to do and I do it well.

1	2	3	4	5

20. It really drives me crazy when someone else gets the credit for something I've done.

1	2	3	4	5

21. I consider myself a good neighbor.

1	2	3	4	5

22. The real problem with poverty in this country is that too many people just don't want to work.

1	2	3	4	5

Now, add all of your answers from the odd-numbered questions. Then add all of your answers from the even-numbered questions. Subtract the even from the odd. Add five points to your score if you've volunteered with an organization doing good work sometime over the past two years.
If you scored . . .

35 to 49 You could be serious community development material. You should actively consider community development as a field.

20 to 35 You could have considerable promise as a community development professional.

0 to 20 Look carefully at the various possibilities in the field. Some of the positions that are not at the core of the field may be a match for you.

0 to –20 A score in this range probably indicates that it isn't likely that you and community development will hit it off.

–20 to –49 Give this book to someone else.

Chapter 4

Getting Started and Getting Ahead

This chapter discusses the paths to take to find your way into the field. As explained later in the chapter, career advancement also takes many of the same paths.

There are basically four ways to get into the community development world. The first is through community service, which can be done either as a volunteer or as an intern in a community organization. The executive director of one of the largest community development groups in the country says that he hardly ever hires someone from outside the field who hasn't demonstrated interest through some form of service. Almost half of people serving as interns are hired into staff jobs in many organizations.

The second path is through training and education. A third is through related work experience, applying a skill or experience from another job to one in this field. And a fourth way is by serendipity—spotting an opportunity and having the courage to grab it.

Finding Your Way through Grassroots Volunteering

You can engage in community service by volunteering for a local, citywide, or national community development organization. You can make your own volunteer opportunity or join an established volunteer program.

Volunteering time, effort, and talent to grassroots efforts to help a neighborhood and the people who live there is rewarding in and of itself and a likely step toward a career in the field. If there's a community development organization active in your neighborhood, you can volunteer to join its members—to swing a hammer to help renovate neighbors' housing, or to serve on the day care center committee, or to work in the food bank. If you understand and feel strongly about the issues facing your neighborhood, you could volunteer to serve on the board of directors of the local community development group. Because community development organizations are neighborhood based and often have trouble finding people who are willing to put in the hours that board service requires, you would probably be welcome. It's a way to learn the nuts and bolts, to prove yourself, and to make sure you really like the work.

It should not be hard to find a local development organization if one exists in your neighborhood. Most community development groups try to make themselves known to neighbors; a leaflet might show up in your mailbox one day, announcing an open meeting or a celebration. If you do not live in a low-income neighborhood but have a strong interest in community development, volunteer to work in one. There is probably no better way to see what it's like to work in the field and, at the same time, to get your foot in the door.

You can find out if a local organization exists in your neighborhood or in one nearby by phoning your city's community development department, looking on the bulletin boards of local community facilities (for example, the Y, the neighborhood health clinic, the public high school), or reading the community newspaper. Look also in the business and real estate sections of the weekend editions of your city's newspapers. Contact some of the national organizations described in appendix D in this book and ask if they have members in your area. One of the most challenging ways to serve a community is a do-it-yourself try at dealing with an unaddressed problem in your community or your city, an approach that requires time, energy, a strong determination, and a personal belief that one individual can make a difference. Many of today's community leaders started this way, by confronting the problems they saw on the neighborhood streets and in their own and their neighbors' lives. These are the men and women who organized their fellow tenants to protest poor housing conditions, or mounted a campaign

calling for the city to help close down drug dealing in the neighborhood park.

"I don't like to be pushed around, and I don't like to see other people pushed around," says Gale Cincotta, a PTA mother of six sons and organizer of her neighbors in Chicago's tough West Side. "We saw that as more and more minorities moved in to our neighborhood, our school system got less and less money; we had forty-five kids in a class, while other neighborhoods had twenty-five. I don't have a college education, but I've run a family with very little money and figured out a budget and paid the mortgage. I did all the research to know what the system had to give you, what public hearings you could go to. If we got a no, we just went and got more people. You don't win all the time. But people who organize feel they have much more control over their lives. It's important not to lose confidence in what you can do. You've got to let the people participate in the democracy," Gale says. She is now the volunteer head of National People's Action, a coalition of grassroots groups that is the force behind the passage of the Community Reinvestment Act and other initiatives that now help to put capital into America's neighborhoods.

If you've been a grassroots activist and eventually decide to apply for a staff job, you'll find that you are among the most welcome of job seekers. Los Sures, a neighborhood development group in Brooklyn that renovates rundown housing and helps the tenants turn their rental buildings into cooperative ownership, hires neighborhood people as organizers. It chooses those who have demonstrated leadership by getting their fellow tenants together for a first meeting, for example.

You can call community development groups and ask, "When is your next community meeting that I can attend?" Or join in their annual paint-a-thon or other activity. After finding out what the group does, tell them, "I'd like to find some way of working with your organization." If the group agrees, you can establish your own volunteer slot. Use the appendices to help locate organizations at the national or local level, including community development corporations, the intermediaries that work with them, government agencies, financing institutions, and organizing and training groups.

You might volunteer to work in one of your church's programs. Churches, synagogues, mosques, meeting houses—thousands of faith-based institutions across the nation, representing all of the country's religious, ethnic, and racial diversity—have gotten actively involved in helping to improve the lives of their neighbors. Bethel New Life, described in chapter 1, is one example. From tutoring programs to job banks to housing loan funds, congregations run community projects that offer the opportunity for experience and service.

Or you can join an organized volunteer program that supports people who are doing hands-on grassroots community service, from housing repairs to counseling youth. For example, Habitat for Humanity, whose most famous volunteers are former president Jimmy and first lady Rosalynn Carter, buys the materials (using funds from a powerful fundraising campaign) and organizes volunteers as construction crews who, under expert supervision, build homes for low-income families in rural and urban localities all across America and overseas. Christmas in April helps communities put together their own volunteer program to repair the homes of low-income, elderly, and disabled homeowner neighbors. These are just two of the many such organized community service programs.

Volunteering—during school years, during summers, and after you're out of school—is a highly effective way to make the contacts and gain the experience that leads to a job. Joe

McNeely, founder of the national Development Training Institute and a veteran practitioner, counsels newcomers: "Volunteer for a while. If you hang around long enough and have skills, you will end up with a job. You can help make your own job or someone will make it for you."

There are programs specifically designed to help neighborhood volunteer activists who need to broaden their academic background or learn specific skills to qualify for a community job. Many of these programs are offered by citywide intermediaries, such as the Atlanta Neighborhood Development Partnership and the St. Paul office of the Local Initiatives Support Corporation. The Non-Profit Housing Association of Northern California targets community people in the northern part of the state, while the Massachusetts Association of Community Development Corporations has three statewide programs that offer training and education to attract residents who have volunteered in their neighborhoods and would like to upgrade their skills for a staff position with a CDC. These programs are described in appendix J.

Entering the Field through an Internship

Internships are different from volunteer work, though both offer opportunities to serve a community and get experience. Internships are usually structured positions within an organization, and the organization commits itself to teaching the intern something about the work in return for the intern's labor. Internships serve as a first, foot-in-the-door job for many young men and women. Most formal internship programs pay a stipend.

Fellowships—discussed later in this chapter—are similar to internships but are almost always affiliated with a university program or a program honoring a person's achievements; they almost always pay the fellow and usually are

Career Story: Hipolito (Paul) Roldan

Executive Director, Hispanic Housing Coalition, Chicago

Like many returning Vietnam War combat veterans, Paul Roldan kicked around a bit, trying to find his way. He worked at a bookstore, and read a lot. Drove a cab for a while in his home base of Brooklyn. Got a taste of the business world at the street level while working for a loan company; he made night calls to collect on loan accounts—work he didn't feel good about.

While working at a good, steady job at a neighborhood manpower center, he got involved as a volunteer with a neighborhood group—involved enough that when a position opened up at the Sunset Park Redevelopment Association, Paul took a job there, working to develop affordable housing with the neighborhood CDC. He went to graduate school at night to get a planning degree, recognizing that the work he was doing was challenging and rewarding but that he needed more skills.

In 1976, his answer to a newspaper ad landed him a job in Chicago as the first employee of the Hispanic Housing Development Corporation. The group had one grant of $145,000 from a state agency. He's still there. Now, a lot of years later, Paul feels a mix of pride and humility at the 1,700 affordable apartments and town homes that

Hispanic Housing has built in several Hispanic communities in Chicago—and the five shopping centers the group has developed.

"This work is in the zone for me. It feels enormously right. I work hard—but there's

a certain imperviousness to the ordinary fatigue that comes with hard work. I never get sick. The work invigorates me," says Paul in describing the personal experience he continues to have as a community development leader.

"Skilled community development people are jugglers. They can go through life in fifteen-minute increments. We need to be broad, and able to handle very different settings, yet technically proficient at what we do. The challenge—the real tough part of the work—is fighting off bureaucracy. Bureaucracy can be the enemy of progress in our communities—whether the bureaucracy is found in governmental agencies or in larger-scale community-based groups," he says.

Paul is a business entrepreneur. In addition to the housing and the shopping centers, Hispanic Housing Development Corporation has formed a property management company that manages over 3,000 housing units, and it has created Tropic Construction Corporation, a residential and commercial contracting company.

In 1988, Paul was the first Puerto Rican to be named a MacArthur Fellow for his work in community development. With part of the cash award he founded the Teresa and Hipolito Roldan Community Development Scholarship Fund to provide financial assistance to Hispanic youth interested in pursuing community development education.

"Get with a CDC or a private developer as early as you can," is Paul's advice to newcomers to the community development field.

meant for people in mid-career.

Recognizing the power of community service as a great path into the field, community groups, networks of CDCs, intermediaries, and foundations have created formal internship programs that offer people grassroots experience. A number of local, statewide, and national community development organizations use the AmeriCorps program, a federally funded national service program that offers money for tuition in return for community service.

These programs create pipelines for people to learn about community development and find permanent jobs in the field. A number of these programs are intended to attract more people of color into the field. There is often a strong emphasis on recruiting people from the neighborhood served by the community development group. Appendix J describes these programs.

Interns in the community development field are not usually asked to do gofer office work, though of course everyone pitches in when necessary. Intern assignments are full of responsibility and challenge, for example: helping residents organize a neighborhood crime watch or a community festival; writing and distributing a newsletter to the hundreds of families in a neighborhood; providing education and counseling to first-time home buying families; assisting in developing affordable housing; assisting in the CDC project that trains local residents to become licensed day care providers; helping to create a community garden; organizing a community cleanup campaign; coordinating after-school programs for local children and youth; getting the word out to families about available child care services and about employment opportunities.

Internships definitely open doors. One local CDC in Detroit found it hired 30 percent of its interns as staff. In the AmeriCorps program sponsored by the national Local Initiatives Support Corporation, 42 percent of the participants found professional positions in the organizations where they had been placed or in other community-based development organizations. In the AmeriCorps-funded program run by the Massachusetts Association of Community Development Organizations, 48 percent of the interns found employment with their host or other community groups.

Darcy Burrell's successful experience with her internship is typical. Selected as an AmeriCorps intern by a CDC in the Boston neighborhood where she grew up, Darcy was assigned to do community organizing. As part of her organizing job for the CDC, she worked with more senior staff to put together training sessions that help residents learn about and take action against neighborhood problems ranging from teenage pregnancy to domestic violence. She also made and distributed flyers for different events and training sessions. She found her associate degree in education helpful in that work. After finishing her AmeriCorps assignment, Darcy successfully leveraged the experience she gained in training to land a job with another nonprofit group as an employment trainer. "It's a great field to go into. In the news everything seems negative, but in the community there's a lot of hope."

Beyond the formal internship programs, almost all national organizations and many local ones, both public and nonprofit, sponsor informal internships from time to time, sometimes with a stipend. It's worth it to contact agencies you might want to work for or organizations active in a part of the community development field you'd like to explore. Ask them if they have internships. Even if they do not, offer to intern for them. Use the appendices to help you find organizations that match your interests. Use the ideas in the last chapter of this guide to identify people and places to contact and ways to network yourself into an internship position.

Amee Olson did just that. After graduating from college, she seized the initiative and volunteered to work as a research

assistant in an informal summer intern position with the
Community Information Exchange, the national information
service for the community development field. She was gladly
accepted. She then located a foundation stipend that supported her over the summer. Amee parlayed that experience
into a paid job for another national nonprofit group, then was
hired by the Community Information Exchange as a full-time
research assistant. In a few years, through networking, she
landed a job in the communications department of LISC and
later won a promotion to assistant program officer.

What makes a good internship?

- one that has specific expectations spelled out for you at
  the beginning;
- institutional support; usually a staff person available to
  provide supervision and to serve as mentor;
- an organization that will include you in its organizational
  activities, so you feel part of the group and can understand
  how it functions;
- substantive, programmatic (minimal gofer) work, with a
  tangible product at the end of the internship so that you
  and the sponsor can feel that something has been accomplished;
- specific and stated in-service training and skill-building
  opportunities.

With these points in mind, if you are offered an internship,
ask for an interview with the highest-level person available
and discuss the position and the organization's commitment
to its interns. Ask for a job description or a written summary
of the assignment. Raise specific questions, such as whether
you will be included in staff meetings; sent to specific training
events; permitted to take part in relevant local activities such
as legislative hearings, press conferences, and tours of the
neighborhood given for visiting bigwigs.

Organizations that offer satisfying internships are "learning organizations," nurturing places that draw the best from
both interns and staff and help them learn and grow. A small
part of your assignment may include routine tasks, but you
should be certain up front that your work will be substantive
and meaningful.

Making the most of the learning opportunity will be up to
you. Be proactive in finding out all you can about the organization, beyond your own assignment. Try to grasp the big picture of the economic and social context in which the
organization works: why it exists, how it meets its challenges.
Try to understand the human element of its work: talk with
other staff members, find out about the other jobs in the organization, talk with the people the organization serves. Read
whatever materials the group has on the organization and the
community development field. Attend whatever events they hold.

Finding the Right Training for Newcomers

As the field has evolved and expanded over the past
three decades, the need for technical skills and knowledge in
each of its aspects has grown. Working to finance housing or
business development requires a grasp of public and private
financial markets, risk analysis, cash-flow analysis,
spreadsheeting, and other demanding subjects. Organizing is
more than setting forth with a handful of leaflets to knock on
doors. Decades of experience lie behind effective organizing
techniques, and there are different schools of organizing, some
confrontational, others consensual, that require different approaches. Training programs are described in appendix K.

An industry has developed to train practitioners in each of
the field's different aspects. There are training workshops
that, over the course of a few days or a week, introduce new-

comers to the general concepts and strategies of community development. Others focus on specific techniques, starting with the basics and offering follow-up courses that teach increasingly sophisticated methods. A few training institutes enroll practitioners for a year or more, covering a curriculum in almost as much depth as a master's degree; often trainees work on a real-life project from their neighborhood as part of the course so they can put their new skills into practice.

Many training programs serve specific constituents, from bankers learning community development lending to people learning how to manage multimillion-dollar portfolios of affordable housing properties, from organizers to the staff of community development credit unions. Dozens of training programs focus on laypeople and clergy who are interested in doing community development from a faith-based perspective.

Training programs are also diverse in where they are located and in the geographic area they cover, from rural to citywide to national.

Because of the broad variety of training programs, newcomers have a chance to acquire skills that will qualify them for employment. However, most of the programs have been created to broaden the capacities of people already working in community development. Knowing this, you might decide that your best strategy would be to get a job in the field, prove your usefulness to your employer, and then ask the organization to send you for training. Your employer might, at that point, give you the time and also pay for your training.

Getting on the Path through Education

If you have your sights set on a job with a nonprofit at the local, regional, or national level, a bachelor's degree in planning, real estate development, or business–public administration makes sense. If you have a specific kind of career in mind, let's say housing development, you might study architecture and design, engineering, real estate development, or finance.

In the public sector, an undergraduate major in economics, public administration, business, or public policy might give you a competitive edge.

If you hope for a career in community development banking, an undergraduate business or finance degree gives you a good background,since that major typically covers economics, banking, accounting, an understanding of what money is, the flow of funds, and credit. Because many banks provide first-rate training, your first step might be to find an entry-level position, then go through the institution's in-house training. After that, you can weigh the benefits of going on for an advanced degree.

Your options are broad. You can study biology and work in a community health clinic. You can major in social work and work with tenants who are organizing to turn their building into a co-op. You can study economics and join an organization working on the policy issues underlying neighborhood problems.

But as the life stories of community developers throughout this guide show, people who have done well in their community development careers count among their majors many that seem totally unrelated—psychology, philosophy, English literature, textile manufacture, biology, music education. If you are still searching for your niche, let your skills and interests decide your major. What matters will be your basic abilities—to think critically, to communicate well, to listen and learn (the qualities of a good liberal arts education); your basic understanding and knowledge of our society—its economic system, its social history, and the oncoming effects of

Career Story: India Pierce Lee

Program Manager, Local Initiatives Support Corporation, Cleveland

"I was born and raised in the Glenville neighborhood of Cleveland, still live and now work there. I grew up wanting to be an airline stewardess. Instead, on one of the jobs I worked during college, I found out about and passed the test for an air traffic controller's job, was given training, and worked as an air traffic controller till then-President Reagan fired the striking controllers. At that point, the minister in my church, who was also the area's councilman, asked me to take over the church's housing construction program. He wanted to get me back into the community. My family had always been active in the community. My sister is a neighborhood planner.

"The church program involved taking vacant lots out of the city's land bank and on that land building nice houses for low-income families who were eligible to get mortgages from the city. Many people thought that no one would buy the houses, but growing up in that neighborhood as I did, knowing lawyers and doctors and professors stayed in Glenville because they wanted to, I knew it could succeed.

"I learned about housing development, about block grants and so on, on the job. I learned the basics of construction, like how the footers go in, and people from the city

helped me understand what studies and tests had to be done for site approval. I learned how to take a house and totally rehab it. It is a wonderful thing when you hear families say how much they liked their house, know-

ing you made a difference in a person's life. Having a vision and seeing that vision become reality.

"From that church program, two years later, a new organization was formed—Mt. Pleasant NOW [Neighborhood Organizing Working Development Corporation], and I became its executive director. We moved out of the church, and eventually the staff grew from four to fourteen. By the time I left, we had developed two hundred new and re-

habbed houses.

"While I was running Mt. Pleasant NOW, I worked with Neighborhood Progress, Inc., and LISC, two intermediaries in the city that offered us a lot of technical assistance. I also went through the Development Training Institute program. These experiences helped me grow, and in turn I helped the staff to grow their capacities.

"I stayed at Mt. Pleasant NOW for seven years. It was one of the greatest experiences of my life and helped shape what I like to do. I loved what I did. When other people said, 'We can't revitalize our neighborhood,' we proved you could. Our housing development included market-rate houses—we bought vacant parkland from the city and built new houses there that sell for a base price of $160,000. We did the first new construction scattered-site housing with Low Income Housing Tax Credits for moderate-income families, with basements, garages, carpets, and 1 1/2 baths, that rent for $350 a month.

"As executive director, I knew every street we built on. I walked door to door and talked to the families about their concerns, to gain their support, over a fifty-block area. I wore out a lot of shoe leather. I knew it was important to build relationships with community people, not just with funders and banks.

"Then I was asked by Mayor White to join the city government as program director for the Empowerment Zone office, and one month later I was made director of the program. In that year, I changed jobs, sold my house, got married, and got promoted!

"I stayed with the city's Empowerment Zone program for 2 1/2 years. During that period, we made $20 million in loans to small businesses, retained over seven hundred jobs, created a center for employment and training, opened a job-match program, and opened a one-stop career center. During that time, I was also selected to take a certificate course in community and economic development at Harvard's Kennedy School of Government, developed for Empowerment Zone directors and paid for by HUD.

"I left to join LISC and its National Equity Fund (which finances affordable housing using Low Income Housing Tax Credits), because as a national organization, it offers opportunities that don't come up that often. My responsibilities here include securing tax credit financing for projects in Cleveland, Lorain, and Youngstown counties. I also provide technical assistance to CDCs and help them manage their properties. I'm expanding CDC initiatives to include economic development, commercial development, and day care and assisting them in increasing market-rate homeownership."

globalization; and your interest in issues affecting low-income communities.

If as an undergraduate you have made the decision to pursue a community development career, you might choose a college that offers internships. For example, the Los Angeles Trade-Technical College's Community Development Technologies Center offers a two-year associate of arts degree, with courses ranging from an introduction to community economic development to community organizing, business law, and strategies to increase community employment. With a grant it won from the U.S. Department of Housing and Urban Development's Hispanic-Serving Institutions Work-Study program, the college offers its students work-study internships in community-building projects.

Read the material in the "Advancing Your Career" section below on qualities to look for in a school and how to find good schools.

Do You Need a Graduate Degree for an Entry-Level Job?

An advanced degree is usually not necessary for your first job, though there are some professions where it may be key to gaining a foot in the door. In a policy institute, for instance, you are likely to need a graduate degree in public policy, management, business, or finance.

More often, the need for an advanced degree depends on the level of the job. You can enter a city or state agency as a housing or community development assistant with a B.A. When you want to advance to a position as program manager, it will be useful to earn a master's in urban planning or public administration or some equivalent discipline. Look at the job descriptions in appendix A to identify which jobs require an advanced degree and the disciplines involved.

If you have an advanced degree and at least three years' experience in some relevant work before you look for a job in community development, you may be able to enter in a somewhat higher-level position. The difference is captured in a statement from Carol Norris, vice president with ICF Consulting, a firm with a substantial number of community development consulting assignments. She explains, "I look primarily for smarts and skills and will hire people at the beginning of their careers. A person with a master's degree always starts as an associate, while non-master's hires with similar experience begin as research assistants or analysts. There is also a difference in pay."

You might consider taking a year or two off after finishing college and get some work experience in the field before enrolling in a graduate program. You might take a series of jobs, some paid, some volunteer, to try out different settings. This experience will help you decide whether you really do enjoy community development work, what kind of work you like best (organizing or policy or number crunching), and what kind of setting you prefer. Do you prefer work at the local level dealing directly with people or work at the level of an intermediary organization? The work experience will help you decide what discipline you'd like to study, or it may show you that technical training is what you need rather than a graduate education.

"Do what and go where you will learn the most. After you've been in the for-profit sector, it may be hard to disengage because of salary. Start in the nonprofit sector, get skills that are useful, learn your job, take advantage of the greater responsibility you'll get in the nonprofit sector than you would in the private sector. This makes you more value-added if you switch sides," advises Stephanie Smith, whose own life story—told in chapter 3—is living proof.

Using Related Work Experience to Get a First Community Development Job

Community development is a remarkably fluid field; that's one of its outstanding characteristics. This makes it possible for people to move from related work into a community development job. Contrast this with other fields, such as medicine or law or teaching, in which entry requires years of new training or education.

Most job seekers will probably come from a position in another nonprofit organization dealing with some aspect of community life. But work in the private, for-profit sector or in government can equip someone with skills related to community development. A person who learned real estate finance at a bank can apply that knowledge within a nonprofit housing organization; a computer techie would be welcome in an organization large enough to need communications and database management know-how; an entrepreneur might be hired to set up a community business to train and employ neighborhood residents. Kathryn Walker, after ten years as a staff accountant and then as comptroller for a commercial real estate company and a law firm, used her knowledge of accounting, business, banking relationships, and real estate development to land a job as chief financial officer as well as chief operating officer for the Kansas City Neighborhood Alliance. (One of its projects, Quality Hill, is profiled in chapter 5.)

Seizing Opportunity If a First Job Comes Your Way

In community development's earlier years, when the field was figuring out what it was, what it did, and how it was different from other economic development and social justice movements, perhaps as many as half the people migrated by chance and happenstance into community development jobs. They brought a useful trade or talent, though they often knew little specifically about the field.

Today, as the field has matured, serendipity is less common, and getting a job usually takes a real and demonstrated interest in the field and serious effort. But there is still an element of chance. The key is to recognize the moments of potential opportunity and capitalize on them. If the opportunities don't exist, make them. There's a definite connection between the kind of people who gravitate to community development and the kind who seize serendipity and turn it their way.

Advancing Your Career

The same strategies that can lead to a first job in the field—volunteer work, internships and fellowships, training and education, and other work experience—are the ones that make it possible to advance your career. There are subtle differences: For career advancement, a person would need to have a long history of volunteering, rather than a year or two, to turn that experience into a senior-level job; and while internships are usually intended for people at an earlier stage of their lives, fellowships often come with advanced studies or as a mid-career reward. But, basically, the paths are alike.

Volunteering as a Launch Pad for Mid-Career Shifts

A surprising number of people in the field, including many who have risen to leadership positions, started out in unrelated work and shifted to community development in mid-career. India Pierce Lee, whose life story is told in this chapter, was an air traffic controller. John Taylor, who now heads a citywide intermediary in Philadelphia, was a policeman for six years. But they, and others who have made mid-career shifts, did have a long history of community volunteering, which both demonstrated their commitment and gave them intimate knowledge of the neighborhood, the CDC, and the workings of community development in

their city.

Taylor explains, "After working as a policeman, I subsequently got a job with a local manufacturing company that always had a representative on the board of directors of the Allegheny West Foundation, the CDC of the area in which the company is located. They asked me to serve on the board; I enjoyed it and got involved in committees and other work there. As the company I worked for shrank, I knew that I would be transferred or laid off. When the CDC's executive director left to set up another program, I was asked to take over, which I did."

John Taylor's story is informative in another way—it underscores the fact that employees who have been downsized out of a job are an important pool of talent for the community development world. The field encourages them to consider making this shift.

Finding Fellowships for Mid-Career Professionals

Many people in the community development world enroll mid-career in graduate programs and in nondegree training programs. Sometimes they secure a fellowship to take time out for a degree; sometimes they study while continuing to work. They are usually motivated by the urge to learn more about what works.

Recognizing how invigorating time to think and new surroundings can be for professionals, a number of institutions have created mid-career fellowship programs. The U.S. Department of Housing and Urban Development, together with Harvard University's Kennedy School, created the Community Builders program for mid-career neighborhood leaders and top HUD staff. The philanthropic world also offers programs for mid-career people to stretch their horizons; these include

Community Story: Colorado Coalition for the Homeless

Housing Homeless Families

When a family has reached the point of homelessness, it means they have lost a livable income, their ability to cope, and the support network of friends or relatives. The state of homelessness and the destructive journey to homelessness often undermine their mental health. Drug or alcohol abuse may be either a cause or a result.

Because of these factors, housing homeless families calls for more than just a decent place to live. To address these problems, the Colorado Coalition for the Homeless develops housing with many human services built right in. In the Renaissance at Loretto Heights, which the Colorado Coalition built in Denver in 1998, twenty-five once homeless families live in apartments side by side with fifty-one working families, who are, in a way, role models. Mixing families also means the homeless are

not isolated in one segregated group.

The Renaissance is a handsome cluster of new townhouses facing a central courtyard on six acres of land. The complex includes a community center with an exercise room, swimming pool, gazebo, and playgrounds.

The homeless families receive a remarkable range of support services—job training, continuing education, help finding employment, child care, mental health counseling, substance abuse

counseling, and health care, all arranged by the Colorado Coalition for the Homeless. The families must agree to develop and abide by a plan designed to help them achieve increased self-sufficiency.

Eleven sources of capital went into the financing for the Renaissance, making it possible for families with incomes as low as 7 percent of the area median income to live there without needing public subsidies. By using Low Income Housing Tax Credits, the Coalition raised $3 million in equity from private investors. The Bank of America provided the construction and permanent mortgage. The county donated the land, HUD gave the project a special grant for the Coalition's program for the homeless, and the local housing authority and the state, among others, also provided funds.

The Colorado Coalition for the Homeless, with a staff of 125, works throughout a five-county Denver metro area to develop affordable housing and provide supportive services for homeless and low-income people. Among the 377 units of housing it has produced are 225 units of transitional or special needs housing. CCH manages 133 rental assistance vouchers for homeless families and individuals. It operates two health clinics for homeless people and public housing residents and runs a mental health outreach and treatment program for the homeless.

the Fannie Mae Foundation Fellowships and the Rockefeller Foundation Next Generation Leadership Fellowship. See appendix J.

Tom Zuniga was one of HUD's first Community Builders. As Tom says, he "got hooked on community development" when, as a novice in a monastery, he was sent for a summer to work as a youth organizer in a Chicago parish in the very neighborhood where Saul Alinsky, the fabled organizer, had organized his first project. "That's when I decided I needed to do something other than be a monk," he says. He left the monastery to earn a master's in policy science from Columbia University. Then he recruited a group of black teachers to run an after-school program in the Bronx, work that took him to a tenement where conditions were so awful he organized a tenant rent strike. The landlord threw up his hands and gave the building to the group. To make a long career short, after many other career moves (including a position in which he financed the development of the Geno Baroni Apartments, an affordable housing project in Washington, D.C.), Tom enrolled in a Virginia seminary and is researching faith-based initiatives during his HUD Community Builders fellowship.

Training to Advance a Career

Training has to occur at the right time—the "aha" factor of higher education, when you learn something you are ready to apply to your own experience. Training can teach technical skills, skills in handling organizational issues, and skills in dealing with people. It can expand your circle of colleagues and networking contacts. These are the benefits that Jack Trawick describes. Jack, the executive director of the Louisville Community Design Center, had worked for three years in that position and seven years altogether in the field when he signed up for the year-long training program offered by the national Development Training Institute. "It was a critical experience for me, an opportunity to understand the root philosophies upon which the modern community development movement is based, and to get to know my classmates, a fabulous group of people from all over the country who became dear friends and helped to sustain me. At DTI, I also began my 'graduate education' in real estate and business finance and development. I learned about the fundamentals of real estate finance including the Low Income Housing Tax Credit program. Up to that point I had thought everything was built with grants.

"Second, I learned about organizational development, that effective organizations are effective because they are well organized. The third thing I learned is the importance of 'soul,' of a spiritual commitment to the work regardless of where that comes from. You have to believe in what you're doing, particularly in this work."

Most training programs are intended for practitioners who have a job in the field. Through training, they often are able to expand the scope of their practice, making their jobs broader and increasingly more interesting. Training can also lead to career advances within their organization or to a higher-level position in a new setting.

Pursuing Education to Move a Career

Going on for an advanced degree may be the right next step for people in a number of different places in their lives. If you have tried without success to land the community development job of your dreams, earning a master's may open new doors for you. If you have worked in an entry-level job, advanced education or training will help you move up or find a job in another part of the field.

The disciplines where you will find community develop-

Community Story: Financing Community Health Centers

Roxbury, Massachusetts

The Community Health Center Fund is the brainchild of Toby Yarmolinsky, public finance officer at Tucker Anthony, a Boston-based regional investment bank.

Community health centers, typically located in economically distressed neighborhoods, serve low-income and uninsured patients. As a result, they operate with non-existent profit margins and cannot build up significant reserves to upgrade and expand their facilities. Nor can they secure loans, given their weak credit profiles and the risk lenders perceive of lending to real estate in lower-income communities. The Community Health Center Fund, now working in several Massachusetts cities, is a model of how to

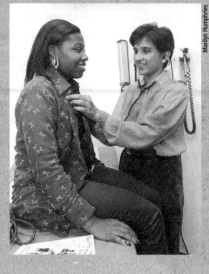

Marilyn Humphries

provide the missing financing.

It does this through a new tax-exempt bond program, in which the bonds issued by the state are backed by a multimillion-dollar guarantee shared by the teaching hospitals and health maintenance organizations in Boston. The first beneficiary of the fund was the Dimock Community Health Center, which used the money to upgrade an old building into a state-of-the-art medical facility and added thirty full-time positions. Founded in 1863 by women physicians, and one of the nation's first community health centers, Dimock provides family-centered primary health care and is the largest employer in its Roxbury neighborhood.

ment explicitly included as part of the curriculum are as diverse as planning, economics, business or public administration (many schools have combined these two disciplines), law, social work, sociology, real estate development, public policy, urban affairs, real estate development, and religion. See appendix B, "Universities and Colleges."

The most common graduate school disciplines among people now working in the field are planning, urban studies, and social work. But there are changes underway. Because the field is growing larger, with more complex transactions and relationships, the increased involvement of private-sector investors, and a greater demand for technical and management skills, future job seekers will do well to study business,

economics, law, and public finance. At the same time, because developing a community means not just bricks and mortar buildings but also all the other aspects of community life—youth development, day care, and education, to name a few—people with social science credentials can also find their way into higher-level jobs in the field.

Community development work always challenges you to keep building your skills: to learn the newest finance techniques or the changing concepts of "community building"; to become more knowledgeable about how to tie the improvement of a rural community or an inner-city neighborhood to the strengths of the regional economy; or to learn how to mount a program to encourage the growth of local micro-enterprises. Or to study economics or business or public administration because the global economy, the corporate world, and the political system are the context within which community development functions.

Experienced community developers also find that further education helps their careers in several ways. Going on for an advanced degree is the path most traveled by practitioners to move upward in their careers or to shift from one sector of community development work to another.

Taking time off for further education is also a fine way for practitioners to take a breather. "One of the things that happens when you work in one place a long time, as I did for the Enterprise Foundation for eleven years, is you find it hard to step back. I needed to get a new view, to change gears," Pat Costigan explains. His advice to fellow practitioners: If you have the financial means, work in an international program for a year; or take a sabbatical; or look into the exchange programs in the government. Pat found that his two years studying for a master's in public administration at Harvard's Kennedy School of Government gave him the "intellectual restimulation" he needed at that point in his life.

Education also makes it possible for someone to make a transition in mid-life from a totally unrelated career into community development. Tony Hall (whose story is told in chapter 5) had years of computer programming and database management experience with for-profit companies; he shifted to the community development field after earning a master's degree from Yale's Nonprofit Management School, where he met colleagues whose networks were instrumental in helping him make that shift.

What Are the Qualities to Look for in a School?

There are a number of factors to consider when choosing a school and a discipline to pursue:

- What schools offer the curriculum and courses you want?
- Which school and discipline are most likely to give you the web of relationships that can help you network for your first and subsequent jobs?
- Which schools offer internship and fellowship programs that might expose you to the people and experiences that will help you decide the next step in your career and give you the skills to achieve that step?
- Which discipline matches your temperament and talents and is most likely to prepare you for the kind of work you think you'll want?
- What financial aid is available?

The good schools for learning community development are those that combine faculty members who have a demonstrated background and interest in some aspect of community development; a relevant curriculum; applied research; internships or fellowships that provide the chance for hands-on field work and professional experience; and a demonstrated connection to communities in their areas.

How Do You Find These Schools?

One way to look for these qualities is to find out whether the school has a community development center or institute attached to the department in which you are considering enrollment. Look for a center that combines academic education, community service (often called service learning), and research. Consult appendix D, which describes schools with undergraduate and graduate curricula relevant to community development, the internships and fellowships they offer, and other features of particular relevance to community development work.

An example of a university internship: The Massachusetts

Institute of Technology's internship program for master's students in the Department of Urban Studies and Planning. Students work ten to fifteen hours per week for a semester on projects in organizations devoted to urban planning, research, or development. They also attend professional development workshops run by the department. Compensation for the semester is about $2,000, or $14 an hour.

In addition to appendix D in this guide, a source of information on schools likely to provide a good community development education is HUD's Office of University Partnerships (OUP). One of that office's programs, the Community Development Work-Study program, provides grants to institutions of higher learning specifically to support community development work-study internships and fellowships for graduate students. Another HUD OUP program, the Community Outreach Partnership Center program, makes grants to institutions of higher learning that partner with their communities in practical community-building initiatives; HUD funding gives these schools the resources to develop relevant curricula and often to fund internships.

Similarly, the Fannie Mae Foundation supports a University-Community Partnership Program that funds selected universities to work with community-based nonprofit housing groups and other partners to expand their ability to develop affordable housing for low- and moderate-income households. These university initiatives offer outstanding opportunities for professors and graduate students to work on real-life neighborhood projects.

Even if you do not apply for an internship or financial aid, use these sources to learn about the educational institutions with a good community development curriculum.

"Over the years, students [who] have attempted to work in the field as they have gone through their academic program discovered what the field expects of them, what their vocational skills are, who the players are in the field of development, and where opportunities for work and entrepreneurship are. Those students who have restricted their work to library research have not been so fortunate," finds Eastern College, a faith-based school outside of Philadelphia with several master's programs in community development.

Adding Community Development to the Agenda of Another Job

Even if a job is not specifically in community development, you may be able to add a community development element to your current work. That might satisfy your interest in community development or offer the experience to land a job later.

You could, for example, work in the world of public finance and spend part of your time on deals that benefit communities, as does Toby Yarmolinsky, public finance officer with Tucker Anthony, a Boston-based regional investment bank. "It was while I was studying at the Yale School of Management that I discovered the world of public finance, which is where finance and community needs meet, both for public and nonprofit organizations," he says. At Tucker Anthony, he finances bonds on behalf of nonprofits and for affordable housing and economic development transactions that are too small to interest Wall Street firms.

Another example is that of Ernie Hughes, a cooperative extension service agent based in Baton Rouge, Louisiana. Over the years, he has so successfully expanded his work agenda that he now coordinates the state's community development trade association. There are similar examples by the score. The point is that a surprising range of professions can be molded to add a community development angle, once you know something about the field.

Chapter 5

Jobs in the Community Development Field

Chapter 2 described the diverse set of organizations in the community development field. This chapter summarizes the wide variety of positions available within those organizations. Jobs are very different in the level and kind of skills needed. The day-to-day experience varies widely from working with people in street and storefront settings to writing policy briefs looking at a computer screen. As in virtually all fields of endeavor, the jobs also differ based on degree of responsibility and on whether they include the management of other employees. The highest-paying jobs in the field are those that require the management of organizations and the people in them.

This chapter describes jobs that require skills specifically related to community development. Figure 5.1 provides a guide to a sample of the jobs in the field and is the structure for this chapter.[1] The next chapter describes what the day-to-day work is like in these jobs, through the personal accounts of people in the field. Appendix A lists job descriptions based on the job categories in Figure 5.1.

Figure 5.1 and the descriptions of jobs in appendix A serve as a shorthand guide to the kinds of jobs in the field and who

has them. The job descriptions are based on actual job openings in 1999 and have been modified to make them clearer and to emphasize the skills and experience needed for each job. However, the community development field does not have a rigid set of job titles and job descriptions. A position called a program manager in one place might be called a community development specialist in another; a job development specialist might be an economic development specialist in another location. Each of the jobs shown in appendix A is as close to an actual job with actual salary as possible. However, salaries very widely, based on the size of the organization, the general cost of living in the metropolitan area where it is located, and whether the job is at the community level or elsewhere in the community development field.

This chapter follows figure 5.1 based on the location of the jobs. But figure 5.1 can be used by looking at the kinds of jobs as well. If you know you want to be a community organizer, for example, figure 5.1 will tell you that those jobs exist at the community level and the intermediary level but are not likely to be found elsewhere.

Where the Community Development Jobs Are

> ● = Most to all groups in category have jobs.
> ◗ = Some groups in category have jobs.
> ○ = Few or no groups in category have jobs.

JOB	LOCATION							
	Com-Based Groups	Intermediary	Government	Capital Provider	Advocacy/ Trade	Academic	Consult	Partners
Executive and Administrative								
Executive Director, Community Program, e.g., Community Development, Neighborhood Housing Services Program	●	○	○	○	◗	○	○	○
Community Manager, Community Program, e.g., Community Development, Neighborhood Housing Services Program	●	○	○	○	◗	○	○	○
Executive Director, Specialty Program, e.g., Economic Development Group, Rural Partnership Business Development Corporation	●	◗	◗	◗	◗	○	○	◗
Executive Director, Public Housing Authority	○	○	●	○	○	○	○	○
Executive Director, Community-wide Housing Partnership	◗	◗	○	○	○	○	○	○
Volunteer Coordinator, Local Housing Group	◗	◗	○	○	◗	○	○	◗
Chief Financial Operator	◗	●	◗	●	◗	○	◗	◗
President, National Intermediary	○	●	○	○	◗	○	○	◗
District Director, e.g., National Intermediary	○	●	○	○	◗	○	○	◗
Department Director, e.g., Intermediary	○	●	○	○	◗	○	○	◗

JOB	LOCATION							
	Com-Based Groups	Intermediary	Government	Capital Provider	Advocacy/ Trade	Academic	Consult	Partners
Community Organizing								
Community Resident Organizer	◖	◖	○	○	◖	○	○	◖
Community Organizer, Public Safety	◖	◖	○	○	◖	○	○	◖
Assistant Director, Organizing Group	◖	◖	○	○	◖	○	○	◖
Construction Related								
Housing Rehabilitation Specialist	●	●	●	●	○	○	○	○
Housing Rehabilitation Manager	●	●	●	●	○	○	○	○
Construction Manager	●	●	●	●	○	○	○	○
Housing Development								
Housing Development Manager	●	◖	●	◖	○	○	◖	○
Housing Development Officer	●	◖	●	◖	○	○	○	○
Housing/Community Developer	●	●	◖	◖	○	○	○	○
Housing Development Specialist	●	●	◖	◖	○	○	○	○
Economic/Workforce Development								
Employment Specialist	◖	◖	◖	○	◖	○	○	○
Community Economic Development Specialist	◖	◖	◖	◖	◖	○	○	○
Director, Economic Development	◖	◖	◖	◖	◖	○	○	○
Storefront Renovation Director	◖	◖	◖	◖	○	○	○	○
Business Consultant	◖	◖	◖	◖	○	○	○	○

JOB	LOCATION							
	Com-Based Groups	Intermediary	Government	Capital Provider	Advocacy/ Trade	Academic	Consult	Partners
Technical Assistance/Training								
Field Service Officer	○	●	○	○	◗	○	○	○
Financial Service Specialist	○	●	◗	◗	○	○	○	○
Director, Specialty Program, e.g., Native American Programs	●	●	●	◗	◗	○	◗	◗
Community Services								
Case Worker	◗	◗	◗	○	○	○	○	◗
Resident Services Coordinator	◗	○	◗	○	○	○	○	◗
Resident Services Manager, Community Service Center	◗	○	◗	○	○	○	○	◗
Coordinator, e.g., Youth, Drug Elimination	◗	◗	◗	○	○	○	○	◗
Homeownership Counselor	◗	◗	◗	◗	○	○	○	○
Program Planner, e.g., Community Service Agency	◗	◗	◗	○	◗	○	○	◗
Real Estate Management								
Assistant Property Manager	◗	◗	◗	○	○	○	○	◗
Property Manager	◗	◗	◗	○	○	○	○	◗
Senior Property Manager	◗	◗	◗	○	○	○	○	◗
Housing and Community Development Management Supervisor	◗	◗	◗	○	○	○	○	◗
Research and Public Policy								
Research Assistant, Public Policy	○	◗	◗	◗	◗	◗	◗	◗
Legislative Assistant	○	◗	◗	◗	◗	◗	◗	◗
Policy Analyst/Legislative Assistant	○	◗	◗	◗	◗	◗	◗	◗

JOB	LOCATION							
	Com-Based Groups	Intermediary	Government	Capital Provider	Advocacy/ Trade	Academic	Consult	Partners
Research and Public Policy, cont.								
Director of Public Policy	O	◐	O	O	◐	◐	O	O
Vice President for Research	O	◐	◐	◐	◐	◐	◐	◐
Capital Provision/Community Lending								
Loan Processor	◐	●	●	●	O	O	O	O
Community Lending Specialist	◐	●	●	●	O	O	O	O
Finance Specialist	◐	●	●	●	O	O	O	O
Senior Loan Officer	◐	●	●	●	O	O	O	O
Senior Finance Specialist	◐	●	●	●	O	O	O	O
Housing and Community Development Officer	O	◐	●	◐	O	O	O	O
Senior Asset Manager	O	◐	◐	◐	O	O	O	◐
Senior Vice President, Real Estate Lending	O	◐	O	◐	O	O	O	O
Portfolio manager/Vice President	O	◐	◐	◐	O	O	O	O
Portfolio Analyst	O	◐	◐	◐	O	O	O	O
Director of Community Development Programs, e.g., Community Foundation	O	◐	◐	◐	◐	◐	◐	◐
Director Regional Office, CDFI	O	◐	◐	◐	O	O	O	O
Planning								
Housing & Community Development Policy Planner	O	◐	◐	O	◐	O	◐	O
Community Planner	O	◐	◐	O	O	O	O	O

JOB	LOCATION							
	Com-Based Groups	Intermediary	Government	Capital Provider	Advocacy/ Trade	Academic	Consult	Partners
Fund Raising Grant Writer	◗	●	●	○	●	●	●	●
Vice President, Development	◗	●	●	○	●	●	●	●

At the Community Level

The bulk of the jobs in community development are at the core of the field—are those closest to the ground, working directly with neighborhood residents and other players in the system, as indicated in figure 2.1, the Map of the Field. Virtually all of the other jobs in the field support the work at the community level, either directly or indirectly.

The jobs at the core of the field involve working for small, mostly nonprofit organizations. As many of the professionals who have shared their stories with us indicate, the jobs at this level are exciting and rewarding—and can be very challenging— "the agony and the ecstasy." Jobs at the community level include:

- executive and managerial positions
- community organizers and community builders
- construction-related positions
- housing development positions
- economic/workforce development
- community services
- real estate management
- capital provision/community lending

- grant writers and fundraisers

Executive and administrative positions, usually called executive director and deputy director, are the equivalent of the chief executive officer and chief operating officer of small companies. These people run the organizations. They are the paid leaders of the community development groups. They report to a volunteer board of directors that sets overall policy. They oversee the staff and are responsible for making and meeting budgets, hiring and firing staff, and seeing to it that objectives are defined and reached. The professionals who succeed in these positions know the content of the work, have management skills, are entrepreneurial, handle both inside and outside (stakeholder) responsibilities, and generally speaking, have the capacity to lead the organizations they run. They have lots of the same skills and traits as small business owners and executives. The executive director of a community-based group works with a wide array of partners: bankers, intermediary staff, neighborhood leaders, and many others. The position requires great versatility. These people lead the way in the development of real estate, working with a broad range of partners; they have the ability to "be the suit,"

Career Story: Bob Giloth

Program Officer, Annie E. Casey Foundation, Baltimore

When Bob Giloth finished college in 1972, he had some vague political ambitions, but he wasn't at all sure what shape they would take. As in the case of many of his contemporaries, his reading led him to City Planning School at the University of Illinois. Bob was influenced by the urban writing of Lewis Mumford and by Jane Jacobs's *The Death and Life of American Cities*.

Immediately after getting his master's in city planning in 1976, Bob was attracted to some Chicago neighborhoods that were fighting a big comprehensive plan from the business community—neighborhoods that were trying to control their own destinies. He took a job in the Pilsen neighborhood in Chicago, with the Eighteenth Street Development Corporation, where he worked for five years. The work he did there involved renovation of older housing in the neighborhood and job training of young adults. That meant he had to negotiate with construction companies, do some rehabilitation with his own crews—that means training them—and figure out how to finance the purchase and rehab of as many homes as possible.

"When I saw the problem I was confronting—learning more about the business every day—I turned to other neighborhood workers who were confronting the same challenges. I quickly realized that we

weren't talking with each other nearly enough. We needed a structured way of getting information exchanged—maybe even saving money by group purchases—things like that." That realization—plus a lot of work—led to the creation of the Chicago Rehab Network, a nonprofit established to help Chicago-area neighborhood groups with their projects. Bob eventually left his job at the neighborhood group and went to work in city government under Mayor Harold Washington, the first black mayor of Chicago.

Bob then went on to Cornell, where he received a Ph.D. in planning—and then followed his wife to Baltimore, where he directed a neighborhood group called SECO, the Southeast Community Organization. After five or so years there, he moved on to become a program officer with the Annie E. Casey Foundation.

"Don't skip over the step of getting your hands dirty" is Bob's advice to a newcomer to the field. "My work with Pilsen and then with local government is the basis for my understanding as a professional of what it takes to get things done at the local level. New people coming into the field need to experience firsthand what it takes to have an impact in a community," he says.

Bob views himself as one of two kinds of people in the field. There are those who take on a neighborhood challenge and stay with it for a long time, people like Mary Nelson and Paul Roldan, who are highlighted elsewhere. These people, says Bob, are challenged to reinvent their organizations and themselves periodically. And there are those like Bob, who stay with a project or an employer for four or five years and then move on, doing their best to take the learning from one job and apply it to the next. In both types, success at community development requires a dynamism and a willingness to adapt.

to work with bankers, investors, and business people—they have to raise money to support the organization—and at the same time they have to be comfortable in the neighborhood, coping successfully with the rough-and-tumble street environment in the community.

Nancy O'Brien, the executive director of the Neighborhood Housing Services Program in Great Falls, Montana, puts it this way: "A good CDC executive director is one who knows how to get dollars. Begging for dollars is not to be underestimated. You need the contribution dollars for your operating support. Somehow you have to be able to talk to the bankers to get them to lend money. You've got to be able to write applications that will get grant funds. You need to be a marketer; you need to sell the program, to convince people to give or lend your program the money it needs to accomplish its mission. This is tough work. You've got to be creative, resourceful, and not give up. It's a demanding job."

Salaries for these positions vary widely depending on the size of the organization, its location, and the experience of the person holding the position. Salary surveys indicate that these jobs can pay from $35,000 to a high well over $100,000 annually.

Community organizers and *community builders* work directly with groups of people to develop agendas for action and to keep an organization strong through outreach and recruitment of new active people who are willing and able to work to improve conditions in the community. Organizers and community builders need solid skills in interpersonal and intergroup dynamics, and in helping people successfully achieve community goals. They must be very savvy politically, since the work they do is often about the redistribution of power. They knock on doors to involve people, get people to meetings to make important strategic decisions on improving the neighborhood, organize tenants to get improved condi-

tions from landlords, or even to convert rental property to cooperative ownership. They organize the community to fight against environmental hazards, to secure jobs for residents, to close down drug dealers—or to mount a community celebration focusing on the positive aspects of the neighborhood.

Community builders often work with select groups in a neighborhood such as youth workers, or senior citizen community workers, helping these groups with special community building needs. Salaries for organizers and community builders at the neighborhood level range from $20,000 to $40,000 annually.

Construction-related positions are closely connected to real estate development. These positions are responsible for the close monitoring of residential and commercial buildings that the neighborhood group is building or rehabbing. Strong knowledge of construction techniques, pricing, and the design process is needed for these positions. In many instances real estate developers may know little about construction techniques themselves. While this is a handicap, it can be overcome by having CDC staff skilled in actual construction and rehabilitation techniques. The construction or rehab specialists work on the construction estimates of a building and, when it is underway, visit the site almost daily to monitor construction, seeing to it that the work is consistent with the plans and that corners are not being cut, and determining when modifications called change orders should be made to the job. A small percentage of people filling these positions are architects or engineers; most have hands-on construction skills, learned on the job or in some earlier position. Jobs in this speciality pay from $40,000 to $90,000 annually.

The *housing development professionals* are at the heart of most CDCs working on real estate development. The pro-

Career Story: Pete Garcia

President, Chicanos Por La Causa, Phoenix

Pete Garcia is the kind of community development professional who puts down roots and stays, working with the same community organization for decades. Pete joined Chicanos Por La Causa, Inc. (CPLC) in 1972 and is today its President and CEO.

"I started as an economic development aide. I had a degree in elementary education," says Pete. "While working for the city of Phoenix as a youth counselor, I moved to CPLC as an economic development aide, packaging business loans to create jobs for the youth I knew needed jobs. It meant I had a lot to learn, and I took classes at night to learn about economic development. I advanced in the organization and by 1975 became director of economic development. I shifted again and in 1979 became social services coordinator, heading a newly created department."

Pete then left for two years, as part of an Intergovernmental Management Training Program sponsored by the U.S. Department of Health and Human Services, where he received a master's in public administration.

Then it was back to Phoenix, where he became president of Valle del Sol, Inc., a nonprofit specializing in drug rehabilitation. Pete turned that organization around, and when CPLC needed a new president, he agreed to take the job. "It was supposed to be temporary—but here I am," he notes.

Under Pete's direction, CPLC has become a statewide community development corporation with forty-seven offices in twenty-three cities, employing over five hundred people.

"One of the things I'm proudest of," says Pete, "is our success in growing people from clerical to management positions within CPLC. The people we have here care about the people we're trying to help—and we've been able to take that concern and combine it with careful in-house development and training. But, more and more, our new hires are people with a college degree. They have more versatility and can move in and out of jobs more readily."

"It's key that community development workers are aware of what's going on. They should read the newspaper, be current, and committed to the mission of the organization. We use people who are skilled in fundraising, in deal making, in finding new opportunities, in managing what we've already developed. With over five hundred employees, we have lots of needs for different kinds of skills."

"I'm still here because I get rewarded every day by seeing how we are changing lives and communities, now across the state. But there's a lot more to be done."

fessionals in these spots are skilled in housing development, either multifamily (rentals) or single-family homes. Their job is to get housing built or rehabilitated, following goals set by the board and executive of the CDC. In these jobs, breadth of community development experience is not nearly as important as skill in the speciality. In addition to technical skills, these positions call for a high degree of interpersonal skills. These people usually report to the CDC executive director, although in a very large CDC the real estate development and management staff make up a department or division, and there are likely to be staff who report to a senior real estate development position.

A housing development professional knows how to get housing built or rehabilitated. That means knowledge of land acquisition and improvement, zoning laws, building codes, and finance approaches. He or she must work effectively with banks, planners, architects, and the other partners critical to getting real estate developed. A housing development specialist knows how to find sites that are suitable for development (or buildings fit for rehabilitation) and negotiate a purchase with an owner. Sometimes the work involves getting the property from local government if it is abandoned and has been acquired through tax sale.

The number of these kinds of housing development positions in a CDC varies with its size, as do the specific program skills needed. The salary range for these positions is not as broad as for the executive and management positions. Salaries vary mostly based on the salary structure of the CDC and the experience level of the person in the job. Expect to see annual salaries in the $50,000 to $70,000 range.

Economic/workforce development specialists work on improvement of the economic conditions in a community and/or on increasing the capacity of people in the community to get jobs in the neighborhood or elsewhere. These community development professionals work with local government, business associations, and others to recruit businesses into neighborhoods, work with existing businesses to help them expand, and may work with staff involved with workforce development to match employers with potential employees. There are far fewer of these positions at the community level than there are real estate development positions, in large part because most CDCs work in residential settings, and because they have only recently moved into the economic development area. These positions usually require knowledge of how business functions and what business needs, as well as knowledge of government programs that help businesses. These professionals may work on big-picture issues—what kinds of businesses can we recruit to our neighborhood—or very specific activities like helping a retail business get a loan or grant to improve a storefront.

Other positions in this category focus on job placement and job training of people in the community. People in these jobs work directly with unemployed and underemployed people to link them to jobs for which they qualify. To do this efficiently, workforce development specialists develop relationships with employers and job training programs. Usually, there is considerable job-readiness training and follow-up with those placed in jobs to coach them through the challenges of being new on a job—or even new to the whole world of work.

Economic/workforce development specialist positions are not as numerous as the housing positions listed above, but CDC involvement in economic development and the workforce is developing. Salaries for these positions are in the $40,000 to $75,000 range.

Community services workers help individuals and families, or groups of people with similar issues in their lives, deal with those issues so they are more able to reach life goals. These workers are often trained in social work, but they come from many backgrounds. These are jobs for people who are most skilled at relating closely to people and who have a disposition toward caregiving of some sort. The specific content of the work varies depending on the community's priorities. In one neighborhood a social service worker might be in-

Community Story: New Quality Hill, Kansas City

Building a Mixed-Income Community

Community development efforts take many forms depending on neighborhood circumstances and what the residents are striving to achieve. Sometimes the goals are aimed at helping low-income people with housing or economic development; in other instances, the goal is to redevelop a neighborhood to bring about a mix of incomes and backgrounds in the people living there.

New Quality Hill in Kansas City, Missouri, is an example of a successful mixed-income development stabilizing a neighborhood that common wisdom held was on a path of decline. Once one of Kansas City's prime residential locations, by the early 1970s Quality Hill had become another distressed neighborhood adjoining a midwestern downtown and struggling to overcome the flight of residents to the suburbs.

Stirred by historic preservation activists bent on saving the architectural heritage of the area, the local business community formed a consortium to provide a low-interest loan pool to help rehabilitate the neighborhood. City government got deeply involved, offering community development funds, tax-exempt bond financing, new streets, sidewalks, lighting, and other public improvements, and a property tax abatement for the first years of the project's life. A Missouri-based developer, McCormack Baron, worked with these local groups and neighborhood residents to develop a vision for a new Quality Hill—a neighborhood with preserved residences, well-designed homes and apartments, safe streets, and some new retail shops. The developer and local business leaders raised more than $11 million from over one hundred local people who agreed to invest in the future of the neighborhood. This combination of private and public resources produced a large-scale, mixed-income community that integrated rehabilitated historic structures with well-designed new buildings.

The development resulted in the construction or rehabilitation of 466 apartments, 30 condominium units, and over 50,000 square feet of commercial space. Red-brick facades on the new structures recall the area's earlier look. Attractive streetscape amenities add charm with decorative street lights, brickwork, and monuments to mark the main entry points into the neighborhood.

But beyond the physical, the neighborhood works well as a place to live. It is more racially integrated than most Kansas City neighborhoods. White-collar workers make up more than 75 percent of the residents. Household income ranges from a very low $1,900 to $180,000 annually. Rents range from $410 to $820 a month, with the very low-income residents receiving a subsidy so that they pay no more than 30 percent of their incomes toward rent.

New Quality Hill exemplifies the complexity of larger-scale community development efforts. First, determined leadership from the community is needed to develop the commitment to bring about a better future for the area. Then, the group must agree on a vision for the neighborhood. In fact, there are many decisions about the future: What kinds of housing? What income groups? What should the neighborhood look like? What are crime reduction and avoidance strategies? And many more. The vision has to be practical enough that it can be financed. Where will the money come from in the form of debt, soft loans, grants, tax credits, abatements, and below-market rate financing. Can investors be persuaded to invest in the project? Can the physical product—houses, commercial space, public improvements—be built within the estimated budget? How do the decisions get made about those physical improvements that the community wants but can't afford? Once the homes are underway, they have to be leased or sold—and in the case of rental properties, the apartments must be well managed so they look attractive for years to come.

Complex, with many players: community organizers, real estate developers, architects and urban design professionals, finance specialists and lenders, construction companies, property managers. It all takes time—but the results are improved neighborhoods and opportunities for people who otherwise might not have them.

Career Story: Tony Hall

High-tech Consultant, New York City

Tony Hall grew up in a segregated rural Mississippi town in the 1960s. A summer fellowship he won from the National Science Foundation during high school sent him to San Diego State University, where he was first exposed to computers and programming. He went to Dartmouth, majored in math, which is where most of the computer courses were taught, and then spent his last college year at the University of California at San Diego. He got a job there at Litton Data. When he returned to Memphis, he studied for an MBA at night, working days at a small software consulting firm. He spent twelve years at that firm and became a partner toward the end.

"At about thirty-five, after I had been doing computer programming for almost fifteen years, I began to think about where to go next, even considering Ph.D. programs in computer science. But I looked at my own personal priorities and basically decided I wanted to realign my career to coincide with my personal interests. I knew I had been very blessed in the opportunities I had had and wanted to give that back to my community, to make that happen for others. I didn't want to do computer programming forever.

"The Yale School of Management master's program attracted me. The program always had a large sector of students who were making career changes and a mix of people from both private and nonprofit sectors. I didn't really know much about the community development field. I didn't know what it meant to go into the nonprofit world. At Yale, the

real learning experience comes from the students and the things they had been doing before they enrolled.

"I knew I wanted to work in disenfranchised communities. The Yale program included lots of guest speakers whose experiences gave me concrete examples of what one could do.

"I learned good management skills. Also, I knew I'd be working in a very political environment, so I took courses in the political science department and as many courses as I could in the economics department, during my first year. One of my professors took an interest in me, introduced me to a program officer at the Annie Casey Foundation, where I was offered a summer position with one of the companies that the foundation funds to provide technical assistance to its CDC grantees. It was a New York–based firm that does lots of work around information systems for nonprofits. I worked there during the summer between my first and second years at Yale and during the school year commuted down there one day a week. At that job, I could use my skills in the computer field and learn more about the world of community-building initiatives.

"After graduation, I worked full-time at that firm on a project for the Comprehensive Community Revitalization Program that funds CDCs in the South Bronx. My project was to create a computer system for nonprofits that offer a multitude of services; the system tracks and coordinates those services and their clients. This was for me a fuller introduction to the issues around improved service coordination for the families. Our entry point was the technology, but we used that as a platform to put questions about service coordination on the table.

"About two years later, I went off to work on my own. That appeals to me because as a consultant you do not do the same thing over and over. The networks are what keep me alive in my consulting work. It was through an Annie Casey Foundation person, for example, that I was introduced to The Enterprise Foundation and got an assign- ment working with them on an employment and workforce development initiative in Baltimore.

"There's a lot of learning we all have to do to integrate the economic, social services, and social fabric components of community.

I've been putting together my own work that will allow me to understand those connections and how strategies can be better created that reinforce those connections."

volved in organizing a self-help group of teenage single moms who are trying to finish their high school education and get a job. In another, the work might involve helping senior citizens through Meals on Wheels programs or some form of adult day care. There is growth in these positions in the areas of job placement specialists, who help unemployed community residents find jobs. Salaries for these positions are in the broad range of $20,000 to $100,000.

Real estate management positions are found in larger CDCs that have developed enough housing and other real estate that they choose to manage the housing themselves and employ real estate management personnel to do that. Real estate managers have the job of seeing to it that the real estate owned by the CDC is well run and is an asset to the community. In the case of housing management, the challenge of the job is often to create a community among the residents living in the housing. But the basic and important responsibilities of this job include: marketing the units or the commercial space, collecting rents, seeing to it that the property is well maintained, completing reports to the owner as well as lenders and regulatory bodies, handling emergencies, and evicting tenants for nonpayment of rent or improper behavior. Within the housing context, all of these responsibilities are of-

ten combined with actions to build a community among the residents. This can involve sponsoring social activities, running programs for children, connecting with community service agencies to provide special programs on site—all in an effort to make the housing an asset to the community and to the people living in it.

Real estate management is one of the entry-level positions in the field. Assistant property managers are often trained on the job and can then advance to higher levels of management or, with training, to other areas of the community development field. Higher levels of real estate management often require certification from a qualified trade group or training program, such as the Institute for Real Estate Management (IREM) or the Consortium for Housing and Asset Management's Nonprofit Housing Management Specialists (NHMS) credential. See appendix A,

Capital provision/community lending positions are found in those community-based groups involved in lending funds to community residents and businesses. Those organizations are not usually CDCs but other community-based institutions like community development financial institutions (CDFIs), community credit unions, and micro-enterprise loan programs. These positions are more fully described in the section on capital providers below.

Grant writers and fundraisers raise money for the work of the CDC or neighborhood groups at the neighborhood level. These positions tend to be filled with good writers and relationship people who can help the organization in its capital needs. The job involves staying current on the organizations that are interested in funding the work of community groups and applying to them for funds through written proposals and follow-up. In some cases, such a position might involve planning and executing fundraising strategies and programs, such as special events, awards dinners, and the like. Most neighborhood groups do not have the budget for a full-time position of this kind. Often the job is handled by the executive director, or by another staff person, or is contracted out to a fundraising consultant. When a full-time position in grant writing or fundraising exists, the salary is generally in the $30,000 to $75,000 range.

These are the major paid job categories at the neighborhood level. Some of the larger CDCs may have jobs in additional service categories or programs—working with the elderly, youth, health care, or volunteers—all of which require specialized skills in the program area. In most cases, the staff reports to the executive director. In larger CDCs, staff are divided by functions, with a senior-level person heading up each group of specialists. Successful CDCs orchestrate the different skills and activities well, so that the work is well coordinated. The additional challenge is for the staff to stay fresh and not become burned out. Successful CDCs—and other community development organizations—take steps to help the staff stay committed to the mission of the organization.

Intermediaries

The jobs at these intermediary organizations are one step removed from the community-based groups. The direct "customers" or clients of the intermediaries are the community-based groups themselves. What usually makes a person a desirable hire for an intermediary is earlier experience with neighborhood-based groups. Salaries for staff at the intermediaries are usually considerably higher than those at the neighborhood level.

The positions at the intermediary level include many of the areas of technical skills found at the community level, but they are wider in range and more highly technical than at the community level. While each intermediary and national organization is organized differently, there is, in general, a separation at this level between jobs that provide technical help and training; jobs that provide capital (loans, equity, and grants); and jobs that raise capital. There is also specialization in the kinds of technical assistance provided: real estate development, construction and rehab, financing, job placement, and others.

The intermediaries also have two other kinds of positions—communications professionals to handle the many newsletters and other forms of information sharing needed in the field, and public policy professionals, whose job involves trying to get governments to develop and improve programs that help the field.

As figure 5.1 shows, the key jobs at an intermediary are:

- executive and management
- community organizing
- construction-related positions
- technical assistance and training
- community services
- real estate management

- research and public policy
- capital provision and community lending
- planning
- fundraising
- public relations and communications

Many of these positions require the same kind of background as those at the community level but with more experience and depth. Perhaps the biggest difference is that intermediaries have staffs actively involved in providing training and technical assistance to community-based groups. The *technical assistance and training positions*— sometimes called program officers or field officers—provide a form of consulting. One field officer from a national intermediary described his job as having three functions: First, a field officer serves as a liaison to some number of organizations in an assigned territory and coordinates services from the intermediary to those groups. Second, the field officer provides technical assistance to those groups, including housing development, organizational development, fundraising, marketing, management systems, and community and business planning. Third, the field officer coordinates interaction among groups at the state and regional levels. Field officers are expected to stay current on "best practices"—those approaches that have been tried and appear to be successful—and, conversely, know what's been tried and failed, so that newer groups learn from the mistakes of others.

There are a considerable number of these positions within the intermediaries—both in headquarters and in field offices and at local intermediaries.

In addition to the technical assistance and training work, activities that occur at the intermediary level but not usually at the community level include: substantial lending and grant making, management of the loans (assets), and policy and research work. There is also a considerable amount of coordination work involved at the intermediary level. Part of the challenge is to get the component parts of an organization working well together, so that the services delivered to the community nonprofits is as user friendly to the CDCs as possible.

The *lending function* at the intermediary level is essentially one of structuring loans to CDCs, evaluating the risks of making the loans, and servicing (collecting) payments on the loans once they are made. The skill needed here is the ability to assess the strengths of the organization borrowing the funds by looking at its financial situation, its leadership, and its operations—and the quality of the real estate or other business venture the group is seeking to finance. These skills are virtually the same as those at financial institutions and CDFIs (see "Capital Providers," later in this chapter). In addition to lending, some intermediaries are conduits for equity between private investors and CDCs developing low-income rental housing. The staff people making these investment decisions are skilled in real estate finance and undertake a careful review of project feasibility and developer capacity before recommending investments.

Because part of their mission is to build the field, the intermediaries are involved with *research and public policy*. Although the staffing in this area is small, most national groups have research and policy staff that work on documenting successes in the field; focusing on legislation and policy issues at the local, state, and federal levels; and providing the results of the research to the field for its use. Recently, these policy staffs have begun to use the Internet to promulgate their work and to communicate quickly with large numbers of people in the field to influence national policy and legislation when necessary. See appendix H for a list of the best Internet sites.

Intermediaries also have staff involved in *public relations and communications,* since getting the word to people outside of the field about what is being accomplished is a key part of the role of the national organizations. Community development jobs in public relations and communication need people skilled in writing, media relations, communication strategies, and event planning and management. In any given week, a communications person might be working on an an-

> *"I like to work on things that others have had trouble doing, leveraging my experience and negotiating deals to make things happen. Figuring out ways to do the things the communities have wanted for years, such as putting the Hunters Point [Los Angeles] shipyard back into productive use."*
>
> *—Kofi Bonner,*
> *Los Angeles Economic Development Department.*

nual report, a newsletter, a public event like a ribbon cutting or ground breaking, and a proposal for funding.

Government

There are thousands of jobs in government—HUD alone employs over 10,000 people, a significant percentage of whom are directly involved in community development (community building is how HUD refers to many of its community development jobs)—and many other positions in housing finance and the administration of the public housing program,

fair housing, and the regulation of the expenditure of HUD funds. HUD has offices throughout the country. For information on HUD's programs and the Community Builders Program, visit www.hud.gov. Vacant positions at HUD are posted on its Web site.

Work at the federal level generally means joining a large agency. Federal workers interact mostly with officials from other levels of government, other federal agencies, and community groups. Jobs at the federal level work best for people who desire positions that are quite clearly defined and have sharper boundaries than jobs in the community development field at the community level or elsewhere. These specialists operate programs such as the Community Development Block Program, a key funding source for community development. The government professionals in these positions see to it that the funds are used creatively and for eligible purposes, and they seek to bring a level of creativity to the uses of the federal funds. Program professionals typically have a set of regulations that their program must follow, and they spend considerable time informing grant recipients on how to meet requirements placed on the programs by Congress.

The federal-level jobs offer more job protection and continuity than most at the community level. It is not unusual for federal employees to stay with the same agency for twenty years or more. Many jobs are unionized and subject to collective bargaining agreements, offering additional protection. Virtually all jobs have excellent fringe benefits.

The jobs at the state level—housing finance agencies (HFAs) and departments of community affairs, for example—are usually technical and finance oriented. An HFA is like a bank in some respects. Jobs at HFAs involve underwriting loans and equity investments, outreach to community groups and housing developers, and oversight of investments that

Career Story: Nicolle Boujaber-Diederichs

Business Manager, Community Development Research Center, New School University, New York City

International community development is another career path. This work can be pursued overseas working hands-on in a community, or through organizations headquartered in the United States. You can also alternate work in the international arena with work in U.S. community development.

Nicolle Boujaber-Diederichs, born in Iowa and now married to a Moroccan, has had assignments in Rabat with the Moroccan Association for Solidarity and Development, creating an evaluation scheme for their microlending program for women; in New York City with Shared Interest, a community development fund for South Africa, raising money to capitalize the organization's social investment fund; and in Wash-

ington, D.C., doing research, including field research in South Africa, to launch a new mutual fund that would invest in low-income micro-enterprises in developing countries.

She has a master's of international affairs from Columbia University; is now, at thirty years old, the business manager for the Community Development Research Center of the New School for Social Research; and is enrolled in New Hampshire College's Ph.D. community economic development program.

"Jobs in the U.S. in international development are very gratifying in terms of the mission, but there's a lot of paper pushing, administration, and fundraising involved," she says. "Working overseas, depending on the type of job you get, you are more likely to be working with communities on the ground. That's gratifying if you are a people person and don't want to be staring at a computer all day but extremely challenging for personal life—for relationships and marriages."

have been made (asset management). Some state HFA workers play an outreach role, working with prospective users (customers) to encourage use of the state's programs. As is true at the intermediaries, these relatively small organizations also offer positions in management, positions generally filled by people who have significant housing and finance experience and who have demonstrated the ability to manage an organization.

State jobs in other program areas require skills in their

particular program concentration. The bulk of the interaction at the state level is with the federal and local governments and with community-based groups. The positions here often concentrate on developing and implementing loan, grant, and technical assistance programs aimed at assisting local governments and neighborhood groups to build and rehabilitate housing, foster economic development, historically preserve important buildings and areas, and address problems of special needs groups.

At the local government level, community development jobs exist in a number of agencies and departments:

- *City planning departments* require people with planning skills to develop plans and development policies for the city or county, usually in conjunction with neighborhood-based groups on plans for neighborhoods. These professionals often view themselves as partners with community groups in carrying out agreed upon plans, and they often must make the tough decisions of allocating scarce locally controlled government funds. The planner's function is to help achieve a shared vision for the future of an area. Doing that means being able to collect and understand data on the area, conduct and participate in meetings to reach consensus on a future for the area, and orchestrate the steps needed to help advance the vision through meetings with other local officials, elected and appointed. Lots of time is spent in meetings with community groups, in discussions about the best planning approach in a particular situation, and in meeting with appointed and elected officials to develop consensus around projects and plans.

- *City and county housing departments* are generally in the business of improving the quantity and quality of the area's privately owned housing stock. These departments work in conjunction with developers (profit and nonprofit) to get housing built and rehabbed and often in partnership with banks to make mortgage and home improvement loans available to home buyers and homeowners. The size of these departments and the kinds of jobs in them vary widely based on the size of the city and the particular emphasis that a mayor or county executive is placing on these community improvement issues. The positions in housing departments vary from policy and planning (advising on what programs ought to be cre-

"Community development lending jobs involve financing and project management. That means analyzing proposals for financial feasibility; using spreadsheets; getting redevelopment reports prepared; evaluating the borrower's track record, capacity and financial condition. Like any real estate lending it's a 'people-property deal.'"

—*Dan Nissenbaum,*
vice president, Chase CDC.

ated or improved), to program management (seeing to it that programs are operated on budget and consistent with the rules for the programs), to rehab and construction (working with homeowners and contractors to see to it that physical improvement work is done in a fair and competent way). The policy work done here and in the planning departments is a mix of analysis, negotiations, and finding approaches that fit within the politics of the situation—the many competing forces that shape a community's future. Considerable work occurs in helping for-profit and nonprofit groups make their projects feasible—and in negotiating approaches that result in the least amount of financial help from the local government, so as to spread scarce public funds across many projects. Quantitative skills and negotiating skills are highly desirable.

- *Public housing authorities (PHAs)* are different from housing departments in that housing authorities are special purpose institutions that develop and own housing for

Community Story: Pioneer Human Services, Seattle

Helping At-risk Youth and Adults Change Their Lives

Seattle, Washington, despite its high quality of life, has a sizeable number of people who were once criminals or addicted to alcohol or drugs. Serving these "at-risk " youth and adults is the mission of Pioneer Human Services. One characteristic that distinguishes Pioneer is that it views its clients as customers. Rather than simply taking care of their needs, it tries to give them "a chance for change," to restore themselves.

Another characteristic that distinguishes Pioneer is its strong entrepreneurial spirit. Pioneer almost entirely supports itself by structuring its array of employment and training programs and housing as nonprofit businesses. By selling services to commercial clients at market rates and furnishing treatment and services under contracts with local, state, and federal agencies, Pioneer earns income to keep its activities going and growing. While in its early years, 75 percent of its funds came from the government, today only 25 percent of its operation is publicly subsidized.

To provide employment and job training to "client customers"—while generating revenue—Pioneer runs businesses that give

those customers extensive and paid job training and work experience. It also provides good alcohol- and drug-free housing for them as well as counseling. Among its businesses, Pioneer manufactures and finishes light metal parts for commercial customers including aircraft, telecommunication, medical, and power management companies. It is Boeing's sole supplier of sheet-metal liners for aircraft fuselages.

Pioneer operates a business that provides assembly, repackaging, and transportation services. Its Food Buying Services is a wholesale business that buys, packages, and sells about 7.2 million pounds of food annually to over four hundred food banks and nonprofit customers in twenty-five states. Another unit operates a 150-seat cafe in Starbucks' corporate headquarters. Its institutional kitchen prepares and delivers over 500,000 meals a year. It runs a full-service printing, mailing, and fulfillment business.

For people with chemical dependency, and those who are also mentally ill, it runs inpatient facilities. It also offers outpatient adolescent and family mental health and chemical dependency services, as well as counseling services onsite in public schools.

Today, Pioneer Human Services serves over 6,000 client customers a year and 1,750 individuals at any given time, employs over 1,000 staff, and operates a budget of over $55 million, of which $2 million is profit, plowed back into its businesses.

low-income people. Because PHAs are public bodies, the housing they own is known as public housing. These authorities vary in size from the many that own and manage a very small number of units to big-city authorities that own and manage thousands of units. The jobs at PHAs are largely in housing management and maintenance, but many housing authorities employ resident services specialists who work with residents of public housing to improve conditions in the public housing communities, and others are employed to develop or redevelop housing communities, calling on many of the same skills used in community development elsewhere. Many authorities also employ development directors who work with architects, planners, and private developers to get public housing built or improved. These positions require a skill set very similar to that required in housing departments.

- *Redevelopment authorities* are special purpose public bodies with the mission of redeveloping areas needing improvement, usually those that have been declared blighted. Redevelopment authorities have staff skilled in acquiring land, improving it, working with developers to build new homes and businesses on the land, and providing technical and financial assistance to developers as needed. These specialized positions often work with community groups when they are developing land owned by the redevelopment authorities. The work can involve planning and executing larger-scale projects, such as downtown improvements, as well as neighborhood efforts. Skills in planning, urban design, deal making, finance, and real estate development are particularly valuable here.

Capital Providers

Community development financial institutions and banks provide investment capital to the community development field. Employed in these institutions are community de-velopment professionals who specialize in lending to community development groups and projects. Some of the staff in these departments have lending backgrounds, but many do not. The lenders have skills in promoting loan products and reviewing the risk elements of loan applications in areas such as affordable housing, small business, micro-enterprises, community facilities, intermediaries and CDCs, and other community development activities.

Many other bank and CDFI employees are not lenders but work on key partnership relationships, promoting loan and other investment products, providing technical help to possible borrowers to see to it that they are more likely to qualify for an investment, and working on product innovations. A considerable number of these workers move into banks and CDFIs from neighborhood-based work, intermediaries, or government.

Bank of America, has a community development group of nearly six hundred, with people from diverse backgrounds: lenders, community activists, former government agency officials, CDC directors, real estate developers, and others. Most banks have community development groups that are considerably smaller.

Some positions in these institutions are primarily relationship and sales positions with the goal of stimulating the use of bank lending products. Other positions involve more technical skills, the abilities to structure deals and assess risks in a loan or pool of loans. The positions are challenging and can be either diverse or specialized depending on the size of the institutions and the management structure. Unlike jobs with community-based groups, the jobs in commercial banks tend to feel more "corporate," since the institutions are for-profit institutions and mix the goals of community development with the goal of making a profit.

Some caution may be needed here in that the commitment

HUD's Community Builders

HUD Secretary Andrew Cuomo reinvigorated HUD's community development activities in dramatic ways, perhaps the most important of which was the Community Builders program, a fellowship program aimed at the "best and the brightest." In his letter of invitation to prospective interns, Secretary Cuomo said, "President Kennedy attracted the best and the brightest into government because he challenged Americans to do more for each other and for their country. In that tradition, the Community Builders Fellowship program offers you an opportunity to give back to your community and, at the same time, enhance your skills and grow in your current profession."

Operated in partnership with Harvard University's John F. Kennedy School of Government, the program educated and trained individuals from diverse professions as Community Builders, who obtained practical, hands-on experience in community and economic development. Many Community Builders then returned to their communities energized with new knowledge and experiences and dedicated to creating better neighborhoods and stronger communities.

The Community Builders Fellowship program offered paid, temporary fellowships for skilled professionals from many walks of life: bankers, school principals, law enforcement officials, directors of nonprofits, social workers, academics, architects, planners, lawyers, economic development experts, health care workers, doctors, nurses, technology specialists, and many other professions.

As part of HUD's new focus on outreach and customer service, Community Builders work out of HUD field offices in up to eighty-one cities across the country and travel into nearby cities and towns to serve as the agency's first point of contact—or, in other words, HUD's front door. Community builders were the first point of contact with the agency for the thousands of people who need HUD's help: home buyers, tenants, community leaders, nonprofits, foundations, mayors, county executives, governors, bankers, real estate agents, business owners, and many, many others.

Community Builders are told to always find answers—they never say, "It's not my job." They are trained in all aspects of the agency and serve as team builders, fostering partnerships both inside and outside of the agency. Community Builders empower communities by providing technical expertise in finance and economic development programs.

---

to the community development lending arena varies among commercial banks. Some see the community development arena as an exciting business venture in which profits can be made by doing good in the community, and those institutions seek to work proactively at the building of community development customers. Senior management of those banks highly value the work of the community development department. In other companies, the community development department is seen as a necessity because the law requires that the bank make community development loans, but there is little enthusiasm within senior management for the community development area. In those situations, employees who take community development seriously may be frustrated by an unwillingness of senior management to be an aggressive lender.

Private foundations are another capital source—grants and low-interest loans—to the community development field. Many foundations employ program staff in the community development field. These jobs require significant expertise in community development, with day-to-day activity involving the review of proposals from participants in the community development field. The challenge of these positions is to be current enough on what is occurring in the field and discerning enough about the quality of requests for funds to make good judgments about the use of scarce foundation funds. Program officers at foundations spend considerable time in reading proposals, discussing the relative merits of one approach compared to another, and monitoring the progress of those organizations that are funded to see to it that the funds are used consistently with the intentions of the funder.

Partners

The jobs available in the category called community development partners also vary widely. The service-delivery partners—those that provide casework and other services to people in need in neighborhoods, need professionals skilled in the particular services being offered: social workers or caseworkers who can work with youth, the elderly, the homeless, AIDS victims, ex-offenders, unemployed people looking for work. They also need managers who can run programs, and they need grant writers who help the organization raise funds to support its activities. The social services field is larger and more fully established than the community development field, with its own network of schools that train social workers and its own professional associations.

The other major partner in community development—the business world—is involved because businesses and corporations see the value of healthy communities. Large corporations are often involved through a community relations professional, with the job of seeing to it that the corporation provides grants (although these are not nearly as large or as widespread as grants from philanthropies) and volunteers to

help with community improvement activities. From a job perspective, the positions here are relatively few, although it is not uncommon to see a community development professional move into a company in a community relations position. Smaller businesses are involved as community partners as well. In the vast majority of these situations, the officers and employees of the company are directly involved, because the company is behaving as a community development participant—and community development professionals are not hired to undertake the work.

Academic Institutions

There are essentially three kinds of community development jobs in academic institutions: teaching, research, and community involvement.

The teaching positions in academic institutions require a high degree of credentials. Most professors and instructors have a Ph.D. or other advanced degree in their field of interest. In recent years there has been a dramatic increase in the number of institutions of higher learning offering community development courses, and this has led to a substantial increase in the number of teaching positions.

The research positions require strong academic and research skills, with researchers often working for and with professors as part of the work of institutes that focus on various elements of the community development field or subject matters that affect community development.

The community involvement jobs have less to do with the fact that the institution is an academic one and more to do with the relationship between the institution and its surrounding community and an effort to build successful partnerships with others to improve the surrounding areas. The skills required for these positions are similar to those found in national intermediaries or larger real estate development

Entry-Level Jobs

Even entry-level jobs are competitive and may call for a few years' experience—which you will have if you've volunteered or worked as an intern.

First jobs in a *neighborhood nonprofit* might be as a community outreach worker or community organizer, responsible for getting the word out to residents about matters important to their lives, such as a new health clinic; for finding out what's on people's minds and being a liaison back to the CDC; for bringing residents together to understand the issues and collectively decide what to do about them.

A training program coordinator would put together workshops for residents. A housing counselor is a specialist, training and counseling families who are first-time home buyers, helping them to understand how to get a mortgage, how to improve their credit history, how to be a wise homeowner.

A job developer, through phone calls, office visits, written announcements, and perhaps Internet communication, would encourage employers in the city to make job vacancies open to neighborhood residents, keep residents up to date about job opportunities, and help in sessions that teach potential job seekers about interview and job search techniques and office behavior.

A development assistant works with the staff who are doing physical or economic de-velopment projects. An assistant property manager works with the staff who manage the organization's properties, making sure the buildings are in good condition, that the residents' needs are met and they know how to care for their apartments, and that crime and property destruction are minimized.

Program associate is the name for the entry-level job participating in one of the organization's special programs, such as after-school activities for neighborhood youth, urban gardens, or the loan fund that finances small business enterprises in the community.

Researchers and writers and fundraisers can be other entry points in larger nonprofits.

Caseworker is a first job, working one-on-one with individuals and families, helping them connect to services they need, such as health care, and to other agencies, such as those that teach English as a second language, and helping residents manage the problems that occur in their lives.

For a *regional or national nonprofit intermediary* (the organizations that exist to provide funding and technical assistance to neighborhood groups), a program assistant or program associate (also called project associate) is the entry-level position, working with more experienced staff in an ongoing program of the organization.

A technical assistance coordinator works with staff who provide training and technical assistance to the neighborhood-based groups in the organization's geographic area. Intermediaries offer first jobs as assistants in research, whose responsibilities include the dissemination of information; computer services; fundraising; monitoring, tracking, and writing about legislation; and communications. In the field offices of intermediaries, the program associate keeps the program work of the office going while other staff are out in the field.

For a *community development financial institution* (which provides equity and loans to community nonprofits and small businesses), a first job might be as a loan associate, preparing the documents used in making loans, monitoring the disbursement and repayment of the loans, assisting loan officers in underwriting new loans, and dealing with potential or current borrowers. Another entry position could be as an associate (also called development assistant), assisting the staff who are responsible for attracting the money that capitalizes the CDFI's investing fund.

For a *bank*, analyst is usually the title for a first position. The job might involve work that is similar in many ways to a CDFI's. The analyst would manage a portfolio of the bank's loans, track its financial reports, prepare loan applications for permanent loan closings, and work with loan officers in underwriting new loans.

In a *city or state agency*, underwriter could be a first position, making loans from the agency's program funds. Planner is another entry-level job. A job as housing and community developer has a wide scope of activities, from working with nonprofit groups on their development projects, to site visits to properties to see what repairs are needed, to putting together public hearings on community issues. These public-sector jobs can be a fine stepping stone anywhere in the field.

If you have an advanced degree and have at least three years' experience in some relevant work, you may be able to enter in a mid-level position. Chapter 7, "Career Paths," describes those jobs. See appendix A for job descriptions.

companies. Only a relatively small number of academic institutions have such positions.

Policy and Advocacy and Trade Associations

Advocacy and trade organizations have as their goals the development of policies and laws that help the community development field and, in the case of trade associations, the well-being of the group's members. These organizations tend to be based in Washington, D.C., or in one location per state where there is a statewide advocacy or trade association. The work in these organizations is not deal related or casework related but is aimed at changing the overall environment in which community development is carried out. Many of the jobs are research and policy oriented. These kinds of jobs deal with concepts and ideas, and the outputs are often position papers and research reports.

Considerable time is spent in advocacy organizations in meetings, strategizing, building support for ideas and legislative approaches, and raising funds for such efforts. The work can have very important outcomes for the community development field. Getting the low-income housing tax credit passed and continued, for example, has meant that commu-

nity groups have a major financing tool available to them that has led to the production of thousands of needed housing units.

The trade associations—groups like the National Congress of Community Economic Development, the trade association for community-based groups involved in community economic development, and NAHRO, a trade association for housing authority board members and staffs—have policy-oriented staffs similar to those in the advocacy organizations. In addition, these organizations provide member services, seeing to it that members are receiving training, technical help, and information to keep current on developments in the field. These positions require strong organizational, writing, and interpersonal skills.

Often called think tanks, research and policy organizations collect and analyze information on a wide range of issues, such as the performance levels of CDCs, the investment practices of churches and banks, and the most effective ways to train unemployed people for living-wage jobs. Positions in research and policy organizations call for an analytic mind and critical thinking, theoretical expertise in a range of topics, computer skills, and high communication capacity.

At the Aspen Institute, a think tank based in Washington, D.C., Tim Walter is a senior associate heading the organization's program that explores the intersection of telecommunications and economic development policy, specifically how rural development practitioners use the Internet to share information and learn from each other. Tim has an MBA from Yale and spent years working in rural places for farm worker advocacy groups, followed by years working for private companies. He sees his job and the Aspen Institute as "a learning intermediary for the field of rural economic development."

Consultants

The community development field is widely served by consultants who provide services to virtually every player in the community development field. The role of the consultant is, essentially, to provide advice, but that often takes the form of carrying out detailed work assignments that may involve many months of sustained work.

There are far more small firms than larger ones, each with its specific niche in the field. While consultants come with varying degrees of experience, generally speaking, those consultants with relatively little experience find themselves in a larger firm, working in a team situation, being directed and mentored by more senior staff. Only after gaining considerable experience is a community development professional likely to succeed as a consultant independently.

Examples of the kind of work undertaken by consultants include:

- organizational development for nonprofits in the field—that is, helping organizations operate more efficiently and effectively (This work, similar to management consulting undertaken by mainstream consulting firms, involves the development of specific systems for organizations and strategic planning assistance.);
- work on particular real estate deals, especially those complex enough that the staff at the community development group cannot handle all of the analysis itself;
- evaluating the effectiveness of particular programs;
- fundraising campaigns.

Virtually every specialized activity in the community development field—except for the management of organizations—can be handled by consultants.

Chapter 5 endnote:

1. Most of the organizations in the field require people with skills that are not unique to the field but are necessary: bookkeeping, clerical skills, accounting, maintenance, etc. These kinds of jobs are not included here. People with these skills who want to work in community development should use this book as an indication of the kinds of places that might be in need of these more general services and skills. There are many instances of people who began work in a community development organization in a support position and learned the field well enough that they moved into a more mainstream community development job.

Chapter 6

A Day in the Life of Community Developers

In this chapter, people who work in different kinds of jobs in different kinds of organizations—from youth organizer to lender to social services provider—describe what their jobs are like.

Most jobs in community development have a lot of contact with people: clients, who could be, for instance, potential buyers of the homes you're building or trainees in your job placement program; colleagues, including neighborhood leaders and office coworkers; and the professionals you work with outside your organization to get a project done, such as bankers and city planners. Most jobs involve the kind of stress that comes with responsibility. Every job has its paperwork, telephoning (and emailing), and staff meetings, though in varying amounts. One day is likely to be fairly different from the next. The work is rarely boring. Most jobs reward you with a high level of fulfillment. Other than these common characteristics, there is a great deal of variation between one kind of job and another.

Youth Organizer

The setting: El Puente Academy for Peace and Justice, Brooklyn, New York

The person: Paula Ximena Rojas-Urritia, age twenty-nine

Career path: Chilean born; researcher, Valparaiso Women's Research and Organizing Project; undergraduate degree, Smith College; community organizer, Holyoke (MA) substance abuse prevention project and fair housing project; staff organizer, Greater Williamsburg (Brooklyn) community-building collaborative; cofounder, volunteer freedom school for young women of color, Brooklyn; project director, El Puente Academy.

The job: Project director

The project: El Puente Academy for Peace and Justice, a public high school in the Williamsburg section of Brooklyn, was chartered to foster a movement for peace, justice, and human rights in the struggling, predominantly Latino neighborhood. The school sponsors a project that has created a community organization for its students to learn and work as community organizers.

"The first piece of my work is to help different groups of the high school students who are becoming community organizers. The groups are at different stages. One is a group that's been working together for two years now on a documentary video so they can tell others about the community vendor project, where the women of the neighborhood organized a weekly outdoor market to sell food and their arts and crafts. I meet with this group twice a week after school to go over their action plans and decision making. My role with this group is to connect resources from adults in the outside world to their work. I helped by bringing them a professional videographer. As an organizer, I consciously try not to take a front role; my role is to work myself out of a job.

"Two other afternoons a week, I meet with another group, one that's relatively new at organizing. My role with this group is more guided. I help them check out the basics of how you get to understand your community, its institutions and players. We talk about the history of people who have made change for themselves, we discuss what is power and the power we each have to make change.

"My coworker with this group, Michael Walkers, is nineteen years old. Michael and I spend a few hours together each week planning what the sessions are going to be like. We take turns writing different training curricula about the analytic skills we use in organizing. We take turns leading the group.

"Since it's young people involved, there's also a youth development component. We play games, getting people analyzing their own self-interest and their achievements, then we make a chart about what made them attain their achievements. We help them connect their own analysis to figuring out what will motivate community people to join in a project.

"I spend a lot of time doing one-on-one as a mentor with a core of about ten high school–age young people. They talk to me over lunch or on the weekend. I'm the first person they call to talk about things like their best friend being run over by a car or trouble with family.

"One evening a week I meet with the group of women who pooled their money to create an outdoor vendors market. They are all Spanish speaking, all older women, who've become a pretty autonomous group. I'm the person who helps them keep on track on how to realize their goals. One of the women takes the lead; I only help move the process along.

"I also teach a class once a week in the high school on community issues. I spend a lot of time as a resource with the teachers in El Puente to work with them, suggesting books or presentations, to make classes in economics, math, and history more relevant to the community.

"Then there's the behind-the-scenes work. I meet with many people in New York to tell them about the El Puente project. I'm on the board of the New York Organizing Support Center, so I spend some time fundraising for that and putting on trainings. There are tons of phone calls, just networking.

"I work also for the Consensus Organizing Institute, as part of their national program. Once a month I travel to another city to help groups doing a youth initiative. I talk on the phone to help other cities think through what they're doing, as they're planning their program. I participate in some meetings with other COI staff.

"People like me tend to take on more than we can handle, and that can be stressful. I know that part of my responsibility is to model what an adult organizer is supposed to be like, that if I'm running around like a chicken without its head and with circles under my eyes, the young people won't want to be an organizer. But I try to be realistic for them, tell them it's a lot of hours, long hours; you take your job home."

Institution-Based Organizer

The setting: The InterValley Project, southern New England

The person: Ken Galdston, age fifty-three

Career path: Master's degree in public and private management, Yale School of Organization and Management; organizer and VISTA intern, Community Action Council, Rose Hill, NC; director, then staff director and lead organizer, Industrial Areas Foundation, Chicago; organizer, Coalition to Save Jobs, Boston; staff director and lead organizer, Naugatuck Valley Project, Waterbury, CT; staff director and lead organizer, Merrimack Valley Project, Lawrence, MA; staff director and organizer, InterValley Project.

The job: Staff director and organizer

The project: The InterValley Project, a network of organizations in southern New England, does institution-based organizing. This kind of organizing seeks to build the power of congregations, tenant groups, labor unions, ethnic groups, and chambers of commerce. The goal is democratic economic development, to see that people have more voice in the decisions that affect them, as a political force, a civil force, or an eco-

nomic force—as workers or consumers. In institution-based organizing, the organizer serves as a coach for institutions' leaders, teaching them how to organize.

The InterValley Project members are regional citizen action organizations that are multi-issue and broad based, combining organizing and development strategies. One member group, the Naugatuck Valley Project based in Waterbury, Connecticut, in the center of one of the nation's oldest industrial regions, shows what institution-based organizing can achieve. To stem the loss of thousands of jobs as plants shut down and businesses moved elsewhere, NVP organized the employees of one company to buy out the company and turn it over to employee ownership, saving 225 jobs. To create new jobs, the Project set up the Valley Care Cooperative to train formerly unemployed and low-income women as home health aides. Currently, NVP is organizing to win the redevelopment of brownfield sites (abandoned factories) to create jobs and clean up the valley's environment.

"The goal of an organizer is to find and develop leaders in the community and teach them how to build a powerful organization which can shape their community's future. As an organizer, in a typical week I would set up five to twenty-five one-on-one meetings with leaders in congregations, labor unions, and community organizations to see if they are interested in becoming active in our work. I'd also meet with several different groups of leaders to help them plan meetings or public actions. I'd help them prepare an agenda and prepare for the meeting itself. Though I might be asked to conduct a brief training for a given meeting, much of my role is to raise questions and serve as a resource.

"For example, if a group decided that they wanted an after-school program in their city, I might help them think through what's at stake here—why such a program is important to the broader community as well as for their kids; who has the power to decide to set up such a program; where they might find examples of the best after-school programs, locally or nationally. I'd help learn how to research the school budget, how to build support for a particular campaign within their own congregation or organization, and find allies in the city.

"Our goal is to be out in the field. We're on the phone to organize meetings, but the office is only a way station to the field.

"A typical week might also include working with leaders and staff months in advance to design one of our periodic leadership training institutes that teach potential community leaders what it means to be a leader, what the sources of our power are, what a community issue is, what a community campaign looks like, and how you recruit other leaders and broaden the diversity of the organization.

"As the director of the network, I'm out two or three days a week visiting the staff and leaders of the organizations in our network, helping them develop strategies to strengthen the leadership, membership, or financial self-sufficiency of their organizations. Or helping to develop campaigns around an issue. Or helping to plan network-wide activities, such as a training workshop. Or identifying resources, from technical assistance and funding to new ways of thinking about the global economy or urban problems."

CDC Executive Director

The setting: East Bay Asian Local Development Corporation, Oakland, CA

The person: Lynette Jung Lee, age 53

Career path: B.A., English literature, University of Hawaii, teaching credential, San Francisco State University; teacher, program for immigrant youth, Chinatown agency's alternative school; case worker, Community Action Agency, Oakland; volunteer and Board member, newly-formed East Bay Asian Local Development Corporation (EBALDC); staff, EABLDC, from volunteer coordinator/secretary to other positions, including assistant director, and interim director; executive director since 1982.

The job: Executive director

The project: EBALDC was formed in 1975 by a group of college students and citizens determined to convert a deteriorated yet historic warehouse in Chinatown into a community facility. Today this warehouse is the Asian Resource Center, housing nonprofit agencies, medical facilities, retail businesses and a state office. EBALDC has expanded over the years beyond Chinatown to work with groups in other Oakland neighborhoods, particularly those serving low-income Asian and Pacific Islanders. The organization has developed 550 units of affordable housing and two child care centers serving 130 children. It also provides homebuyer training.

Among its economic development activities, EBALDC developed neighborhood retail and office spaces totaling 85,700 square feet, organized merchant associations, and revitalized the historic Swans Market into a mixed use building including housing, a food vendors market, and office and retail space. It operates a revolving microloan fund and provides business planning and marketing assistance to the small businesses located in its commercial properties. It also operates a pilot Individual Development Account (IDA) project offering funds to match savings made by low-income families for homeownership or education or to start a business.

"Now that our organization has grown to about seventy-five staff, my job is very different from when I became the executive director eighteen years ago. At that time, there were two of us, and I did almost anything that the other staff person didn't do.

"One of the main ways I spend my time these days is in meetings. I am in meetings with staff and EBALDC partners, usually other nonprofits. This week, for example, staff and I met with the executive director of another nonprofit development organization called Affordable Housing Associates, to talk about the division of roles and responsibilities for a joint venture project. On another day I got together with a staffer from our local computer street academy, which, like EBALDC, is part of an East Oakland committee working on an Annie E. Casey Foundation initiative intended to strengthen families and neighborhoods.

"Staff and I also made a presentation this week to the Oakland Housing Authority, on one of their projects we're competing to develop. I met with other Housing Authority folks to break through a logjam for a housing project that will use housing vouchers. I took comp time to meet with a national officer from the United Methodist Church's community development program. I also met with a city council member to talk about issues in her district. On top of those meetings, I also participate in the staff meetings we hold by department and with individuals. And I give a good chunk of my time to the strategic planning process our organization is going through now, which involves both board and staff. I try very consciously to be a booster for our staff, which is now large, to make each person feel part of a team and important to our work.

"My time is also given to providing free technical assistance to other, smaller nonprofits, and to an affordable housing coalition in the city of Alameda. That is in addition to serving on a number of task forces and boards, such as the Oakland Sharing the Vision board for the city of Oakland and, just recently, the Housing Task Force. I also speak out on advocacy issues, giving formal testimony most often before the city council or a council committee.

"Fundraising and public relations are also part of my job. I write proposals, talk with program officers of foundations and officers of bank community development departments. I take many people out for a tour of our neighborhoods and our projects so they can get a better sense of the work we're doing. Our annual fundraiser is coming up, for which I make calls for sponsorships to banks, businesses, and other supporters who will purchase tables. Board work is also very important. I spend time meeting and orienting new Board members, as well as meeting with long-time board members to understand how they feel about changes in the organization and within the board.

"My job also involves national functions. I'm on the board of the Development Leadership Network and on the committee that deals with CDC issues such as how we can strengthen organizational sustainability.

"The Fannie Mae Foundation awarded me one of their James A. Johnson Fellowships this year. I'm using that opportunity to look at how I can apply my real estate experience to look in the Lower San Antonio neighborhood where schools are terribly overcrowded for sites that could be developed into new small schools.

"Being an executive director of a community development corporation often requires long hours but it has a richness and diversity of experience, as well as provides 'soul satisfying' opportunities, that are hard to match."

CDC Housing Developer

The setting: Rural Opportunities, Inc., (ROI) Rochester, New York

The person: Jay Golden, age sixty-two

Career path: MA in urban planning, University of Chicago; further training through numerous technical seminars; urban planner for a large for-profit company developing regional shopping malls and subdivisions; housing development officer, Illinois housing finance authority; vice president for a company developing government-financed multifamily housing; consultant in affordable housing development and property management to nonprofit housing development organizations; director of real estate development, Rural Opportunities, Inc.

The job: Director of real estate development

The project: Rural Opportunities, Inc., is a rural, community-based development organization founded in 1969 to help farmworkers, who are among the most disadvantaged people in the nation. Headquartered in Rochester, New York, ROI's work now stretches across six states—from New York to Michigan. It is a member of the Neighborhood Reinvestment Corporation's network and a certified community development financial institution. To improve housing conditions, ROI builds and manages multifamily housing, renovates run-down properties, and counsels low-income families about how to qualify for and obtain a mortgage so they can buy a home. Through its CDFI, it makes loans to small businesses and gives them business advice. It trains over six thousand migrant and seasonal farmworkers every year in vocational skills so they can compete for more stable jobs. While farmworkers earned $5,289 on average before training, their earnings leapt to $14,851 after training. The organization operates migrant Head Start centers for children. It offers services ranging from health education, emergency shelter and transportation, to lessons in English as a second language. With an annual budget of over $24 million, ROI now has a staff of 250.

"I have my own projects and I also work with our staff on their projects. I have technical oversight over all the projects done by ROI no matter where they're done. My job is to look at the proposed projects to spot all the holes in the fabric through which a problem could pass and engage my staff people in a conversation about sewing up that hole.

"I travel about twelve times a year. This week I visited our office in Alliance, Ohio, to look at two potential project sites and to meet with a funding source and our staff housing director for Ohio. We talked about how, why here, when, in what way, what will be the nature of our relationship with this vendor or that.

"Every morning, I stay in touch with other staff, in Rochester or outside, through email and phone calls. Then I may work on renewing the option agreement for Xyz project this week; at that point I have to think about whether or not the project is expected to survive and whether therefore I should extend the agreement, which typically involves the expenditure of money. If I decide not to go forward, I'll call or write the seller and tell them. If it does go forward, I have to know that there is a source of funds behind this check. If there isn't, it's my job to find the money, either from the revolving loan fund we administer or other sources we know, such as United Way, and discuss this with the ROI financing unit.

"I might get a telephone call from an architect. He says to me, 'Jay, this old building we're renovating, its building envelope will not show an R value of greater than R8 and the state energy code requires 16, so we're going to have wreck the inside, remove it, put in new studs and put in new wiring, and

that means there's going to be increased costs.' I get the builder on the phone, go over the same dialogue with him so the builder can know what's coming before he ever sees it in drawings.

"Because the Federal Home Loan Bank funding round for the Affordable Housing Program is coming, I call one of the several institutions in Rochester to invite them to tell me whether or not they're interested in working with us to get those funds and negotiate some benefit for them in return for the AHP application.

"In the evening I may go to a hearing for a planning board to talk about a project I hope they will approve in their community. I'll bring along the architect who created the site plan, the building footprint plan, and floor plan for the building. In this public forum, we tell them what we propose, with neighbors and reporters there, and they'll ask me all sorts of questions. I want them to know we're on one team, that we'll take as many of their suggestions into account as my project and financing can support. So a major part of what a developer does is to create and sustain relationships that are mutually beneficial. Without relationships, you are nowhere.

"It's been quite gratifying. I get a real big charge out of seeing what other people have done with their abilities."

CDC Construction Site Manager

The setting: Homes In Partnership, Inc., Apopka, Florida

The person: Reuben Herrera, age thirty-six

Career path: Farm worker, then construction worker, moving from entry-level construction work at Homes In Partnership to current position.

The job: Construction site supervisor

The project: Homes In Partnership works throughout five rural counties in central Florida, producing single-family homes for low- and very low-income families. Since its founding in 1975, it has built or renovated over two thousand affordable homes. Most are built in small subdivisions; in most cases, the homeowner families give hundreds of hours participating in the building of the homes, and those hours count as their "sweat equity" downpayment on the home.

"Homes In Partnership helped me build a house for my family—we have three kids—and then I got a job with the company.

"I'm responsible for scheduling and overseeing the work of each individual subcontractor—the masons, carpenters, dry wall crew—and to make sure their work progresses in stages. I look over the specifications and can answer any of the questions they raise. Every day I go through every home. Right now I've got about seven homes under construction. I know enough about each trade to know what's good workmanship and how it should be done, though carpentry is my specialty. If there are mistakes, I get the subcontractors back out there to correct it. After I approve their work, they can submit their charges to the office.

"I also check the quality of the materials used in each home to make sure we're not getting something I wouldn't want in my own house. If it's bad, it goes back. We send back any material left over, and the homeowners get credit on their account.

"We use a few different models for the homes, but they all have three bedrooms, two baths, a one-car garage, central air conditioning and heat, wall-to-wall carpets, and appliances.

"We work in five different counties. Everything has to be built according to the code for each county and each city. I have to know the inspectors, know how to communicate with them, what they want, how they want it done, and gain their trust at the same time.

"Saturdays I work with homeowners, who do the clean-up after each of the subcontractors is finished. I also show them how to lay sod and paint their houses. The family has to give 840 hours to their homebuilding or until the job is completed. I have to speak Spanish with about 50 percent of the families.

"My workday starts about 7 A.M. and goes until 5 or 5:30, though sometimes it's over by 2:30 in the afternoon. We budget forty hours a week, which includes Saturdays, but we put in more hours than that."

Citywide Intermediary President

The setting: Neighborhood Progress, Inc., Cleveland

The person: Eric Hoddersen, age fifty-three

Career path: Graduate certificate in non-profit management, Case Western Reserve University; founding director, Union-Miles Development Corporation; director, neighborhood investment program for an organization of Cleveland's corporate executives; president, Neighborhood Progress, Inc.

The job: President

The project: Cleveland is a city with many CDCs and a long history of CDC accomplishment. NPI's core mission is to provide multiyear operating support and technical assistance for community organizations, based on competitive applications. Quantum Leap, another NPI program, strengthens the human capital of the community development field in Cleveland through board recruitment, leadership and staff development, and systems development. NPI also designs pilot programs and does research for specific purposes, for example, brownfields remediation, clean building technology, and the financing of child care facilities.

NPI also has two subsidiaries. One is the New Village Corporation, a real estate development division, which partners with CDCs in larger projects, mostly market-rate housing and retail projects. The other is a $10 million community development financial institution, Village Capital Corporation, from which NPI makes loans and grants and provides predevelopment financing for projects.

"This week was an external one. I started the day today at 8 A.M. meeting with the external relations staff of BP, the multinational oil company. BP is leaving their headquarters in Cleveland, and they want to talk about a legacy project, a $1 million fund they will leave behind for economic development. They're considering Village Capital as a home for the money. We're trying to pull together a proposal for them before they leave town at the end of this month. Another proposal I'm working on is one to the local Cleveland Foundation to fund a 'main street' neighborhood commercial revitalization program that NPI could offer to its CDCs.

"When I came back to the office, I finalized plans for my trip to Lexington, Kentucky, to speak at a one-day conference on regional development and the role of neighborhoods. And I set up a visit to Chicago with a CDC director and his councilperson where we will look at some factory buildings successfully converted into apartments, as a possible model for Cleveland. Then I had a meeting with the executive director of a community organization we fund, to go over her year's development agenda and talk about political and strategic issues. This was one of the series of quarterly meetings I hold with the directors of the fourteen CDCs we fund.

"After lunch at my desk, I went to a two-hour meeting with the new director of the Mandell Foundation (a local foundation that gives NPI half a million dollars annually), to talk about a range of things, including long-term funding for NPI and questions of how we work with the local universities.

"When I got back, I took a call from a CDC director to talk about negotiations to acquire a shopping center that Village Capital is involved in. Then I met with a consultant from HUD's public housing division. Tonight I and one of my staff people are going out to dinner with a banker we're trying to recruit.

"During the week I'll also meet with other citywide organizations—the Cleveland Housing Network and the local Enterprise Foundation—to talk about coordinating our initiatives.

"Last week I spent time in negotiations with a bank over their neighborhood investment plans, in an interview with a reporter doing an article on a real estate project, interviewing someone about a job, meeting with LISC, meeting with a child care facilities committee, in another councilman meeting, and discussing with staff a job description for an asset manager. And I spent time orienting the new director for Quantum Leap. On the average, I probably work fifty-five hours a week.

"I probably spend 30 to 40 percent of my time with my staff. I have an open door; people can come in and talk about problems or ideas. As time for a board meeting comes closer, I spend time preparing for that. With our operating budget at $5 million and various contractual arrangements, I spend a lot of time signing checks."

City Agency Housing and Community Development Director

The setting: City of Portland, Oregon, Bureau of Housing and Community Development

The person: Steve Rudman, age forty-six

Career path: Master of Arts in public policy studies, Claremont Graduate School; policy analyst and trainer, Grantsmanship Center; codirector, RAIN Community Resource Center; executive director, Southeast Uplift Neighborhood Program; resource development manager, then program manager, Planning and Resource Development, Portland Bureau of Community Development; now bureau director.

The job: Director

The project: The Bureau of Housing and Community Development's annual budget is $35 million (90 percent from HUD, 10 percent from local, general funds). It does not run any direct programs. Ninety percent of its work is accomplished through about 150 contracts, all of them with nonprofit organizations. The city's community development industry has grown very quickly. Ten years ago, Portland had two CDC-type organizations; now there are over twenty. They own over three thousand units of housing, as well as commercial and day care properties.

The bureau operates seven different program areas, from affordable housing to programs for people who are homeless or have special needs; workforce development; youth employment; commercial development; neighborhood infrastructure improvements—improving streets, parks, and other public facilities in lower-income neighborhoods; public safety; and neighborhood revitalization—hiring organizers to help citizens identify needs and come up with plans, then providing operating support to groups to carry out those plans.

"We try to use our office to put systems of support in place for CDCs. We work with intermediaries here—the Neighborhood Partnership Fund and the Enterprise Foundation—to support their programs of technical assistance, to fund CDCs' operations and to make loans for CDC projects. Community folks need to have the tools; then they can succeed in tackling the problems the larger sectors have been unable to do.

"We keep a small staff of twenty very motivated people who are responsible for the soup to nuts in their area. They deal with all aspects of grant administration and contract monitoring and also strategic policy planning in their area.

"I manage the staff, deal with the city council and other funders, and with the problems that rise up to my level. I put a lot of time into keeping the big picture in mind—the strategy and policy planning, and building constituencies for those strategies and policies. One of the downsides is that I don't have enough time to spend with the community folks myself.

"This week I took part in meetings with Portland's big downtown property owners; with one of the Portland commissioners to talk about his interest in employer-assisted housing and assistance to small and minority businesses; and with the nonprofit organization that's designing a citywide one-stop employment center to link jobs with job seekers.

"I also met with congressional office staff to begin to look at how we could encourage companies such as Intel to support employer-assisted housing. I also talked with my staff about new investment products we could design to tap the new wealth of this region.

"I try to do something educational as often as possible as part of my work. So the Portland commissioner who has oversight over initiatives for the homeless and I did a workshop on that subject for a Portland State University course.

"I met with the group that's evaluating our policies, including our policy about rent-to-income ratios and long-term affordability—Portland now requires a property to remain affordable housing for sixty years.

"I also went to a meeting on lead-based paint, because Portland's housing stock is of the age where there's a lot of lead. The state is planning to test ten thousand kids on Medicaid, and we're exploring contracting and liability issues if we use folks from the Urban League or AmeriCorps program in a lead paint hazard control initiative.

"Then I met with people from the local food bank to discuss some innovative ways to use Community Development Block Grant funds in financing a loan to construct their new warehouse.

"In Portland a big part of the agenda is about working regionally, around planning, resource development, and brownfields. We hold regular citywide meetings with environmental justice folks, big developers, property owners, and city bureau directors to focus on brownfields. We're trying to work on this creatively, with my department especially looking at our responsibility in the lower-income areas of town.

"Seven people directly report to me; every other week I spend forty-five minutes with each of them. The other week we have a staff meeting. And I have weekly meetings with my commissioner.

"What I love about the job is that it deals with such a wide range of issues. I've had to learn to be a manager. Doing the work isn't hard, but managing is harder than I thought, a continual struggle. I have high expectations of myself and people who work for me, but I trust and give them lots of responsibility.

"As an appointed official, I think you can get more indepth into issues and programs, where an elected official has more breadth. In this kind of work, the challenge is to build a better society."

Bank Community Development Lender

The setting: LaSalle National Bank, Chicago

The person: Kristin Faust, age forty

Career path: Undergraduate degree, double major in political science and philosophy, Brown University; one year of a two-year MBA, Carnegie Mellon University; master's in city and regional planning, Harvard University John F. Kennedy School of Government; presidential management intern, U.S. Securities and Exchange Commission; assistant vice president, Neighborhood Lending Division, First National Bank of Chicago; senior vice president, Community Development, LaSalle National Bank. (Since this interview, Kristin Faust became the chief deputy treasurer, state of California.)

The job: Senior vice president, Community Development Lending Department

The project: The Community Development Lending Department markets community development loan products inside the bank and to communities throughout Chicago. In the seven years since the department was started, it has generated $150 million in loans and created or preserved 6,000 units of affordable housing and 500,000 square feet of commercial space, while attaining profitability.

"I started Monday morning with a 7:30 A.M. breakfast for a nonprofit board I'm on, then I came into the bank and went to a loan committee meeting where I presented a loan for approval. After that, I had a meeting with another bank in town, a foreign-owned bank (foreign banks are also subject to some extent to Community Reinvestment Act [CRA] regulations) that needs to do more community reinvestment and had asked me for ideas about how to get involved. Like most community development lenders, I'm glad to share information; that's part of educating and spreading the word, even to our own competition.

"Then I went out on a customer call. The customers in this case were some professional people, doctors and such, who on the side buy troubled apartment buildings and turn them around. We discussed ideas they had for their next deal. At 5:30 P.M., our bank hosted a reception for one of the city's nonprofit groups.

"In the course of the day yesterday, I also probably spoke by phone with about forty different people. Several of the calls were from existing customers, and there were several cold calls from people wanting to ask about loans. Our bank is hosting a big community forum on the future of public housing. I'm putting together a panel and video, so I was on the phone today with the commissioner of housing about that forum.

"Our bank is part of the Chicago Housing Partnership committee, made up of all the lenders that fund and finance tax credit deals. I'm the chair of the subcommittee looking into what the actual expenses are in doing a tax credit deal. I spent all morning putting together a memo on this, using the data my staff had collected, and adding my recommendations about the issue. We faxed the memo to the subcommittee members to get their response. I've been taking calls all day on their feedback.

"Being a good lender is being a good detective or a puzzle solver. You're given information by your applicant, your appraiser, the applicant's general contractor, and then you have to go and verify these facts yourself. Can the building really get $650 a month for a two-bedroom apartment? Does the borrower understand real estate is not an 'armchair' investment? At the same time, you have to encourage and inspire your applicant because doing rehab in a low- or moderate-income neighborhood isn't easy.

"As a person in a management job, part of my day sometimes includes interviewing people for a staff position. I have a monthly 7:30 A.M. meeting with my boss. I participate in the CRA oversight committee of the board of directors of the bank, which meets quarterly.

"Sometimes you're out there advocating for community development lending, you find it's not always a shared philosophy within the bank as a whole, particularly when you get involved in policy. Sometimes you'll find you want to take positions that your bank doesn't, and you have to figure out how do you try to educate your own senior management to see it your way. That's a theme, whether you're trying to educate your management about a specific loan all the way up to an overall policy issue where you would take an aggressive stance and upset the mayor.

"I took on a lot of extras because I have the energy to do them, but it helps make people view the bank as a leader."

Bank Community Development Team Leader

The setting: SunTrust Bank, Washington, D.C.

The person: Terri Copeland Devaney, age thirty-eight

Career path: BA in economics, University of Pennsylvania, MBA in international finance, George Washington University. Worked for various CDCs on business development projects and community organizing. Management Consultant at PriceWaterhouseCoopers; Market Leader for Community Development Lending at National Cooperative Bank (NCB).

The job: Senior vice president, Community Development Department

The project: The community development department is the bank's link to community-based organizations and regional governments involved in community development activities. The community development department provides technical assistance to individuals, small businesses and nonprofit organizations in areas such as business development, financial literacy, credit counseling, home buyer education and organizational development. In addition to an on-staff Housing Counseling expert, the department has an on-site Small Business Resource Center that offers one-on-one counseling as well as group workshops for current or aspiring entrepreneurs. At the end of 1999, the Resource Center was utilized by 340 existing and/or aspiring small business owners.

"What is a typical day for me? There is no such animal. As senior vice president of Community Development, my role is to ensure that the community knows that SunTrust is not just a bank in the community but that we are a member of the community. Usually about 25 percent of my time is on the phone, 40 percent in meetings and 25 percent in preparation for meetings doing research and material review, and 10 percent for impromptu staff meetings and other administrative management-related activities. Here is a rundown of a recent Thursday.

I went to Covenant House–Washington, a nonprofit organization providing comprehensive services to at-risk youth in D.C. to discuss their New Building Campaign. As chair of their Fundraising Committee for the Banking Sector, I am working with other area banks to raise the rest of the money needed to complete the project.

I spent many hours on the phone finalizing logistics for our quarterly Community Leaders Forum, which provides community development practitioners an opportunity to meet with peers, experts, and local government officials involved in community development. In developing topics for the forum, I drafted invite letters, talked with government staff about scheduling, and coordinated other logistics.

I am a member of the Steering and Executive Committees of the D.C. Community Development Support Collaborative, which provides grants for operations and technical assistance to nonprofit community development corporations in the District of Columbia. As a member, I went to a board meeting hosted by the National Capital Revitalization Corporation (NCRC), a newly created, quasi-public/private entity charged with implementing neighborhood revitalization in the District of Columbia. I presented a statement from the Collaborative on how we could assist the NCRC in learning more about how

CDCs can assist in their neighborhood development efforts. On behalf of the CDSC, I invited the NCRC Board to attend one of our Steering Committee meetings and also offered to arrange neighborhood tours for them. That meeting was from 6:30 p.m. to 8:30 p.m.

I also manage the bank's local TV campaigns. These campaigns (Camp-4-Kids and School Supplies 9 are two examples) are done in conjunction with two local TV stations and focus on underprivileged children. In 1999, School Supplies 9 collected 31,000 pounds of supplies, which helped almost 43,000 children across the Greater Washington region. I also manage the SunTrust Cares Committee, which is the bank's vehicle to promote and support staff volunteerism. Last year the SunTrust, Greater Washington staff amassed over 30,000 volunteer hours in the Greater Washington region."

"The hours can be long but the rewards are great. I work with a range of people—from grassroots community leaders to high-ranking government officials—all of whom have a great deal to contribute. It keeps the job exciting."

"While the position does not include lending, having a financial background definitely helps, as does a graduate degree. You also need to be a people person and have a broad view of the community. It was the thing I found most attractive about this position. While I enjoyed working as a lender for several years, I was at a point in my career where I wanted a change. I was ready for a nonlending role that required more contact with the people in my community."

Community Development Financial Institution President

The setting: Northeast Entrepreneur Fund, Inc., Virginia, Minnesota

The person: Mary Mathews, age fifty-one

Career path: BS in textiles and clothing, Iowa State University; small business management training, Hibbing Technical Institute and WEDCO Enterprise Development Institute; economic development training, National Development Council; representative, zipper manufacturing company; owner and manager, fabric sales company; managing director, Hibbing Chamber Business Development Corporation; executive vice president, Hibbing Chamber of Commerce; president, Northeast Entrepreneur Fund.

The job: President

The project: In rural northeastern Minnesota in the 1980s, the iron mining industry, once a major employer in the region, declined as America's steel market closed down; twenty thousand people lost their jobs. Northeast Ventures was founded to counter the area's reliance on external economic influences by promoting economic self-sufficiency. One affiliate is a venture capital company that makes equity investments in homegrown companies. The other, Northeast Entrepreneur Fund, nurtures the start-up and growth of small businesses through loans and guidance. It is an award-winning community development financial institution. It has helped start, stabilize, or expand over 390 businesses; and the businesses' survival rate is remarkably high.

"Though we're a private nonprofit organization, we operate using sound business principles. We seek to be a role model for the business owners we work with. Our mission is to assist unemployed and underemployed people, most of whom are low-income women. We are a staff of nine. Three of the staff are business consultants, and these are all people who have been business owners themselves. There are two people making the loans (one a manager, the other a loan assistant); a vice president, who runs the day-to-day operations and does some consulting and training; a finance director; an administrative assistant; and myself.

"When a prospective business person calls for information, we invite them to attend one of the business planning sessions we hold three or four times a month. At these sessions, we introduce people to the services of the Entrepreneur Fund and begin to talk about what it means to be a business owner and the skills needed, and we encourage them to go through our twelve-hour business planning course.

"We have three offices throughout the 20,000-square-mile region we cover, with a business consultant in each. If you're a business consultant, you spend three to four days in the office, and one to two days a week on site visits or traveling to one of six other locations throughout the region where we meet with clients.

"In our one-on-one sessions, which generally last an hour, we work with the business owner to develop an action plan of the steps they will take next and what they will accomplish before the next session with us. This way, through the individual plans, we help them figure out what they need to learn next . . . what the potential is for their business. We never tell anyone, 'This business isn't going to work.' We ask them, 'What makes you think this is going to work?' and guide them to the actions they need to take to make the decision for themselves.

"Our training and technical assistance functions take up two-thirds of our budget. Some customers, through our help, learn how to manage with what they already have. Others find that after they're prepared by their work with us, they can go to a bank for financing. The balance, we finance.

"Our business consultants help the customer put together a business plan, which can be taken to a bank, or to us, or for us to finance together with another source. Our loan fund manager does the loan functions, reviewing the business plan, taking it to the investment committee of our board of directors, then preparing the documents and disbursing the funds. He does the ongoing tracking of the loan activity, may have conversations with the borrower if they're delinquent, and does annual monitoring.

"As the president, I have a number of policy and representational responsibilities. Last Thursday and Friday I spent in Washington, D.C., in a meeting for an evaluation project run by the Aspen Institute. Fifteen of us from micro-enterprise organizations are working together to figure out how to measure and communicate outcomes and impact. We identify performance measures, then test them in our own organizations. I also met with our congressional representatives.

"As president, my corporate responsibilities today center on planning, finance, fundraising, advocacy, and organizational growth and development. I'm active in several local, state, and national associations. For the first five years, while I was building up the organization, I also worked as a business consultant. My job has changed dramatically as we have grown.

"Entrepreneurship is about personal as well as business development. We help people learn to be business owners and to grow as their business grows. It gives me a deep sense of satisfaction to watch the changes that occur in people as they become more confident as they meet their personal and business goals. Often they can buy their first house, or they are building other assets. To see the transformation that occurs in people, to see that and share in that, is really remarkable and very rewarding."

Federal Reserve Bank Community Affairs Director

The setting: Federal Reserve Bank of Atlanta

The person: Courtney Dufries, age forty

Career path: Undergraduate degree in economics, Georgia State University; during college worked part-time as a bank clerk; bank examiner, then applications examiner, then community affairs officer, Federal Reserve Bank of Atlanta.

The job: Director of Community Affairs

The project: The Federal Reserve Bank of Atlanta is one of the district banks in the Federal Reserve system. The Federal Reserve system was founded to ensure the stability of the American economy by influencing the flow of money and credit in the nation's economy and by supervising the activities of certain state-chartered banks and bank holding companies. It is also charged with writing regulations for the major consumer credit laws, including the Community Reinvestment Act. Each district bank has a community affairs function to promote community development and fair lending throughout the district.

"This is at least a fifty-hour-a-week job. My typical day includes a lot of time spent on the telephone, providing technical assistance for, among others, public housing authorities, colleges and universities, city planners, and nonprofits. That means helping them to identify what their issues are, then identifying what types of financial institutions and products could be used to address their issues. For example, we provide documentation of various models of loan programs that have worked well in other cities, such as the articles of incorporation and underwriting criteria of a community loan fund, or helping people develop new models, modifying other products to meet local needs. We also publish a newsletter to share problems and success stories.

"The rest of my work includes preparing conferences and speeches, and conducting research for the technical assistance I give. There's a lot of networking.

"Fifty percent of my work involves travel time. You've got to be in the community and meet the people face to face. Often you are called on to understand and explain very sensitive issues. There are times when some people are angry or upset, or have misperceptions, around race issues and discrimination, for instance, and you have to understand where they're coming from and figure out how you can get beyond that anger. You can't do that from a desk. My kind of job is built on a strong base of knowledge of how banks operate and a strong link to the financial institutions."

For-profit Development Company Investment Banker

The setting: Urban America, L.P., Manhattan

The person: Richmond McCoy, age forty-five

Career Path: Undergraduate degree in American literature, Fordham University; twenty years of international and national real estate development experience; founded McCoy Realty Group and Development Corporation, which represented the nation's two largest pension funds, Teachers Insurance Annuity Association and California Public Employees Retirement System; served as the asset property manager and development manager on new projects.

The job: President and chief executive officer

The project: Urban America is a private real estate investment company focused exclusively on developing income-producing commercial real estate in major urban inner-city markets.

Twenty-five percent of Urban America's partnerships are with CDCs and faith-based groups. Urban America works with these groups in acquiring their idle commercial properties, providing CDCs cash flow from purchase, management control of the property, joint venture ownership, an increase in employment for the community, and reintroduction of services long missing from the community.

"As President and CEO, I spend a lot of time communicating with investors and board members, working on the real estate acquisitions, managing our image in the press, and providing direction to the staff. About 50 percent of my time is travel. Doing the acquisitions takes a lot of financial analysis, legal analysis, negotiation, and tenacity. These real estate deals don't happen overnight, and the development projects can take one to three years to complete. My days are packed with meetings. Yesterday I was in Miami and met with some local real estate developers about a joint development, then I met with a bank that may become an investor with Urban America on some other projects and discussed our investments strategies and interests in Florida; then I met with the local government agency that oversees CRA activities for Miami and we toured parcels of land with an eye toward future development, and then we all went to lunch and talked about the Miami real estate market. Between all these meetings I fielded ten phone calls and half a dozen e-mails through my phone. Most of my days run like this, meeting investors, looking at possible projects, and so on.

"This is the first time in thirty years that the economic impacts of real estate development is focused on the urban market. We saw a tremendous need for real estate professionals to assist community development and faith-based organizations in maximizing their real estate investment opportunities.

Our objective is to earn profit but also to stimulate economic development in communities. Seventy-five percent of all vendors we work with reflect the diversity of the community and that keeps profits and jobs in the community. We are also bringing goods and services to communities that may have been without supermarkets and other retail operations for fifteen years or more."

Tenant Services Case Manager

The setting: Mid-Bronx Senior Citizens Council

The person: Roxie Perez Lohuis, age twenty-eight

Career Path: Undergraduate degree in psychology, University of Miami; case manager, Mid-Bronx Senior Citizens Council for five years. (Since this interview, Roxie Perez Lohuis has become an outreach counselor for the Lupus Foundation.)

The job: Case manager

The project: When Mid-Bronx Senior Citizens Council (MBSCO) was formed in 1970, its mission was advocating on behalf of an elderly population left behind in deteriorated areas of the South Bronx. Since then, with a workforce now numbering 235 employees, it has become a housing developer and manager of 1,580 units of housing and has taken on the role of providing services to homeless and other low-income families as well as seniors. It runs a child health campaign, providing immunization and lead screening; is developing a primary health care center; and operates a Head Start center. It runs job training and vocational counseling and job readiness programs. It started three local businesses, including a catering enterprise that delivers food for several thousand people daily. MBSCC also runs an anticrime program, a youth program, adult education programs, and a large-scale open space project.

"As a case manager, I spend most of my time just talking with people, getting them to express exactly what their problem is, what do they think their next step should be, asking them, have you tried this or that? We don't get paid abundantly, but I can't see sitting in a chair for $100 an hour listening to someone talk about her troubles and when the timer goes off, you say, 'Goodbye, I'll see you next week.' Here, I'll talk to anyone who needs me for as long as I have to.

"Sometimes it will be a parent coming in to ask me about summer programs for their child, or a person who's worried because they didn't get their Medicaid reimbursement. If it's a parent who needs to find day care so they can go to work, I'll go through my child care provider list, match the person's zip code to the providers who are the nearest to them. Then I'll give them advice about how to interview the provider, that they should inspect the provider's apartment and should stay a couple of hours to observe what the day care is like. Another part of my work involves screening low-income families who are applying to rent apartments in buildings that Mid-Bronx Seniors owns.

"Over half of my cases are about an income-related issue. Then I try to help them, for instance, to get into a job training program, or to take a course in English as a second language, to get their GED, or to find employment. Mid-Bronx has a job resource center, and I can refer people to that. If the day care provider hasn't gotten paid, or if the rent check has not reached the landlord, I track down what went wrong and try to fix it.

"There are some people who sabotage themselves because they have a victim mentality; they feel that the world has wronged them, and they can't rise above that. I try to motivate them enough to have them take the first step. I tell them, we're not going to make the phone call for them. We'll help them, but they have to take initiative.

"The tenants who live in Mid-Bronx's building know that the organization has a social service component, because they get letters from Mid-Bronx with my name on it. So they come in when they need help. Because our office is located right in the middle of the neighborhood, people will walk in and ask for help. Theoretically, I could be dealing with between 300 to 350 tenants.

"I also have responsibility for tenant leadership development. This part is not so much focused on social services but on a transferring of skills, so in a way you're working yourself out of a job, you're leaving something in place once the funding is gone. Each staff member staffs a different neighborhood committee. At first we are the point person for the committee, but we gradually transfer our skills to them and then just look in every so often. Last year three neighborhood groups planted twelve trees on three different blocks; that involved securing permits from the community board, working with the parks department, scheduling the event, managing the funds now available. It was beautiful, something lasting, one of the best things I've ever been involved in.

"In a typical week, I'll be working on one or two different projects, setting up meetings, scheduling time for the neighborhood person to come in to work with me. Meanwhile, clients are coming into my office. I take care of them. I write memos, then I follow up the next day with a phone call to see what progress has been made. Then I work some more on my project. We hold our meetings in the evening because the neighborhood people are working people. We have at least one evening meeting a week, sometimes two or three.

"I'm a people person, and this work is very rewarding and satisfying—not all the cases, but most. It's not the type of rewards that show up in a bank statement, it's the type of reward that nourishes the soul."

Employment Services Job Networker

The setting: DenverWorks

The person: Debbie Montavon, age forty-six

Career path: Bachelor's in music education, Ouachita Baptist University, Arkadelphia, AR; graduate studies toward master's in social work, University of Denver; support services coordinator, Hope Communities, Inc.; music instructor; pianist accompanist; program director, Hope Communities, Inc.; founder and executive director, DenverWorks.

The job: Executive director and job networker

The project: DenverWorks, founded and run by Debbie Montavon, links unemployed people to job openings, providing the job counseling and person-to-person relationship that helps long-time unemployed people find and stay in their jobs.

"I was working for a nonprofit housing organization in an impoverished section of Denver. People kept walking in the door saying, 'What I really need is a job.' I knew of businesses that needed workers, so I'd give them the card for that business, but no one ever used those cards. I realized that people usually need a personal relationship to help them through difficult times. My dream was DenverWorks, which would be a real relational way to hook people up with the many jobs that there are in Denver. I started DenverWorks over three years ago.

"We now have a staff of five, including a Spanish-speaking staffer. We have one foot in the business world, the other in the social services world; we chose staff who can relate to both. Our organization is not very large, but we see about 275 people a year. Each of us who is a client worker works with sixty to one hundred people a year one-on-one to help them find jobs.

"We start people with a crash course in job preparedness—helping them with interviewing skills, inventorying their interests, teaching them goal setting and professional behavior, that they need to show up for work every day on time, how to dress. We're in an office building; this is a business; it's very professional.

"Most of our clients are very low income. A lot of them are coming off welfare; others have criminal backgrounds. They often lack the little pieces—a car to get to work or clothes to wear—that they need to get and keep a job. After they've gone to the workshop, we meet with them individually, give them three to five job leads, and try to help them with those little pieces that are missing. Mostly we do that by referring them to other organizations. We also have about seventy-five volunteers we can bring in for different pieces, such as if someone needed help getting a driver's license or getting tutoring.

"We place about one-third of the people; for another one-third, what we've done for them gets them off dead center, and they get a job on their own. The last one-third don't get a job. They won't come to appointments, they make excuses, they're just not ready. We've done the statistics over the three years of our history; that's the way it works out.

"This week included meeting with several social service agencies—an alternative high school, a low-income health organization, and a mentoring organization—to tighten up our collaboration to attract national funding. I spent time writing proposals to foundations. I did some more writing on the twice-yearly newsletter we put out. The administrative part of my work, including staff supervision and fundraising, takes about 50 percent of my time.

"I taught about half of the workshop on Wednesday and did a lot of client work this week. My load is about forty-five people—it's like putting on a detective hat; each person has a different situation. A client I saw this week doesn't even have a place to live; I had to figure out, is her first need a job or a place to live? It takes weeks or months to get a placement. Every client is such a challenge. It's so hard, but it's so encouraging.

"Another huge piece of my week is the employer visits we make every Thursday. We put on our suits and heels and beat the streets. Sometimes we make appointments; other times we choose a section of the city, take some material with us, knock on doors, and introduce ourselves. Employers are happy to talk with us because we are another employee source, and we don't charge any fees. Another way we find jobs is that we hold a bosses' breakfast once a month. Denver is such a great place right now. Even McDonalds pays $9 an hour. So we don't even have to look at low minimum-wage jobs. But it still takes the right fit on both sides and the human relationships."

Chapter 7

Career Paths

This chapter will give you a flavor of career patterns and potential, with the caveat that in this field your life will create its own destiny.

What does the future hold for someone who has entered the community development field? Where can you go once you've gotten a job and some experience? Almost anywhere. You can move up, around, and out of the field. It is a springboard to a diverse and challenging future. As every one of the life stories in this guide demonstrates, neat, predictable career ladders seldom exist in the community development world. There are observable patterns rather than ladders.

Entrepreneurship, risk taking, and drive characterize the great majority of people who feel comfortable in community development. They don't mind and, in fact, welcome change, so the very limitless and unknown nature of the world of career opportunities appeals to them.

Community development careers advance in two different ways. One way is to stay within the same sector—for instance, working with neighborhood nonprofits or for the government—and move up or around, within the same institution or from one group to another. The other way is to advance by changing from one sector to another—for instance, from a neighborhood nonprofit to an intermediary.

Staying within One Sector of Community Development

You might join the staff of an organization where you could progress from one job to another and find your work comfortable and fulfilling enough to stay for twenty years. That happens for many practitioners, though at least as often people work for more than one organization over the span of their careers.

Community-Based Organizations

Community-based organizations are hospitable to career moves up the ladder within the organization. From construction worker to housing developer. From maintenance staff to property manager.

As institutions founded on the principle of local democratic control, community development groups usually go the extra mile to encourage personal growth and hiring from within the neighborhood. But even a long-established, very sizeable organization probably has not arranged its positions into a formal career ladder. "It is not so much a career ladder as career moves customized to the individual," says Michael Rochford about the way his CDC encourages growth. He is the associate director of the Brooklyn-based St. Nicholas Housing

Career Story: Willie Jones

Senior Manager, The Community Builders, Boston

The beginnings of Willie Jones's work that led to his becoming a successful community development professional were in the civil rights movement, when he was involved with voter registration drives in the South. A Better Chance scholarship took him from his southern roots to the University of New Hampshire. From there he became a VISTA volunteer and then a youth worker for a settlement house.

Willie then turned to union organizing—part of a conscious strategy on his part to win more places for blacks in labor unions. He drove a bus in Washington, D.C., for three years, then moved to Detroit as a worker in the United Auto Workers. It was in Detroit that his skills as a union organizer first crossed into community development, as he served on the board of a community improvement group. This work, and that which soon followed, caused Willie to develop a "love of community organizing and families in neighborhoods."

While in Boston, Willie picked up a degree in engineering at Northeastern University, so that as he became increasingly involved in community development, he had some hard skills—he knew organizing

and could read architectural and engineering plans. After working with the dean for minority recruitment at the university, Willie moved into community development full-time when he joined The Community

© Jaymes Leavitt

Builders, a Boston-based community development nonprofit that has developed low- and moderate-income housing throughout the Northeast and the Midwest.

"Community development is different from real estate development," says Willie. "Community developers must have an affin-

ity for advocacy. They need to be very energized to improve a neighborhood, not only develop a real estate product. The talent needed for community development is higher than in private business. Community development professionals need to be risk takers and run organizations that are lean and mean. We must understand what has gone wrong in a neighborhood and set out to fix it."

Now at a senior management level, Willie has the responsibility for hiring substantial numbers of people into the firm. What does he look for? "I judge a candidate on four criteria: (1) a demonstrated commitment to community—as a volunteer, or in some way, actively engaged in working to improve the world; (2) is the person a 'play maker,' able to move things along and get them done; (3) quantitative skills, including computer spreadsheet skills, and (4) character and passion. I want people working for me who feel passionately about life. When I find a person with those skills, I can train them on the specifics of what must get done on the job."

"Community development is an expanding field. Cities are becoming healthier again, in part because of the work of the community development field," says Willie.

Preservation Corporation, which has an administrative staff of 250 and in all employs about 1,300 people (a good number of whom work in either the home care program or the family and youth services program). St. Nick staffers who start in an entry-level job can move upward into increasingly more responsible and complex positions. St. Nick calls these positions project manager, senior project manager, assistant director, and program director; other CDCs would have a similar range of jobs, also without prearranging these positions as a set of steps on a ladder.

Large CDCs such as St. Nick are rarer than small and moderate-sized groups, which may not have the resources to hire and groom entry-level employees. One highly acclaimed Boston-based CDC hires only practitioners with experience and technical skills, gained either as a staffer elsewhere or as a volunteer. But there is an advantage to working for a smaller organization: each person performs a range of duties. That exposure and experience make the job more varied and tend to build strong skills more quickly.

Within larger CDCs, the more senior positions are executive director (or the title might be president) and assistant director (also called deputy director or vice president). A program manager is in charge of a specific program of the organization, such as upgrading spaces for local businesses or boosting the ability of unemployed residents to find jobs. Organizations that operate programs to expand the economic well-being of their community by assisting local small businesses would have one or more business counselors on their staff. This position helps the business owners to understand and improve their business plans, strategies, and management; and develops and presents workshops dealing with small business problems. A staffer can advance from working as an assistant in these programs to managing the program

and its staff and budget.

From the financial management side, CDC bookkeepers and accountants can move into chief financial officer positions in a large organization with complex deals and affiliated or subsidiary organizations, managing a staff of six or seven people.

From organizing, you can progress into a position managing programs or as head of an organizing team. Many organizers, with their strong ties to the neighborhood, interpersonal skills, and understanding of power relationships, have become executive directors, after learning enough about the technical details of development to know the right questions to ask and the right people inside and outside an organization to deal with.

For one reason or another, practitioners often make career progress by moving from one CDC to another.

Intermediaries

Intermediaries are usually large enough to have a relatively good-sized staff, with jobs of increasing levels of responsibility and complexity. You can join their ranks and be promoted or move from one intermediary to another.

Within the largest intermediary, the Local Initiatives Support Corporation, there are jobs of increasing responsibility. In the headquarters staff, the progressive positions would be: program or project assistant, assistant program officer, program officer, senior program officer, program director, senior program director, and vice president. The program assistant is an entry-level job, calling for an undergraduate degree and about three years of work experience. The assistant program officer would require three to five years of community development experience, and a graduate degree would add to a candidate's competitiveness. There are also

Career Story: Abdul Rasheed

President, North Carolina Community Development Initiative, Raleigh

Abdul Rasheed grew up in the small, segregated rural town of Henderson, North Carolina. "In eleventh grade—this was 1965—I was selected as one of a number of young folk asked to go across town to integrate one of the majority schools. In my first meeting with my counselor, when I told him I had a desire to take college preparatory classes, he told me I had no chance of going on to college. That was a defining moment in my life: I knew I was going to show him I could go to college." Abdul attended Elizabeth City State University, a historically black university in North Carolina, and majored in business. He then went on, with a fellowship, to Trenton State College for a master's of education. "I worked for a couple of summers in Schuyler Homes, a famous public housing project in East Orange, New Jersey, where my wife is from. It was there that I saw how the children and families lived in that discouraging environment, and that left a strong impression on me.

"When I came back to North Carolina, I knew I wanted to do something in poor communities. I took a job with the community college system in its community mental health program. I managed the internships of graduating seniors who were trying

to understand what was the mentality and psychology that keeps people from thriving and having healthy communities.

"Through an economic development project I worked on with John Wheeler, founder of Mechanics Farmers Bank, the state's largest minority bank, I learned the importance of strong political skills and the need for our communities to build assets and wealth we could control, to define our own life and aspirations. Around that time, in about 1976, I became active in the Muslim community. I saw how the Islamic tradition stresses trying to do something for yourself, through jobs and the ownership

of assets, with a strong spiritual base and a lot of discipline for self-improvement.

"In the late seventies, I took a position as a community educator for Legal Services of North Carolina. I worked with lawyers and paralegals to develop larger strategies to help people take more responsibility not to allow their communities to deteriorate. Through my work there, I met folks from the National Economic Development and Law Center and got acquainted with their community economic development approach. I began as a community educator to integrate a community economic development strategy into every legal services office, to create affordable housing, jobs, and institutions such as credit unions that help individuals in low-wealth communities rise above those circumstances.

"I received a fellowship from HUD to go to New Hampshire College's Community Economic Development Master's Program, and it was there that I gained the competencies and skills to really understand how you put together deals and build a community economic development system in local communities.

"When I finished my degree, I went back to Legal Services but focused my work on building development capacity in low-income communities. I created the North Carolina state association for CDCs and became its first president. During my four years in that job, we created new pools of revenue to support community development deals, a community economic development studies program to train folks in concepts and skills and values, and a project linking churches to community development.

"After that, I secured a major appropriation from the state general assembly to support CDC work, and in 1994, we set up a corporation, the North Carolina Community Development Initiative, as the mechanism to use the statewide funding. I became its president. It is the only such organization in the nation included in the base budget of a state. Since 1994, we've funneled $20 million into communities.

"What I love is to watch a family that never had the opportunity to own a home enter that house for the first time, to witness the joy and pride that goes with that kind of transaction. To watch communities overcome hurdles and gain wealth that they control that transforms them individually as families and transforms them as communities in terms of their aspirations for the future."

specialized jobs, such as in training and research, working from assistant up through assistant director (or manager) and then director.

Intermediaries often maintain field offices as well as a headquarters operation. In these outposts, in the Neighborhood Reinvestment Corporation, for example, you might start as an assistant, move on to field service officer, to senior field service officer (a position with increasing grade levels), then to associate director and district director. The job descriptions in appendix A offer a sense of how the work changes from one position to the next.

Smaller intermediaries can also offer career promise for people who seize all chances to learn and grow. Mia Ford, with a BA in music education, started out as office manager of the Texas Development Institute, an intermediary organization responsible for training and advising community development groups across the state. She says, "I hung in there with the institute for more than seven years through some rough times and good times and took every opportunity to learn," going to conferences, reading newsletters and publications, listening to the trainers hired to give courses at the institute's workshops, and learning from the specialized expertise of consultants. Mia is now the executive director of the institute.

Capital Providers
Banks, Investment Financing Companies, and Regulatory Agencies

Once you are hired by a lending institution, they "train you like crazy," through their in-house credit training programs. After the training, you have to go where the bank needs you first; once you prove yourself, you network your way into the community development lending side. Once you are in a community development position, as an assistant or second vice president, your responsibilities might be underwriting, closing, and portfolio management, and you might work as part of a team providing financing for community development projects in a specific geographic area. These jobs also involve a good deal of outreach to communities, to understand and create products to meet their financial needs.

Or, after you get training and experience in the commercial credit department of a bank, you may be able to work your way up to a commercial loan officer job and then into investment banking.

You might also enter the world of investment finance directly, with a master's in real estate development or business, as a portfolio analyst or asset manager (monitoring the way the portfolio of properties that the fund invested in is performing, to make sure they stay in good condition, remain fully rented, and continue to house people with limited incomes). Moving up in a career comes with talent, hard work, and experience—no different in this than in other settings. After an entry-level job, the next position might be as an assistant vice president, then vice president and portfolio manager, then senior vice president. At some point, in some firms, senior staff have the opportunity to become a part owner or partner.

The agencies that regulate banks—the Federal Reserve Banks, the Comptroller of the Currency, the Federal Home Loan Banks—also offer a growing number of community development careers. With some relevant finance background, you might get hired as a bank examiner (reviewing information from banks to assess the safety and soundness of each financial institution's capital, management ability, earnings, and assets) and later move into the agency's community affairs departments.

Community Development Financial Institutions

Community development financial institutions, as described throughout this guide, exist to provide funds to projects that mainstream institutions would consider too risky or too small. Within CDFIs, from an entry-level job as a loan associate involved in administrative work such as closing loans and monitoring portfolios, the next position would be as a loan officer, who analyzes and underwrites loans and gives technical assistance where needed to the groups that receive loans from the CDFI. That job might call for a bachelor's degree and three years' experience in community development finance. The loan officer (which could also be called a loan underwriter) makes the decisions about which potential borrowers will receive loans from the fund, by underwriting (that is, analyzing) the loan application and applicant organization); closes and monitors loans; and provides technical assistance to potential nonprofit and small business borrowers.

In larger organizations there might be a senior loan officer or fund director position, supervising a team and managing several different lending programs, and participating in the policy decisions of the organization. For example, the Women's Opportunities Resource Center in Philadelphia promotes social and economic self-sufficiency for economically disadvantaged women and their families. The Center's loan fund manager oversees a fund that provides early-stage financing and related business assistance to women starting up their own small ("micro") enterprises. That is a high-level job where an advanced degree in business would be desirable and management and relevant work experience a requirement.

A fundraising director (often called development director), whose job is securing grants for operations and securing loans and grants for capital, is another senior-level position of importance, since giving out money is the bread and butter of the CDFI. For this job, you would need not only a fundraising background and enthusiasm, but also knowledge and experience in development finance.

The chief financial officer (often called comptroller) is

Community Story: Astoria, Oregon

Restoring the Economy and Environment of a Small Town

When the plywood mill closed down, it left the small town of Astoria, Oregon (population 10,000), faced with two great problems. One was environmental: The former owners left the eighteen-acre site heavily polluted with PCBs and petroleum products. Not only was the paper mill adjacent to the mouth of the Columbia River, with its waterfowl and salmon, but an intertidal mill pond on the site was carrying the toxics out with the tide twice a day and into the river.

The other problem was economic, since the mill had been the town's major employer. But the economic and the environmental were intertwined, since the site is immediately next to the town's central business district and its pollution became a bar-

rier to any kind of capital investment in putting the land to new use. The Small Business Administration held a mortgage of $3.2 million on the property, and various other lenders and suppliers held another $3 million of loan.

The community came together and created a vision for that part of downtown, which they called Astoria Gateway. They planned a new municipal swimming pool and a branch of the Oregon State University devoted to research into new seafood products and more. But the problem remained of how to finance the old paper mill's environmental cleanup.

Shorebank Enterprise Pacific, working with the Gateway committee, provided the solution. Shorebank Enterprise Pacific is a nonprofit organization that provides lending and business assistance to small, natural resource–based enterprises in the Pacific Northwest. It is a sister organization to Shorebank Pacific, a development bank that is an affiliate of

the Shorebank Corporation of Chicago.

The organization made a $750,000 loan commitment to cover half the cleanup cost, and, with that commitment, the town was

able to attract funding and participation from the Oregon Department of Environmental Quality to pay the other half. Once the site was cleaned up, it was sold to a private company, which will develop housing and commercial space on the land. The cleanup also supported the nearby construction of an aquatic center; and it made possible the development of a seven-screen movie theatre, a new medical specialist building with sixty high-paying jobs, and a bank with additional new jobs. And a source of visible contamination of a sensitive estuary has been removed. This project is a vivid example of how risk-based capital can influence both ecology and economy.

another senior-level staffer. In addition to controlling the CDFI's finances, this person is usually also in charge of the administrative staff. In small organizations, this is a one-person office; in larger organizations, the CFO would have an accounting staff to manage. You might secure that position with perhaps five or more years of experience, plus two in a supervisory role. A graduate degree in accounting, finance, or business would help.

Career progress can come from moving upward within a CDFI if it's large enough or from moving from one CDFI to an-

other. Usually, the larger the CDFI, the more qualifications it will require in its hiring.

Government

Of all the circles that make up the community development world, government is probably the most likely place for a person to enter and stay, moving up in government jobs, either within one large agency or in different agencies within a city, or shifting from local to state to federal levels. This may be so because government agencies tend to be large enough

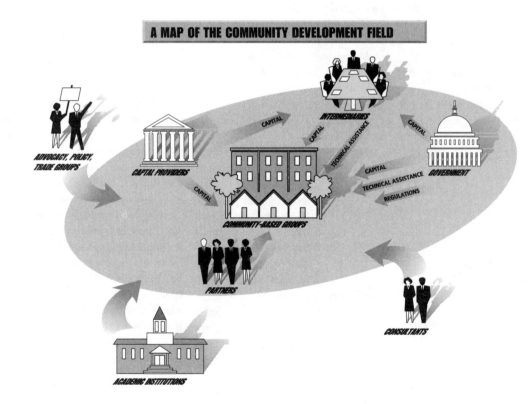

A MAP OF THE COMMUNITY DEVELOPMENT FIELD

ADVOCACY, POLICY, TRADE GROUPS

CAPITAL PROVIDERS

INTERMEDIARIES

CAPITAL

CAPITAL

TECHNICAL ASSISTANCE

CAPITAL

CAPITAL

TECHNICAL ASSISTANCE

REGULATIONS

GOVERNMENT

COMMUNITY-BASED GROUPS

PARTNERS

CONSULTANTS

ACADEMIC INSTITUTIONS

to offer advancement from within; because a public-sector job offers more security than other sectors do; or because the organizational skills are especially applicable from one government agency to the next. Moving from agency to agency, getting new challenges by changing jobs and the job setting, is the path that William Bostic chose. His career also shows the power of networking. A native of Chattanooga, Tennessee, he first worked on the city's Model Cities program. Then, after serving in the Vietnam War, he went to the University of Pittsburgh for a master's in urban and regional planning, stressing community and economic development in his course work. "I feel very strongly that regional plans cannot be effective without the input of people affected by the plans," he explains. After graduation, he found a job as senior planner at a regional planning commission. Then a former colleague asked him to join the governor's administration. From there he moved into the state's Department of Community Affairs, rising from chief of legislative relations up the line to the position of deputy secretary, second in command to the department head. With the change of governors, he moved (again, through a colleague's contacts) to a federal job in the U.S. Department of Education and then back again to state government (after writing position papers for the man who won the gubernatorial race that year), as secretary of the Department of Community Affairs. Today he serves as executive director of the Pennsylvania Housing Finance Agency.

Human Services

It's most likely for people to enter the human services in a position that deals directly with clients and over the years get promoted to administrative and management positions within the same organization or in another organization. For example, someone starting out as a case manager might rise to the job of coordinator of a program and then become manager of a division or of all social service programs.

But, since organizations that once might have focused solely on social and human services now take on explicit "development" roles, people who start as case managers or in other roles providing services can work their way into development jobs. Judy Jarmon, whose career is described in chapter 8, began as a block captain, moved up the career ladder, and is now head of employment services for a St. Louis settlement house that does housing and other types of development. Another economic development example: while day care and in-home health care were once thought of as helping services, today community groups have realized (what for-profit companies knew) the business potential in these lines of work. A good number of community groups have organized worker-owned cooperatives for child care providers and in-home health care aides.

Many community development groups and service-providing agencies hire people with a social work background as community organizers or outreach workers. There is, in fact, a whole subset of the social work world—"macro" social science as contrasted with the "micro" focus on individuals and families—dedicated to community practice and social change, with its own trade association (see appendix B, the "Social Work" section).

Or you can remain in social or human services work and add physical or economic development elements to it. In Pittsburgh, Phil Pappas founded and runs a remarkable nonprofit organization, Community Human Services, which offers health care clinics and an emergency shelter for the homeless, in-home support services for families, and after-school programs for children and youth, to mention just a few of its projects. He has also steered his organization through about

$8 million worth of housing development. This includes the creation of a single-room-occupancy building in downtown Pittsburgh where the tenants, very poor men and women, many once homeless, now run the building's restaurant while others have organized themselves into an outreach team to visit and help their brethren still on the streets. Phil, who has a master's in education, has run a school for delinquent kids, has founded a CDC, and is a poet and artist.

Moving Up among the Sectors

Community developers often move from one sector of the field to another and yet another, practicing community development from different perspectives and taking on increasingly higher-level positions. They progress from the nonprofit to the for-profit or public sector and vice versa. There are certain bundles of skills and experiences—for example, know-how of the housing development process, or financing, or policy making, or public-private relationship building—that make movement from one circle to another more likely. This is one of the observable patterns of career growth in the field.

The only constraint might be in moving from one city to another, because each place has its particular history, politics, and economics. It is doable, but it means you have to learn and adapt. Often people seem either to stay in or return to their hometown, as they progress from one job to the next.

From a Community Group to an Intermediary

A common career move is for people who have worked and done well at the grassroots level, in a CDC or other kind of community-based nonprofit development group, to be hired away by an intermediary. In fact, CDCs often lament that they are the training ground from which their best people are lured away.

The neighborhood-level group is a great place to gain hands-on experience, technical skills, and a direct understanding of the assets and problems that make up a community. Organizer seems to be the job that builds the best base for all kinds of work afterward. It gives people a solid feel for the texture of the problems and issues of a community and how to find solutions, as Mark Jahr, a LISC program officer, puts it.

In an intermediary, people can generalize from the practical skills, experience, and understanding they've gathered from grassroots practice and provide assistance to less experienced groups in the intermediary's network. Both the neighborhood-based organization and the intermediary are focused on "deals" involving tangible physical or economic development projects. In the neighborhood group, practitioners gain knowledge about the regulations, procedures, and politics involved in putting together deals and can generalize from that experience.

After working in a grassroots group, practitioners usually move to an intermediary at the citywide level. Subsequently, people often go on to work for an intermediary that is regional or national in its focus and take on increasingly more senior roles. After a summer's internship with a housing advocacy group, John Berdes got a VISTA volunteer job as the community coordinator for a one-person neighborhood citizens group in Seattle. Then he joined the staff of the neighborhood's housing development corporation and soon became its executive director. The Local Initiatives Support Corporation hired him for a regional position as program director for the Puget Sound area and then for a nationwide job as senior program director, helping to develop a new LISC program for rural communities. A few years ago, he joined one of the rare institutions that bridge the community development and environmental worlds, Shorebank Enterprise Pacific. Astoria, Oregon, whose story is told in this chapter, is one rural community that John helped to rebuild.

But CDC to local intermediary is not the only move from a community-level position. Any combination or permutation is possible. CDC practitioners can go on to jobs in banks, community development financial institutions, foundations, the government at any level, and consulting firms.

Interchanges between Intermediaries and Government

Another frequent move is from an intermediary to a city or state agency, often to a senior position, such as director of housing finance or head of the community development department. Afterward, people often go back to the same or another intermediary, to a job with greater responsibility. The reverse also happens: people move from government to intermediary. Roy Priest came from a post as deputy assistant secretary for economic development at HUD to head the National Congress for Community Economic Development, the trade association for CDCs and sponsor of this career guide.

Skills and experience in policy analysis, in abstracting to a "macro," broad-scale picture, in coping with the more rigid organizational life of a large institution or bureaucracy, and in formal communications lay the groundwork for this set of career moves. In the intermediary and in the government, people acquire a clear understanding of the roles of each sector, what the government can and cannot do. They grow smarter about politics, including the politics of scarcity, gaining the ability to judge which allocations of funds are likely to have a stronger outcome than others and how to leverage money well. These positions share the ability to attract partners at a macro level, invent new approaches, and if needed get legislation enacted that will help those approaches become realities.

Moving between Government and Academe

People who have attained senior positions in government often join the world of academe. Then frequently they return to a next higher position in a government agency, maybe leaving that to head a nonprofit or a think tank.

The bundle of skills and experiences that makes these moves possible—policy making, institution building, the wide vision, the understanding of politics—are similar to those behind switches between government and intermediaries. This kind of work is the most abstract in the community development field, functioning in the world of ideas and ideology. People who can translate themselves from government to academe and back are able to generalize about broad trends and recognize the macro directions in which the field and the world are heading. They look from an outside vantage point into the field, to produce policy documents or legislation that pushes community development forward.

To and from Jobs That Provide Capital

Practitioners from grassroots groups have also frequently gone to work for banks and for community development financial institutions. They might come from positions such as housing or economic development project manager or executive director.

People who provide capital to community development projects—working for for-profit banks or nonprofit financial institutions—also do not stay put. People who have worked in banks move into CDFIs and CDCs, and vice versa. People who move from one institution to another have the ability to assess risks and to find ways to mitigate them, in order to advance the community development goal while protecting the investor's capital. This ability requires good analytic skills and a facility with numbers.

Another set of skills involves the ability to use capital efficiently, which means understanding capital markets and being able to size up one investment compared to others. These jobs also require good people skills. And they have in common the ability to develop and manage a system and the technology to get investments made efficiently.

Many people also switch between the circle of capital and the circle of government. These moves are illustrated by the life stories of two people in this guide, Kristin Faust, who left her long-time, cutting-edge position as senior vice president for community development lending for the Chicago-based LaSalle National Bank for a government post as deputy treasurer for the state of California, and Stephanie Smith, who left her job as deputy commissioner of the Federal Housing Administration at HUD to join the Bank of America's mortgage unit as national manager for community lending.

Another career pattern is the progression from a financial institution to a position in an intermediary.

Defined most broadly, these jobs—whether in a CDFI or bank or CDC or intermediary or public agency—share a vision and the ability to devise ways to get capital into underserved communities.

Community Story: South Carolina

Raising Capital for Community Development

The Capital Access Group, headquartered in Washington, D.C., is a company that Paul Pryde founded after he had spent years in earlier versions of what would today be called community development financial institutions. In his earlier work, he tried first to help create jobs by providing capital to inner-city areas—"a development finance laboratory," he calls it. "I got tired, realizing even if we do a wonderful job, it won't make a difference. The problem is too massive. Then I began thinking about all the assets that cities have and saw that if they could recycle these assets through a secondary market vehicle, they could get a lot more money to lend. The Mott Foundation gave me money to try to sell that idea, and finally I got a contract from HUD. The idea is to create a vehicle that pools individual community and economic loans into a package and turns them into securities that investors can purchase through Wall Street [this is what Fannie Mae does for the housing market]. Then the lenders get a great part of their money back and can recycle those funds into new loans. This could go on perpetually, and community developers won't be so dependent on philanthropy or public money." Capital Access Group's first success was to pool and sell to the MacArthur Foundation a portfolio of community development loans that the state of South Carolina had made to its rural small and minority-owned businesses.

"What drives me is the belief that the only solutions that have legs are positive-sum solutions. Many different economic interests have to be satisfied. I want to see people employed and achieve their dreams. I also want to make a good living doing this, though it hasn't happened yet," Pryde says. Since its first deal with South Carolina, his firm has arranged several other portfolio sales, raising over $7 billion in additional capital for business lending pools across the country.

Sometimes out of Human Services

People with a social services background usually remain in a line of work that continues to make use of their enjoyment of and skills in person-to-person relationships, their familiarity with national and local social services programs, their understanding of the human, individual aspects of neighborhood life. But career moves to the physical and economic development side of community development can and do happen. A social worker, Sheila Crowley, now heads one of the nation's most powerful groups advocating for policies in support of affordable housing, the National Low Income Housing Coalition. With a master's in social work, Sheila worked for a YWCA on battered women's issues and then directed a large homeless service organization. There she gained insight into the underlying policy issues of housing and then explored those issues further through Ph.D. work and helped found a CDC in Richmond, Virginia. It was her combination of hands-on and theoretical background in both policy and housing that made her the right person to head the Coalition.

Furthermore, now that groups developing and managing housing realize that their tenants will be better tenants if they are supported by social services, those organizations are hiring or training social service workers as property managers.

Moving Up by Growing Your Own

Growing the Small Organization

Being entrepreneurial, a remarkable number of today's community development leaders started their careers in a small organization, stayed and helped it grow and grew within it. That is how it happens that Nancy O'Brien now finds herself at the head of the Neighborhood Housing Services (NHS) of Great Falls, Montana, a housing organization that now works across the state and is in fact a model for how to do development in sparsely populated rural areas. When Nancy, a well-regarded Junior League volunteer, took over the failing organization, it had only six weeks left of funding and faced a major crisis. There were 350 boarded-up houses on blocks scattered throughout two neighborhoods, and she was the staff of one, backed by her board of directors and volunteers. Since then, the NHS has removed or repaired those dilapidated houses, built 80 new homes, painted or in other ways rehabilitated hundreds of other houses, lowered the vacancy rate from 16 percent to less than 2 percent, and increased the area's yearly tax revenues to over $4.5 million.

As this tale illustrates, grassroots community groups by their very nature often start out small and grow based on and because of the capacities of their staff, especially if, as was the case in Great Falls, they work in a friendly environment. Similarly, community development financial institutions often start with a founding executive director and secretary who slowly build up board and staff. Mary Mathews, president of the Northeast Entrepreneur Fund in Minnesota, an organization that funds micro-enterprises, says of her work, "It's my first successful start-up. I enjoyed the start-up, but because we're expanding and growing, there's still opportunity, and I like that part."

Growing a small outfit to a larger one happens in the for-profit world as well, as in the case when a bank's one-person community development lending unit expands into a good-sized division.

Starting a New Company

Many people who have spent years in organizations of one sort or another, whether CDC, bank, intermediary, foundation, or government, eventually decide to go out on their own,

Career Story: From the Community to Elected Office

There's a direct connection between community work and politics. Community development professionals learn the political process, figure out how to use it for the benefit of the community in which they are working, and, in so doing, often get to know their elected officials personally and get a good picture of how politics works.

Sometimes they decide to run for election and become the elected representative. Community developers are in the U.S. Senate, Congress, and statehouses; are on city councils, and are mayors. Here are the stories of three out of the many.

Marcy Kaptur, now serving her ninth term as a member of Congress representing Ohio's Ninth Congressional District, which takes in her home town of Toledo, started her life in a working-class Polish-American family that was politically active. Marcy worked in Vice President Hubert Humphrey's presidential campaign in 1968. Her interest turned to urban planning. With a master's degree in urban planning, she worked for Toledo's planning depart-

Rep. Marcy Kaptur

ment until she met up with Monsignor Geno Baroni, who recruited her to join his National Center for Urban Ethnic Affairs. She worked in a number of cities, spending a year in Chicago developing a plan for the Riverwest Neighborhood. Her report on the Chicago work made it into the hands of the top domestic policy advisor for President Carter, and she was asked to join the White House staff. She worked on the development of key neighborhood legislation, including the Community Reinvestment Act. Marcy went to MIT to get a Ph.D. in city planning, but her studies were interrupted when the folks back in Toledo asked her to run for Congress. She accepted the offer of the local Democrats. Her election victory in 1982 was the national upset of the year. Her community development work now takes the form of getting legislation introduced and passed that

Mayor Tom Murphy

will help community groups.

Tom Murphy, returning in 1972 from a two-year stint in the Peace Corps in South America, went to work for a large Pittsburgh-based corporation. But what he was looking for was a job working at the community level, using the skills he had developed in the Peace Corps. Through persistent networking, he found a position as executive director of a community group in a racially changing neighborhood, Perry-Hilltop. Tom led the way with innovation in the neighborhood. When it became clear that the real estate agents in the neighborhood were steering people racially, Tom worked to get his community group into the real estate sales business— in addition to the youth work, the block clubs, and the other activities that were underway. After six years, Tom decided he wanted to represent the area directly, and he led a grassroots campaign to become the elected state delegate from the area. He won and stayed in that position for over a decade before successfully running for mayor of Pittsburgh. Now in his second term, Mayor Tom Murphy continues to challenge community groups to live unto their ideals and make a difference in their neighborhoods.

Barbara Mikulski, now a United States Senator from Maryland, grew up in a working-class section of Baltimore, where her parents ran a small grocery store. The neighborhood, already in decline in the early 1960s, was threatened with devastation when a sixteen-lane interstate highway was planned that would cut through the area. A social worker by training, Barbara was instrumental in forming SECO, the Southeast Community Organization, which successfully fought off the highway, and now, more than thirty years later, she is still working on community improvement. She went from community activist to city council to U.S. Congress to the United States Senate, where she has served Maryland since 1986. Senator Mikulski has maintained her interest in community development by becoming the ranking member of the Senate committee that appropriates funds for HUD. Over the past five years she has been instrumental in creating the HOPE VI program, the program most responsible for the transition of dysfunctional public housing projects into successful mixed-income communities.

Sen. Barbara Mikulski

as consultants working alone or through their own start-up company. This usually happens not because of downsizing or some other factor over which they have no control, but instead as an affirmative career move.

The founders structure their organizations either as non-profit institutions—for example, the nonprofit City Year, which helps localities place youth as volunteers in community improvement programs—or as for-profit companies, such as the Capital Access Group, which uses the financial markets to raise capital to lend to community businesses.

🏠 Up and (Not Quite) Out of the Community Development Field

There seems to be no limit to the places and positions that people have moved to (though not yet the presidency) after working in the community development world. They have become mayors and assistant mayors of cities of all sizes. A number have been elected to the U.S. Congress, and one community organizer, Barbara Mikulski, is now a U.S. Senator (D-MD). Local politics and state legislatures are places to find another remarkably large number of people with community backgrounds.

Community practitioners can move to the world of other nonprofit institutions. Greg Berman, who worked as a special assistant to the executive director of a community development group and then won a Coro Fellowship, is now at thirty-two years old, the deputy director and one of the founders of a national center that creates new prototypes to enable courts to be involved in solving neighborhood problems. Others have gone on to the United Way, the Save the Children Federation, and university positions. The president of Island Press, the nonprofit company that published this guide, is Chuck

Savitt, who in his first job after college worked for the Save the Children Federation on programs that moved the organization from being a child welfare agency to encompassing community development as well. Chuck subsequently moved on to a position with a national public policy institute, running a division that helped community-based cooperatives expand to include community development finance. Now, as head of Island Press, he advocates for incorporating community-based organizing and development into environmentalists' thinking about sustainable development.

In the business world, a transition to the local chamber of commerce is not uncommon. Others have gone on to head for-profit companies.

In the world of philanthropy, community developers have risen to top positions in foundations: for example, Frank Thomas (chapter 1) became the president of the Ford Foundation. Many have served as foundation program officers.

Dig down a bit into the work these people do, however, and you'll find that though their titles and institutions may seem to fall outside the community development field, the people bring community development to their jobs. They continue throughout their working lives to advocate, to create new programs, and to act on initiatives that are definitely part of the community development vision.

Chapter 8

Landing That First Job

If you have come this far in reading this guide and find community development appealing, here are some suggestions to get you started.

 ## Organizing Yourself for a Job Search

Deciding What You Want out of Life
Read!

First, send away for some (usually free) back issues of newsletters from different kinds of institutions (CDCs, foundations, banks, and so forth) from different parts of the country. Look at various Web sites. (Information about newsletters and Web sites is included in appendices D, E, F, and G.) These sources carry news about innovative strategies, trends, and legislation and stories about people working in community development. They give you short case examples of how XYZ group renovated Mr. and Mrs. Senior's rundown house, how they brought business back to Main Street, how Ms. Volunteer organized a Cinco de Mayo celebration, how the P.C. investment firm found a new way to finance affordable housing. By reading some of these newsletters, you can explore the field and its different parts more thoroughly. If and when you go for a job interview, the people who interview you will be impressed that you know something about their field.

Second, read the books recommended in the first chapter—books that give a broad view of the world. Read the books recommended in appendix L—books that focus on community development as a practice. Read the books described at the end of this chapter.

As you read, think about whether the actual work sounds interesting, whether you can see yourself doing it, and whether the issues that community development deals with are issues that you would like to be immersed in from day to day.

Step Back and Look at the Big Picture

Now is the time to step back and think creatively and critically about what you truly want out of life and about your career as part of that. What type of lifestyle do you want? What are the issues you care about? What ways of working on those issues will make you happy?[1]

Don't launch yourself immediately into the concrete details of a job search. Back up first to understand what makes you tick. What does the way you've lived your life so far—your hobbies, extracurricular activities, past work, or volunteer experience—tell you about yourself? Go back to the self-assessment test in chapter 2. Look at the areas where you scored low and try an introspection of your values and attitudes.

Given your personality, try to determine what part of the field would make the best match for you. You can work in community development directly with people or with physical structures or with finance. You can do research and writing. You can work on policy issues. Which appeals to you most? Or, since you could actually start in one mode and eventually switch to another, where would you like to start? Even if your score on chapter 3's self-assessment quiz wasn't high, don't rule out community development entirely; just try to find a place in the field that accommodates who you are.

Are you somewhat risk averse? There are settings in the community development field that will be more comfortable for you than others. Do you want a formal, organized atmosphere, or are you happier with a causal, nonhierarchical setting? There are places in the community development world that fit each of these descriptions.

If you're not sure just which discipline—economics, business, social work, sociology, law, planning, architecture, urban affairs, religion, the environment, or some other approach—will hold the strongest appeal for you, try what Paul Brophy, coauthor of this guide, did in his early years: Walk through the stacks of the periodicals section of your local library; take down a few journals from several topics to browse through; repeat this process until you have covered all the topics. Include journals of commentary and opinion.

Create a Game Plan

Before you launch into action, create a game plan for experimentation and soul searching. Figure out which pieces of your self-knowledge and experience are already in place and which pieces you need to acquire. If you have already worked directly with people, then you might want to experiment with a position dealing in policy issues. Try to develop a logical se-

quence for testing your different interests and different practice settings while building the skills and credentials to land the job of your dreams. The suggestions that follow can be incorporated into your game plan.[1]

Get Experience

Get involved in doing something relevant to community development. First, that's the only way for you to find out whether you really like this field. Second, it's the key to getting a job. If you are still in school, you can test your interest areas and work settings during the school year and over the summers. Work for free if you can, or split your summers between paid and nonpaid work if money is an issue. If you are out of school and have a job, there are evenings and weekends for getting experience.

- Work for a professor or for a group on campus.
- Volunteer or get an internship.
- Join a neighborhood or block association.
- Create your own initiative.

Reread chapter 4, "Getting Started and Getting Ahead." Getting out and doing real work is absolutely the most important step you can take.

Developing the Skills and Materials for the Job Search

Prepare a Writing Sample

A number of prospective employers or places where you want to intern will want to see a writing sample. If you don't have one, now would be the time to create one. Write a paper or article for school, or for a school or community newspaper; choose a topic that is relevant to community development. Or use your volunteer work as the opportunity to write something such as a brochure for the organization or one of its programs, or a press release, or an article for its newsletter. A report on a topic that you researched on behalf of the organization would be a particularly impressive writing sample.

Prepare a Résumé

Get one ready. Refer to the many résumé-writing guides available through the public library or high school or college career counseling offices. See "Recommended Career Guides" below.

Learn to Use Technology

In addition to general familiarity with computers for writing, email, and Internet-based research, jobs in the community development field often call for more specific technical skills such as spreadsheet or database management. Look at the sample job descriptions in appendix A to see what technological knowledge is required or desirable for the kind of job you'd like to land. Take steps to learn that technology.

Think about Training and Higher Education

Go back and look at chapter 3, "Choosing Community Development as a Career," and chapter 4, "Getting Started." If you want to work in a certain setting or type of job—let's say, as a housing developer—and you lack some of the skills or knowledge that the job requires, get training or an advanced degree to enhance your eligibility. Use your informational interviews as another way to find out whether and where your background needs to be strengthened.

Most practitioners advise gaining the "hard" financial and analytic skills required in development. "Without financing, nothing is going to happen. You don't have to do the numbers, but you have to understand what works financially and

what doesn't and how you put together a package that works," says one expert, stating a view that others would agree with fully.[2] If there is a college or university in your area that offers courses you might want to enroll in, get permission to audit a few. Talk with some of the students to find out what they are learning. Talk with the professors about what they're teaching. Ask for the syllabus; review the texts. These steps will give you a feel for whether you will do well in this kind of curriculum and whether it will hold your interest.

Finding That First Job
Network

Whether it's for your first job or for moving from one position to the next, success is almost always the result of intentional and informal networking and friendships.

Networking means developing long-term relationships in order to gather information, gain exposure, and learn about unadvertised job opportunities; it means meeting and staying in touch with the people who can help you find a job and give you support as you pursue your career. Networking is a skill that anyone can and should acquire.

Dan Nissenbaum, now a community development lender, said, "I never had to get a job by responding to an ad. I did a lot of networking and internships, and they were critical. I've had the advantage of trying different sectors, from working for a national trade association to a policy institute, an investment bank, and now a local bank."

Finding People for Networking

Talk to teachers and professors in classes you've taken or programs you might consider enrolling in. Contact people who have come to your campus for presentations.

Contact alumni. Read your college or graduate school

Career Story: Judy Jarmon

Director of Employment Services, Grace Hill Neighborhood Services, St. Louis

Judy Jarmon, who lives in one of the Grace Hill neighborhoods, began working for Grace Hill Neighborhood Services as a volunteer in the late 1960s, helping neighbors collaborate to identify and resolve community problems. Then she moved into a paid position as block captain in which her job was to visit every house in an assigned five- to six-block area every two weeks and connect families to resources they might need. She subsequently took a job with Grace Hill doing clerical work. "I hated that work with a passion, but the director at that time saw something in me and moved me to the family services department. That was my niche. I loved working with people.

"From there my next assignment was as a vocational counselor. I supervised a team of intake people and job developers and also worked one-on-one with program participants to help each develop her individual employment plan.

"In the early 1980s when social service dollars had begun to dry up, we tried to figure out how neighbors could meet each other's needs, and created a program called MORE (Member Organized Resource Exchange). In running the pilot, it made me completely rethink how to do so-

cial services. We conducted training not in a formal setting but anywhere you could find tables and chairs. We didn't keep case files. My role became one as an enabler, to train neighbors and to tell them what I knew and transfer that knowledge to the

neighbor, who became the actual service provider. Whereas before I had provided resources for food or housing or day care, now neighbors did that for each other from their homes. It continues to this day.

"In 1986 I was promoted to the position as director of the MORE linkage network, which has staff and neighbors in ten neighborhoods.

"After that, I managed the residential treatment program for homeless, chemically dependent women and children, a program Grace Hill operated in collaboration with Washington University for two

years. My job was to link the women to Grace Hill's health system and to refer the children for assessment to our centers.

"Then I managed our parenting program, Co-Care, which helps parents develop parenting skills. The program organizes group sessions, brings in professionals, sets up mutual support group where parents share their problems with each other and help each other overcome them. The people who emerge as leaders go on to run the child care program.

"In July 1998, I moved over to head up the employment program and to develop a case management program for women moving from welfare to work. The state division of Family Services refers people to us, under contract, and we provide them comprehensive services, from job training to day care assistance to helping them cope with problems in their daily lives. Another part of the employment division is Structured Assistance, where Grace Hill provides shelter and case management for women who are homeless, and we also work under a contract from the city's employment agency to provide employment services for the hardest-to-place welfare recipients.

"This work is not for everybody. You need to like working with people, to be able to deal with people who have many problems and not get caught up in their personal problems. We all have our niche. It's been a good career for me."

alumni magazine to identify people who work in jobs or places that sound appealing. You'll be pleasantly surprised at how responsive most of them will be when you approach them for information and networking help.

Attend functions and events. Community development conferences are held in many cities in the course of a year. Community development organizations in your city may hold events, such as a house-opening tour, to celebrate and publicize their accomplishments. Local universities or citywide intermediaries or advocacy groups may sponsor forums to discuss issues and policies. By attending these events, you can learn a great deal, and you can meet people and make personal contacts. Find out about these functions and events by getting on the mailing lists of the city community development department or the citywide community development intermediary, and by reading the local news section of newspapers. Read the community newspaper(s) in your city; they are a gold mine of articles and announcements about activities that the major newspapers fail to cover. (If you move to a new city at some point in your life, read the community newspaper to learn all about your new hometown.)

If you can afford the travel expense, attend conferences in nearby cities. Conferences and workshops are announced all the time in national and regional community development newsletters.

Join one or more professional associations. Association newsletters and conferences will give you information about a sector, ground you in the language of the field, and help you find networking contacts. (Hint: The National Congress for Community Economic Development, sponsor of this guide, offers a Student Membership; read about NCCED in appendix D.)

Set Up Informational Interviews

Informational interviews (a highly effective approach made popular by the career guide *What Color Is Your Parachute?*) are conversations with people who are doing the work you think you would like to do. They are not job requests in disguise. You are interviewing these people for their personal perspectives about the field in general, their current jobs, and even jobs they've held in the past. They may also be able to offer advice for your job search.

Here are some pointers about informational interviews:[3]

- To set up an informational meeting, send a letter and follow-up with a phone call. Reassure the person that you are not looking for a job interview (and mean it). Tell your prospective interviewee that you want only twenty to thirty minutes of his or her time and do not stay longer.
- Come to the meeting with relevant questions. It's a meeting you asked for, so it's up to you to focus the conversation. Research the person and organization in advance. Prepare key questions in advance.
- If you can't meet in person, a telephone interview or email exchange can accomplish many of the same goals.
- At the conclusion of each interview, ask if the person you're meeting with can refer you to other people who may be helpful.
- Send a thank-you note.
- Stay in touch if the person seemed responsive to you. You never know when a vacancy may arise or how that person may help your career over time. One hotshot networker further suggests sending out a card or a letter every so often to influential contacts; protective of her time, she has preformatted notes on her computer for this purpose.

Career Story: Barbara Hall

Technical Assistance Coordinator, Atlanta Neighborhood Development Partnership, Atlanta

Internships and fellowships were turning points for Barbara Hall, giving her the chance to get development experience and to meet the people who would later, by coincidence, lead to a new life and a new job. She tells her story as follows:

"I was born in 1967 in the Philippines to an African-American father, who was in the navy, and a Filipina mother, but our family moved to Hawaii, and I grew up there. I studied nights and worked a full-time job during the day while I was going to San Francisco State University and graduated with a double major in anthropology and history.

"During my last year in college, when I was trying to decide what to do, I took an interdisciplinary course in urban studies, taught by a planner for the city of San Francisco. He got me an unpaid internship with the city.

"I went on for a graduate degree in city and urban planning at UC Berkeley, with a fellowship, while also working part-time at two jobs and the city internship. Then I took a leave of absence and had a baby son. I moved to a supportive housing project, got welfare and food stamps, used community health clinics, and went back to Berkeley

and finished my course work requirements.

"During graduate school, through the same teacher's contacts, I found one summer internship with an economic development agency studying job creation in a

predominantly Filippino neighborhood and another internship with a leading Bay Area community development group.

"One day in the economic development agency's office I happened to hear about the Nonprofit Housing Association of

Northern California's new fellowship program for people of color from a person visiting the office who was on the program's committee. I applied and was chosen. As the fellowship's full-time, on-the-job experience, I worked for three organizations engaged in a wide array of economic development activities including housing. In the morning, first I had to walk my son to day care, then take the train across the bay to their offices. After graduating from the fellowship program, I took a job as development assistant with a Bay Area CDC.

"When I decided to move to Georgia, where two of my sisters live, through my Bay Area contacts I found out about a job with an Atlanta nonprofit. They flew me across the country for an interview; a director of that nonprofit in turn referred me to the Atlanta Neighborhood Development Project (the citywide neighborhood development intermediary) and its CEO, Hattie Dorsey. ANDP created a job for me and hired me that day.

"Since then, I've become coordinator of ANDP's technical assistance program and the program manager for a fellowship and internship program in community economic development. In these two roles, I work to assist CDCs build their capacity while growing the community development field in Atlanta."

Find a Mentor

A mentor can help you with networking, can provide information, and can give you advice that guides your professional development and your ability to lead and to take risks. A mentor can give you encouragement and support and instill confidence in your abilities.

You do not have to wait for someone to stretch out a helping hand. You can make a positive effort to find a mentor. Terri Montague, who is now senior vice president at Boston Financial, a firm that invests in and manages affordable housing, says, "I've been very intentional about developing mentoring relationships. I've been able to find people within my firm in high positions who let me take them out to lunch to talk about business and issues."

But mentors enrich more than just your professional growth. "Outside my work, I have found personal mentors," Montague adds. "One is an eighty-year-old lady in Georgia who takes in young mothers and helps them get back on their feet; another is a pastor in Chicago in a difficult community. People such as those bring insight to me as a person." A mentor can share his or her dream and insights in life and thus widen your own perspective and vision.

You will find potential mentors through your volunteer work or through networking or through academic programs or by meeting someone at a training session or professional association conference.

In seeking out a mentor, you should look for someone who works in and knows the community development field and has the ability to abstract, to think and talk about specific initiatives and neighborhood issues in a larger context. You are also looking for the kind of person who is a natural connector of people.

In approaching someone, you might ask for limited assistance rather than a long-term and open-ended relationship. You might ask, for example, if the person would be willing to meet with you for an hour every two weeks for a span of a few months. You might offer in turn to do volunteer work for your mentor's organization or project. Prepare in advance for your mentor meetings, deciding what you want to talk about. If that interaction turns out well, the mentor might be willing to extend the relationship.

Ways of Looking for That First Job

You will probably need persistence to get a job; this is not a particularly easy field to enter, though this guide does its best to help. Jobs are indeed advertised, and in appendix J you'll find the places where the ads are most likely to be found. The community development field is so diverse that whatever your background, talents, or personality, you can find a first job that matches. A literature major can join a fundraising staff or do policy work. A person with a social work or psychology background would suit a position in organizing or human services. Someone with computer skills can do database management for any kind of program. Again, these jobs might be entry points, not end points, since, as the previous chapters have shown, there is great fluidity in the community development world.

When you start to focus on your job search, develop a strategy for the ways you will go after the jobs you want.

Use the appendices in this guide, first, to select the kind of work and the kind of organization you think you'd prefer; and second, to create a list of potential employers. If you want to work in a particular geographic location, call national trade associations and regional and national groups, associations representing government employees, and associations representing capital providers. Ask for names and addresses of their member organizations in those locations.

Target fifteen to twenty potential employers; research each one by asking them to send you background information about their work. Do networking and informational interviewing with as many organizations as you can.

Don't rely on job announcements; go after the places you want to work.

Be persistent; don't give up.

If you can, get a job and then go on for training or advanced education. The point is that the sooner you work, the better, because it will clarify who you are and whether in fact this field is right for you.

Taking Charge of Your Career Growth

Don't leave it up to your employer organization to take care of your career. Especially in the community development world, where organizations are relatively small, with scant time and resources, people should create their own plans to manage and develop their careers. As you work in your first and successive jobs, figure out the areas where you want and need to develop your skills and knowledge. Be proactive in finding training workshops, conferences, discussion groups, and educational programs you want to attend. Get your supervisors to understand your commitment to the organization and your desire to learn and grow.

Networking is not a skill or practice to put aside once you are in a job. It should become part of your working life. It will be the way to hear about new opportunities for yourself as well as for your organization and your community.

In the twenty-first century, workers will do better if they pursue what experts call lifetime learning, and this is what career management is all about. You could call it developing the developers.

 ### What a First Job Is Like and How They Landed It

Mark Gaines is assistant land manager in the Dallas Division of Ryland Homes, a for-profit housing development company. He oversees the development of residential lots all across the Dallas–Fort Worth metropolitan area. In his latest project, he is collaborating with a newly formed nonprofit, the African American Pastors' Coalition, in building 284 market-rate homes in a long-ignored section of Dallas. As part of this project, he administers a mentoring program that helps African-American builders to produce about one-fifth of the homes. Mark also organizes community outreach efforts for Ryland; one is a program that teaches building skills to young people from lower-income families. Mark's 1998 graduate degree in urban planning is from Ball State University in Muncie, Indiana.

Mark used networking and informational interviews to find his job with Ryland. "I had many jobs before this one, but none of them were what I wanted to do. I want eventually to be a developer building housing, from low-cost to high-end homes. I looked for companies where I thought I could learn those skills. I targeted those companies, found people in them, phoned them and said, "Hello, I'm a student. You're in the kind of company where I want to be." I asked them for informational interviews. Finally, through those interviews and networking, a gentleman doing community revitalization work told me about Ryland Homes and its management-training program and gave me a contact there. I called his contact, and eventually with his help created an internship with them—they had never had an intern before. From there they hired me, and now I'm going through their management-training program."

Manuel Maysonet, twenty-six, is a program officer for the National Equity Fund (the arm of the Local Initiative Support Corporation that uses Low Income Housing Tax Credits to finance affordable housing across the country). His job is to implement many of the president's initiatives and special projects, to work with the NEF research and development team, and to get the word out about the Low Income Housing Tax Credit program and other new initiatives by participating in CDC conferences across the country. He has a master's degree in city and regional planning from Rutgers University.

Manny is one of the people whose success in landing a dream job forms the basis for the suggestions in this chapter. During his first year of graduate studies at Rutgers, in 1997, he was selected to participate in the first class of the National Congress for Community Economic Development's Emerging Leaders program, which he had seen announced on the department's bulletin board at Rutgers.

Over the summer, he attended the annual meeting of the National Council of La Raza, held that year in his hometown of Chicago. There he introduced himself to the human resources manager of the National Equity Fund, which is headquartered in Chicago; he gave her a copy of his résumé and kept in touch with her over the second year of his studies. That year, networking through a professor's contacts, he says, "I essentially created my own internship" at the LISC Newark office, where he helped screen funding proposals from the CDCs throughout New Jersey. Through frequent review of LISC's on-line job bulletin board, he found out about the job opening in the NEF office that spring, applied for it, and landed it.

Straight out of college, with a major in business administration and a minor in marketing, **Mia Baytop** was hired by a mortgage company, Norwest Mortgage, as a credit counselor in a nationwide program to increase homeownership among ethnic populations, immigrants, low- to moderate-income families, and first-time home buyers. Her job was to help these people to understand the home-buying process and repair any credit problems they might have. Once hired, Mia took advantage of Norwest Mortgage's professional development program to enroll in various community development training workshops, including the National Congress for Community Economic Development's Emerging Leaders program. She says, "That's where I realized how broad the field was and got pulled into it as a career." Recently, she was promoted to a higher-level position as community development analyst, and, while working, she is studying for an MBA full-time, evenings and Saturdays, at Morgan State University in Baltimore.

Jonathan Weinstein works as an associate at Bay Area Economics, a Berkeley-based for-profit consulting firm whose clients are generally public agencies such as city housing authorities and redevelopment agencies. The company does economic analysis and strategic planning. "In the year I've been here, I have helped a city prepare its application for federal empowerment zone designation, worked with a city's housing author-

Community Story:

Urban Agriculture and Gardening across the Country

Urban agriculture, a growing practice in inner-city neighborhoods, transforms trash-covered lots into attractive green spaces while producing jobs, income, and healthy food. City farming is blooming as a new community economic development business line.

In Chicago, where there are 80,000 vacant lots, community groups make use of some for the production of goat cheese for sale; others are trying aquaculture, raising high-profit fish for local markets. In California, a company founded by inner-city youth produces valued-added items—salsa and salad dressing—from vegetables it grows and sells those products nationwide.

A large biosphere greenhouse, sponsored by the Dudley Street Neighborhood Initiative (DSNI), a resident-controlled group, is under construction in the Roxbury neighborhood of Boston. This project represents a concrete example of DSNI's dedication to sustainable community economic development. (It is also a concrete example of community empowerment: the way the space would be used was decided by vote at an an-

nual meeting of the residents.) The 10,000-square-foot, $750,000 commercial green-house, built on the site of a former garage, will grow high-end crops for sale to city restaurants and markets. Profits will be plowed into a Community Development Account and tapped to support community-based programs such as youth activities.

In New York City, community gardeners plant vegetables and flower beds in reclaimed lots and forgotten neighborhood parcels, 113 of them rescued with financial help from Bette Midler.

Isles, Inc., a Trenton, New Jersey–based CDC, had its origins in the community garden movement. While broadening its scope over the years to include affordable housing, job training, and environmental education, Isles has continued its garden activity. It manages a market farm and greenhouse on a community college site outside the city, where it trains and employs neighborhood residents while producing high-quality food for Trenton's inner-city areas.

ity to assess its properties and develop a strategy to improve the quality of the housing and attract mixed-income renters, and several other development projects. It's been a great experience, not just to do the work but to learn about the various cities and meet with the city officials, community representatives, and teams of specialists—for example, architects, engineers, and planners with twenty years of experience in community development—and learn from them."

As part of his MBA at UC Berkeley's Haas School of Business, Jonathan took courses in real estate and economic development, as well as classes in the School of City and Regional Planning. He found out about Bay Area Economics because a professor and a lecturer were associated with the company, and the firm had earlier hired a number of students from the City Planning school. He made a strong campaign for a job there: met with staff several times just to talk, sent them postcards, and visited their Maryland office when he was on the East Coast during an NCCED conference. "I put a lot of effort into showing my interest and staying in touch," he says. Jonathan is a graduate of NCCED's Emerging Leaders program.

Michael Barr got his job as deputy assistant secretary for community policy at the U.S. Treasury Department through networking and experience with issues related to community development. A Rhodes scholar and Supreme Court law clerk, he had gotten interested in public policy while working in a legal clinic at Yale Law School on the problems homeless people face in securing housing. He explains, "Secretary Rubin's chief of staff was a friend of mine from school and from a political campaign. She suggested I come over and talk with him about the new policy position he was creating at Treasury." After an interview, Secretary Rubin offered Barr the job. Barr says, "This is fabulous work. It gives meaning to one's life."

An internship through her university turned into a job for twenty-six-year-old **Shelia Slemp**. As part of her course work in the Community-Based Social Work Pilot Program offered by Case Western Reserve University's Mandel School, Shelia was placed in an internship with Neighborhood Progress, Inc. (NPI), Cleveland's citywide intermediary that funds and assists CDCs. "I worked with one of NPI's vice presidents to design a human capital program that would link student interns to the area's CDCs in order to increase the number of people with skills and interest in CDC jobs. I organized the CDCs in a collaborative effort to design the program proposal, and when it was funded, NPI hired me as project manager. I work twenty-five hours a week and set my own hours, which enables me to continue my studies. My direct supervisor, a former educator, has a vested interest in my education and is a great mentor."

 Recommended Career Guides

These guides cover careers, job hunting, networking, and career management:

Career Intelligence: The 12 New Rules for Work and Life Success, Barbara Moses (Berrett-Koehler Publishers, 1998). Sets forth key principles of life-long career management:

craft your own future; know and leverage your strengths; identify your core values; and be honest about your identity and abilities.

Damn Good Résumé Guide, Yana Parker (Ten Speed Press [800-841-2665], 1990).

Do What You Are: Discover the Perfect Career for You Through the Secrets of Personality Type, Paul D. Tieger and Barbara Barron-Tieger (Little, Brown, 1995, 2nd edition). A guide to charting one's personality type and temperament, it also suggests the most satisfying careers for each type, "doing what you love and what feels natural."

Do What You Love, The Money Will Follow, M. Sinetar (Doubleday Dell [212-354-6500], 1987). Another guide to discovering the career that suits your talents and personality.

Don't Use a Résumé . . . Use a Qualifications Brief, Richard Lathrop (Ten Speed Press [800-841-2665], 1990).

Harvard Law School's Public Interest Job Search Guide. (Harvard Law School, Public Interest Advising Office, Cambridge, MA [617-495-3108], 1998). Includes top-notch advice for anyone, not just lawyers, about résumés, networking, interviewing, job searches.

Non-Profits & Education Job Finder, Daniel Lauber (Planning/Communications [708-366-5200, www.jobfindersonline.com], 1997). Advice on using the Internet for job searches, along with comprehensive lists of associations, directories, and Web sites.

What Color Is Your Parachute? 2000 Richard Nelson Bolles (Ten Speed Press [800-841-2665], 2000).

Chapter 8 endnotes:

1. This material is based on an unpublished memorandum written by Alexa Shabecoff, "Counseling Harvard Law Students and Alumni/ae" (Cambridge, MA: Office of Public Interest Advising, Harvard Law School, 1999).

2. *Community Development Career Guide* (New York: Local Initiatives Support Corporation, 1998), p. 7.

3. Some of these pointers were drawn from two excellent career guides: *Public Interest Job Search Guide,* Stacy M. DeBroff (Cambridge, MA: Harvard Law School, 1998, 8th edition) and *The Complete Guide to Environmental Careers,* Bill Sharp (Washington, DC: Island Press, 1993).

This appendix contains a sampling of job descriptions in the community development field. These descriptions follow the categories listed in Figure 5.1 and have been taken from actual job postings in 1999. The content has been modified in two key ways: First, the identity of the employer has been deleted. Second, the salary ranges are based both on the actual job posting and on informal surveys of other similar jobs, so a fuller range can be shown.

Salaries range based on size of organization, region of the country, and whether the position is in the nonprofit, government, or private for-profit sector.

For a list of places where jobs are routinely posted, see appendix I.

Executive and Management Positions

Executive Director, Community Development Corporation

The executive director oversees the overall direction of the neighborhood-based community development corporation and reports to the board of directors.

Responsibilities:

- Oversee and manage the staff and the overall day-to-day operations of the community development corporation.
- Create and execute a set of programs aimed at improving the community.
- Work with the board in setting the direction and policies for the organization.
- Collaborate with the organization's treasurer and financial manager to establish a budget for each fiscal year.
- Actively promote the goals and image of the corporation within the community.
- Secure funding for the operations and for special projects.

Skills:

- extensive knowledge of community-based organizations and/or a strong background in nonprofit housing development;
- excellent verbal and written communication skills;
- capability to organize and provide direction for working groups;
- ability to do financial analysis and planning;
- ability to work independently and exercise excellent judgment.

Experience:

Advanced degree in the appropriate field and a minimum of three years' experience in a related field.

Employers: Community development corporations
Salary range: $35,000–$150,000

Executive Director, Neighborhood Housing Services Program

The executive director will manage housing revitalization and neighborhood reinvestment programs under the direction of a diverse board of directors.

Responsibilities:

- Work with all partners in the development and execution of housing development, rehabilitation, and lending programs.
- Recommend to the board new initiatives aimed at neighborhood improvement.
- Oversee the staff and operations of the program.
- Raise funds, as needed, for program operations.

Skills: To be able to work cooperatively and build effective working relationships with government, businesses, neighborhood residents, and nonprofit board. Be a self-starter with entrepreneurial spirit.

Experience: Five years of experience in two or more of the following fields is required: grant writing, nonprofit management, housing rehabilitation, finance, real estate, community organizing, organizational development, urban and regional planning. Managerial and supervisory experience in the field is preferred.

Education: Bachelor's degree
Employers: CDCs, Neighborhood Housing Services programs
Salary range: $50,000–$65,000

Program Manager, Community Development Corporation

Develops affordable housing; responsible for entire development process, including land acquisition, financing, construction, marketing, and project lease-up.

Responsibilities:

- Obtain zoning and building permit approvals for project development. Become familiar with licensing requirements and determine sources of funds available to pay for shelter and/or care in an assisted living facility.
- Prepare applications and financing proposals, negotiate loan terms with lenders and ensure that all loan and subsidy requirements are satisfied.
- Select and contract with architects, engineers, and general contractors familiar with local and state requirements for assisted living facilities.

- Oversee the design phase of the development process to ensure that all local and state requirements are satisfied.
- Prepare the project development, operating, and leasing budgets.
- Identify and contract with agencies to manage each facility and provide supportive services to the residents.
- Prepare, package, and make presentations for approval of projects to the CDC board of directors.
- Coordinate the preparation and review of loan closing documents with financing institutions, attorneys, engineers, architects, and contractors.
- Work closely with construction manager to monitor construction and compliance with plans and specifications; work with all parties to reach construction completion on budget and on schedule. Attend on-site owner's meetings. Process construction draws and change orders.
- Prepare application for and obtain the license to operate each project.
- Ensure that all permanent loan requirements are met and appropriate documents are executed.
- Identify and facilitate ties between each project and local community organizations and work with property manager to establish annual budgets.
- Oversee project management to ensure compliance with all local and state requirements and ensure that quality services are being provided to the residents.

Skills: Proficiency in use of spreadsheet software, problem-solving skills, excellent oral and written communication skills, and the ability to confidently and effectively market to and work with individuals at all levels of company and client organizations.

Experience: Three years of related experience providing the ability to evaluate the feasibility of a real estate development as well as knowledge of development, construction, and financing of affordable housing.

Education: Bachelor's degree; graduate degree in finance or planning preferred.

Employers: National intermediaries and large nonprofit and for-profit developers.

Salary range: $35,000–$65,000

Program Manager, Neighborhood Housing Services Program

The program manager handles the affairs of the organization and directs selected aspects of the organization's programs.

Responsibilities:
- Manage day-to-day affairs of neighborhood-based office.
- Develop and implement loan products for home ownership.

- Support training and leadership development for neighborhood residents.
- Market neighborhood to prospective home buyers and the real estate industry.
- Work in partnership with community advisory board composed of neighborhood leaders, appointed and elected government officials, lenders, and real estate professionals.

Skills: Strength in administration, event planning, office management, operations and service delivery, program development, staff team supervision, and volunteer development and management.

Experience: Experienced in real estate, mortgage lending, community development initiatives, or neighborhood revitalization.

Education: College-level course work and any equivalent combination of education, experience, and training.

Employers: CDCs, Neighborhood Housing Services (NHSs)

Salary range: $35,000–$50,000

Executive Director, Economic Development Group

The executive director is responsible for overseeing all of the activities of the CDC, and especially seeing to it that the lending operations of the corporation are carried out consistent with board direction.

Responsibilities:
- Program initiation and implementation consistent with direction from the board of directors.
- Comprehensive management of program operations, budget and finance, personnel, systems development, and maintenance.
- Management of numerous revolving loan funds.
- Supervision of an eleven-member staff.

Skills: Excellent written and oral communication skills and the ability to network with other resources.

Experience: Five years' experience managing community development programs. Two years of lending, preferably commercial lending.

Education: Bachelor's degree in public administration, finance, human services, or related field; master's degree preferred.

Employers: Community development corporations

Salary range: Starting $35,000–$75,000

Executive Director, Rural Partnership

Responsible for the implementation of the rural area program's ten-year plan to produce one thousand low-income housing units.

Responsibilities:

- Create partnerships with local churches, nonprofit groups, towns, and/ or builders on specific housing projects.
- Assess the feasibility of specific real estate development projects and housing programs.
- Broker financial and technical resources.
- Package project financing and set and monitor housing development standards.
- Facilitate community participation in the development process and facilitate the community, organizational, and social support networks for each project.
- Facilitate fundraising for the organization budget.
- Manage and direct all operations, including hiring, firing, and supervising staff, consultants, and subcontractors.
- Represent the organization to the general public (e.g., through the media), the government (local, state, and federal), local communities, and others.
- Encourage and monitor lender agreements of participants in a possible area-wide benevolent loan fund and facilitate marketing efforts to potential depositors.
- Provide training to board and committee leaders.

Skills:

- ability to deal with the public, government officials, and officials of religious institutions;
- ability to write well and speak publicly and work with a board of directors;
- ability to conduct leadership development training.

Experience: At least five years' experience in all aspects of real estate development including project feasibility analysis, design, financial packaging, construction, sales/lease-up. Familiarity with low-income communities and nonprofit housing developers. Experience in fundraising and financial packaging. At least five years' experience in managing an organization as an executive director or key manager.

Education: Prefer a graduate degree in law, business, architecture, planning, or administration.

Employers: Rural housing partnership organizations and national intermediaries

Salary range: $50,000–$75,000.

Executive Director, Business Development Corporation

The executive director is responsible for the strategic growth and general operations of all divisions. Staff management and development is of primary importance, as is integrating the development, finance, technology, and environmental goals and objectives with those of the corporation as a whole.

Responsibilities:

- Evaluate and analyze current business development activities; identify product and service niches; define priorities and allocate resources; refine and execute business plan to reflect market knowledge and experience.
- Oversee management of small business incubators; identify and evaluate need for additional acquisition or management opportunities.
- In conjunction with senior management, oversee execution of programs project and operating fund (grant, program-related investments, and/or loan) acquisition. Define and develop fee-generating product and service lines within the organization.
- Integrate corporate development objectives. Define and track development outputs.
- Oversee budget, staffing; align resources with primary objectives; provide appropriate reports, analyses, and outcome measures to the board, regulatory agencies, and governing bodies of the organization.

Skills:

- ability to manage, mentor, and motivate professional staff and entrepreneurs;
- ability to analyze and evaluate step-by-step approaches to clients' business growth and teach staff to do the same;
- excellent communication skills; the ability to listen and articulate well; adept at all levels of input and delivery.

Experience:

- Prior experience in nonbank lending; prefer mezzanine/subordinated debt and early-stage financing of emerging companies.
- Prior experience operating a business, developing business plans, market analysis, and financial projections. Strategic creative planner with strong history of preparing and implementing business strategies.
- Hands-on approach to creating momentum in growth of organization.

Education: Combination of higher education and experience in appropriate fields.

Employers: Banks, CDFIs, large economic development–oriented CDCs, business incubator programs

Salary range: $70,000–$95,000

Executive Director, Public Housing Authority

The executive director reports to a five-member commission and oversees all operations of the housing authority.

Responsibilities: Administration of 185 public housing units and 1,200 Section 8 vouchers and certificates, an annual budget of approximately $5 million for all programs, and supervision of seventeen employees.

Skills: Budgeting, accounting, personnel, maintenance, communication, and public relations.

Experience: Grant writing and affordable housing.

Education: Bachelor's degree or equivalent and/or experience in public housing administration. A Public Housing Management certification is required or the ability to obtain same within one year of hiring.

Employers: Public housing authorities

Salary range: $45,000 and up, depending on size of Authority

Executive Director, Community-wide Housing Partnership

Lead position managing operations of a nonprofit corporation. Establish the office, hire and manage initial staff of housing partnership representing the city, county, local lenders, and civic leadership. Manage operating funds to finance, purchase, and/or rehab several hundred housing units per year.

Responsibilities:

- Oversee implementation of strategic housing plan for the area in cooperation with city planning and development, redevelopment, housing authority officials, and other representatives.
- Develop partnership policies and procedures for personnel and fiscal management, working with staff and outside consultants, and oversee these functions.
- Obtain commitments and/or set-asides of grants, loans, and other financing sources for partnership projects, partners, and operations. This may include equity sources for syndicated rental projects.
- Manage short-term investments of partnership funds.
- Recommend to the board and/or approve direct loans and grants from partnership funds; sit on loan committee for partnership funds. Propose policies and procedures for loans and grant processing with lending consortium.
- Directly provide local nonprofit housing groups with assistance in strategic planning, accounting, fundraising, and other aspects of organization development to increase effectiveness of locally based organizations.
- Arrange organizational development, capacity building, and project technical assistance to local nonprofit groups engaged in housing production.
- Also do proposal writing, public speaking, real estate project development and management, proforma analyses.

Skills: Exceptional skills in writing, public speaking, finance, fundraising, entrepreneurial thinking, transferring knowledge and skills and relating to clients, as well as local leadership. Concern for disadvantaged individuals and dedication to creating positive solutions to housing problems.

Experience: Hands-on experience with low-income housing development in distressed urban environments; public and private financing, including syndication; successfully managing a growing organization. Minimum of five years of increasingly responsible positions as a project manager or executive director of a similar organization involved in producing affordable housing.

Employers: City and county housing partnership organizations, national intermediary field offices

Salary range: $40,000–$90,000

Volunteer Coordinator, Local Housing Group

The volunteer coordinator helps potential volunteers find an appropriate role in helping the organization accomplish its goals.

Responsibilities:

- Assist people who contact the organization by providing them telephone, email, and written information about volunteer possibilities.
- Help facilitate a decision-making process whereby an inquirer is able to make an informed commitment to participate as a volunteer. Must be able to answer questions about programs, philosophy, expectations, and logistics. Determine needs for volunteers within various committees and projects.
- Must be able to qualify prospective volunteers and assist them in finding suitable volunteer positions.
- Assist prospective volunteers in overcoming obstacles to their participation with the organization.

Skills:

- extremely high degree of self-motivation and desire to excel;
- strong communication skills on the telephone, through e-mail, and in writing;
- ability to pay attention to accuracy of details communicated to others within the organization and throughout the community;
- demonstrated proficiency in sales techniques and time management; sales or recruiting experience preferable;
- customer service skills;
- ability to use computer word-processing and database programs;
- flexibility, creativity, and the ability to work as part of a team.

Experience: Two years' related experience.
Education: Bachelor's degree
Salary range: $25,000–$40,000

Chief Financial Officer

The chief financial officer has primary responsibility for the organization's financial management, information management, human resources, information systems, and operations. The CFO works closely with the executive director, other senior managers, and the board treasurer. The CFO supervises a staff of four.

Responsibilities:

- cash, portfolio, and risk management;
- budgeting and financial planning;
- accounting systems, policies, and procedures;
- vendor selection and management;
- human resources management;
- financial reporting to the executive director, the board of directors, funders, investors, and staff;
- information management;
- information systems management;
- office systems and operations;
- technical assistance to and/or training for internal and external clients;
- staff supervision.

Skills:

- financial planning;
- cash management;
- budgeting with annual budgets greater than $2 million/year;
- bookkeeping and accounting, including use of accounting software;
- policy development and implementation;
- organizational leadership;
- staff supervision;
- human resource management (including hiring);
- information systems management;
- vendor relations;
- public speaking and presentations;
- training;
- knowledge of nonprofit accounting (highly beneficial);
- strong written and verbal communication skills;
- computer literacy, particularly with accounting software and spreadsheet applications;
- database experience;
- excellent interpersonal skills and the ability to work well with diverse

constituents;
- ability to set priorities effectively, delegate effectively, and work independently.

Experience:

- At least five years' professional experience as a CFO, comptroller, and/or senior financial officer is required. Experience working at a financial institution is highly beneficial.
- At least two years supervising employees is required.
- Experience analyzing and managing portfolio risk is highly beneficial.
- Human resources management and underwriting experience is valuable.

Education: Bachelor's degree in accounting, finance, and/or business administration is required; graduate degree in one or more of these fields is highly beneficial.
Employers: Large CDCs, national intermediaries, CDFIs
Salary range: $75,000–$120,000

President, National Intermediary

The president and chief executive officer has all of the authority and accountability traditionally vested in a chief executive.

Responsibilities:

- Define, and propose for board approval, the strategic directions and goals of the corporation.
- Translate board-approved strategic goals into carefully designed and well-coordinated programs to achieve them.
- Lead the mobilization of the financial and human resources necessary to carry out these programs.
- Manage all elements of the corporation's programmatic and advocacy activities.
- Serve as the corporation's primary public representative and spokesperson to the many constituencies critical to its success.
- Recognize the broad scope of both the problems addressed and the remedies offered, and to incorporate all concerns and objectives into a balanced approach.

Skills:

- Commitment to the needs and interests of the very disadvantaged and to the mission of the organization, together with the skills to communicate these convictions compellingly to diverse audiences and constituencies.
- Highly developed leadership and entrepreneurial qualities, as demonstrated by successful service in a leadership role in some form of endeavor aimed at least partially at public benefit.

- Capacity to work with board and staff to develop a common strategic vision for the organization.
- Ability to recognize talent, to retain and motivate an extremely talented staff, and to recruit individuals of equal quality as the need arises.
- Creativity, courage, perseverance, and other qualities necessary to raise funds. Experience at raising money from foundations, corporations, and/or governments is preferred but not required.
- Political savvy, acquired and applied without strident partisanship, along with understanding of how government works at both policy and operating levels.
- Advanced management capability and people skills suited to a far-flung organization composed of free and highly articulate spirits who are best led by persuasion and example, not by exertion of hierarchical authority. Managerial experience should include responsibility as head of an organization or unit of some kind.
- Personal stature, preferably in a field directly relevant to some aspect of the organization's mission, together with a broad network of influential contacts.
- A high level of emotional security and centered self-esteem, with manageable needs for ego satisfaction, a talent for listening, and readiness, after due consultation, to make timely and clearly expressed decisions.
- High energy, a large appetite for hard work, and a vibrant sense of humor.

Experience: Substantial experience in community development, leading an organization, and management capacity.

Employers: National intermediaries

Salary range: $145,000–$200,000+

District Director, National Intermediary

The district director provides senior-level management and leadership within one of nine districts in the United States. As the senior manager representing the corporation in the state district, the district director exercises considerable independence and sets the standard for a district staff team whose mission is to strategically provide technical assistance and financial resources to member organizations in order to advance capacities and achieve positive impact within the communities being served.

The district director is responsible for management functions that include administration of district personnel and resources; communication of corporate programs, policies, and procedures; expansion of district network affiliate groups; technical assistance and grants to affiliate organizations; and evaluation of district field operations. The district director is a member of a corporate management team and reports to the director of field operations.

Responsibilities:

- In consultation with the director of field operations, hires, trains, promotes, and terminates district staff in accordance with personnel policy and procedure; supervises, motivates, and sets an example for a diverse staff who must function as an effective team.
- Serves as the organization's chief spokesperson within the district. Hosts, moderates, and facilitates events for the purpose of planning and partnership development; recognition, training, and education; and resource development.
- Assesses leadership and resources available to undertake new partnership development; and, based on private- and public-sector commitment, recommends the allocation of corporate resources.
- Provides funding and technical assistance by evaluating the status and effectiveness of each affiliate organization; establishes district priorities for technical assistance and grant funding, the outcome of which is to further motivate and reward productive organizations and challenge or leverage cost-effective change in less productive organizations.
- Conducts annual evaluation of district staff; based on performance, promotes and/or terminates employees in consultation with the director of field operations.

Skills:

- Excellent interpersonal communication skills, both oral and written; ability to foster trust, credibility, and cohesive teamwork among people with diverse talents, backgrounds, and perspectives.
- Conversant in the ways, means, and sources through which public-sector and private-sector financing can be utilized to positively impact local communities.
- A career of professional achievement characterized by creative and proactive leadership, which is collaborative and facilitative yet decisive toward results.
- Exceptional skills in strategic planning and organization development with the capacity to discern how to motivate and leverage positive change through the investment of human and financial resources.
- Entrepreneurial spirit with a capacity to exercise independent judgment and sound decision making in the midst of diverse and complex organizational environments.
- Keen analytical and problem-solving skills.

Experience:

- At least five to seven years of demonstrated achievement working with nonprofit, community-based initiatives, a portion of which encompasses hands-on, grassroots experience.

- At least eight to ten years of diverse public- and/or private-sector management with responsibility for personnel hiring, supervision, and development; contract and budgetary management; resource development; program development and/or provision of technical assistance; and program evaluation.
- Belief in community empowerment and the ability of people to improve the quality of their lives and the communities in which they live.

Education: Bachelor's degree; master's degree preferred.
Employers: National intermediaries
Salary range: $70,000–$95,000

Department Director, Intermediary

The department director of an intermediary is responsible for the oversight, management, and administration of three divisions comprising the Management and Organizational Development Division; the Planning, Design & Development Division; and the Community Assets Division. This is a senior management position that reports directly to the vice president of program services.

Responsibilities:
- Prepare the yearly budget and work plan for the department, including projections of program priorities and staffing and travel requirements to carry out the work plan. Maintain the budget and sign off on department expenditures.
- Hire and assign staff responsible for carrying out program priorities.
- Directly supervise the work of the directors of the three divisions, as well as the administrative staff; serve as second-line supervisor to rest of staff. Arrange for the professional development of department staff, including in-house training and outside courses.
- Review and approve written reports issued and contracts entered into by department staff.
- Represent the intermediary to outside organizations and local governments, through both speeches and written reports and articles.
- Serve on selected nonprofit boards of directors.

Skills:
- familiarity with housing and small-business finances;
- strong writing and communications skills; the ability to interface with all sectors and effectively represent the organization's goals and mission;
- strong managerial and supervisory skills;
- computer literacy and word-processing skills;
- strong organizational skills and ability to perform tasks with limited supervision.

Experience: Seven years of experience in housing or community development work, preferably in government, nonprofit organization, or community or local development corporation; graduate degree can substitute for two years of experience. Experience providing effective training and technical assistance. Fundraising experience. Experience working with a board of directors and/or advisory committee.
Education: Bachelor's degree
Employers: National intermediaries
Salary range: $75,000–$115,000

Community Organizing

Community Resident Organizer

The community organizer is responsible for developing an active resident organization in housing developments. This position reports directly to the executive director.

Responsibilities:
- Develop and implement a plan for organizing residents and developing leadership on sites. Conduct outreach to all existing residents using a variety of methods including door knocking, house meetings, flyers, informational meetings, and others.
- Provide information on the roles and responsibilities of residents and the future development plans for their housing. Build excitement and enthusiasm among the residents. Identify and recruit potential leadership and develop resident leadership skills and abilities. Form on-site cluster groups of residents, an on-site steering committee, and on-site committees to work on specific issues of concern to residents; identify vehicles for resident involvement on the site.
- Identify, recruit, and organize potential residents from community organizations and social service and religious centers to participate in the development. Work with residents to encourage and maintain their commitment, motivation, and participation in activities supporting a self-help style of management; initiate planning with residents for appropriate activities to enhance the above. Facilitate a cooperative working relationship among the residents and the property management staff to ensure a smooth transition from rental housing to ownership.

Skills:
- demonstrated ability to organize people to build a strong, diverse community, to develop strong leadership, and to maintain commitment and involvement at high levels over time;
- proven ability to work with large groups of diverse people;
- demonstrated ability in leadership development and resident training;

ability to design and execute a development and training program;
- ability to establish and support grassroots committees and activities;
- excellent oral and written communication skills;
- ability to develop and maintain cooperative relationships with prospective residents, community leadership, government officials, and members of the business community;
- demonstrated commitment to supporting long-term affordable housing for community residents.

Experience: Two years' demonstrated experience in the field of community organizing or equivalent in motivating and involving people.
Education: College degree preferred.
Employers: CDCs, mutual housing associations
Salary range: $20,000–$35,000

Community Organizer, Public Safety

The community organizer helps people in selected neighborhoods build community, increase public safety, decrease violence, and strengthen school improvement efforts.
Responsibilities:
- Conduct outreach and education programs.
- Develop and maintain coalitions of residents, merchants, and service providers.
- Help neighborhood groups to plan, set goals, and carry out strategies that prevent crime and promote public safety.

Skills: Work independently; communicate well, both in writing and orally; keep consistent records; and be computer literate. Must be able to respond with sensitivity to issues of ethnicity, gender, and sexuality.
Experience: Experience in neighborhood-level organizing and coalition building, and in developing positive, trusting relationships with neighborhood activists, residents, and public officials.
Education: Bachelor's degree preferred.
Employers: CDCs, community-based organizations
Salary range: $25,000–$35,000

Assistant Director, National Organizing Group

The assistant director assists the executive director with strategy, program, and management responsibilities.
Responsibilities: Assist the executive director in all aspects of operations of the organization, including personnel management, fundraising, and service delivery.
Skills: Strong collaborative and communication skills are critical, as is

a deep commitment to the idea that empowered people make a stronger society.
Experience: A seasoned understanding of grassroots community organizing and effective management skills; experience obtaining foundation support for social change.
Education: College degree
Employers: National organizing groups
Salary range: $45,000–$55,000

Construction-Related

Housing Rehabilitation Specialist

The housing rehabilitation specialist inspects properties to determine the feasibility of rehabbing them for residential use in a cost-effective manner; prepares specifications and cost estimates for residential rehab and identifies, hires, and manages contractors.
Responsibilities:
- Inspect potential properties and complete an inspection check list.
- Recommend the feasibility and cost of rehabbing potential properties.
- Prepare plans, detailed rehab specifications, and cost estimates using effective cost-cutting techniques.
- Prepare and maintain an active list of potential contractors who have been interviewed and prequalified to do work for the program.
- Prepare bid packages, bid jobs, review and select successful bidders.
- Prepare a construction budget and draw schedule.
- Do program/draw inspections with contractors to monitor schedule performance and authorize draw releases.
- Authorize change orders.
- Facilitate the obtaining of a Certificate of Occupancy and the final close-out and turnover of the property.
- Interface with city and state building inspection staff.

Skills:
- basic knowledge of all phases of housing rehab;
- ability to be innovative in developing and using effective cost-cutting methods;
- good oral and written communication skills;
- ability to supervise contractors.

Experience: Hands-on residential construction experience preferable.
Employers: CDCs, large nonprofit housing producers, local government
Salary range: $32,000–$60,000

Housing Rehabilitation Program Manager

This position is responsible for managing a number of programs that offer loans and grants for various construction activities, including: home repairs, emergency home repairs, weatherization, renovation of investor-owned properties, and acquisition and rehabilitation of homes for sale. The housing rehabilitation manager is responsible for staff supervision and training, quality assurance, meeting production schedules, and federal compliance with regard to construction management activities.

Responsibilities:

- Maintain, and redesign as needed, the construction management systems for city-funded housing programs, including construction standards, inspection procedures, work specifications, cost estimation, contractor recruitment and qualifying, bidding, contract awards, construction monitoring, dispute resolution, federal compliance procedures, and other construction systems that may be required by the city's programs.
- Prepare and present written proposals for projects requiring City Council approval.
- Hire, train, motivate, and supervise rehabilitation technicians. Spot-check work in progress. Plan and oversee training and certification programs. Assign and supervise licensed design professionals as required, or coordinate with design professionals hired by applicants. Recruit construction contractors. Assure compliance with policies for recruiting women- and minority-owned businesses. Make final determinations on contractor qualifications and on sanctions for poor performance.
- Perform (or assure performance of) rehabilitation scope and cost evaluation of properties that may be acquired with city funding. Perform last-resort resolution functions for disputes among owners, contractors, and/or rehabilitation technicians. Oversee job accounting and other record keeping and reporting of construction activities. Perform other duties as required.

Skills:

- Must have the skills required to perform the tasks above, including excellent analytical, writing, public speaking, and personnel management skills.
- Knowledge of the federal housing program requirements is highly desirable.

Experience: A minimum of three years' experience performing inspections and making specifications and cost estimates for housing rehabilitation projects. Experience with both single- and multi-family projects is highly desirable. A minimum of two years' experience supervising a minimum of three specification writers or cost estimators. A minimum of one

year's experience being responsible for creating or maintaining construction management systems for a firm or agency. If any of the above experience was concurrent, a minimum of five years' total experience in the above categories is required. Experience with construction management of federally funded housing rehabilitation is highly desirable.

Education: A bachelor's degree from an accredited college or university is required. A degree or major in a construction-related field is highly desirable. Years in excess of three years' full-time supervisory and management work can be substituted for years of college education.

Employers: Local governments, state governments, large CDCs
Salary range: $45,000–$55,000

Construction Manager

The construction manager operates the construction division within the organization, including the establishment and maintenance of proper procedures and lines of communication. Conducts construction-related development on acquired buildings on a prioritized basis. Develops and implements construction standards, specifications, and methods of contracting.

Responsibilities:

- Develop organizational flow chart showing relationships between construction division and property management, tenants, contractors, administration and bookkeeping, and executive director's office. Develop procedures and lines of reporting and communication to flesh out flow chart. Develop budget for operations.
- Develop preliminary estimates on buildings, including hard and direct soft costs. Help establish construction budget. Develop preliminary project master schedule. Research design parameters based on requirements of local codes, mod rehab, and historic review. Identify areas for negotiation and/or waiver. Develop standards and outline specifications compatible with budgets. Conduct value engineering for alternative material and methods.
- Develop detailed specifications and prepare bid documents. Prepare preliminary construction estimates and construction schedules. Conduct bidding. Coordinate with government agencies and utilities. Coordinate insurance requirements. Award contracts. Develop detailed construction schedules. Develop cost control system, reporting procedures, contingency and "damage control." Monitor construction work by contractors, attend project meetings, approve draws and change orders, and keep accounts of construction expenditures. Coordinate with property management for relocations and marketing. Prepare punch list and follow-up. Perform post-construction evaluation.

Skills:

- extensive knowledge of major construction systems;
- strong business acumen;
- ability to interrelate with personnel at all levels of development activity: public and private, contractors and workers, buyers and renters;
- ability to interpret, advise, support, and train on behalf of staff, community, agencies, and clients;
- strong organizational and supervisory skills;
- assertive, determined personality with ability to anticipate problems and master details;
- good oral and written communication skills.

Experience: Five years' construction experience with a minimum of two years as a construction manager with similar responsibilities and duties.

Education: College degree preferred.

Employers: CDCs, large nonprofit housing developers, local governments

Salary range: $40,000–$90,000

Housing Development

Housing Development Manager

This position is responsible for the management and coordination of development projects that may involve a variety of uses and require a high degree of coordination among city departments, sources of public and private financing, development entities, neighborhood organizations, and city officials. This position requires strong analytical and problem-solving abilities, as well as working knowledge of the programs of the city; local lenders; nonprofit organizations; federal, state, and local sources. This position involves the preparation of correspondence, reports and evaluation documents, budgets, and schedules; the monitoring of project budgets and compliance issues; and the formulation of recommendations concerning development policies, programs, and projects. Work is performed independently under general direction. Extensive contacts both within and outside of the city are required to perform the responsibilities of this position.

Responsibilities:

- Conceptual development of project initiatives including research and analysis related to land acquisition issues, financing options, budgets, and planning issues.
- Preparation of grant applications and planning documentation including redevelopment proposals.
- Meeting with developers to discuss their development plans, as well as

financing alternatives available to meet those plans.

- Assisting in the preparation of requests for proposals for various development sites and evaluating responses.
- Reviewing and analyzing developments with regard to financial, market, and physical feasibility and assessing the capability of development teams.
- Coordinating site visits by the city's staff.
- Coordination of project implementation activities related to all aspects of the development process from planning through financing to construction.
- Maintain development budgets and schedules.
- Assemble and present, as necessary, completed packages to the city's loan committees.
- Represent the city at banking, finance, development, and other meetings, conferences, seminars, etc., as required by city needs or interests.

Skills:

- competency with analytical skills;
- ability to write clearly;
- ability to speak effectively;
- ability to establish and maintain effective working relations with officials in both the public and private sectors;
- an understanding of design and real estate development and financing.

Experience: Five years in real estate development.

Education: Master's degree in business administration, public administration, or a related field. A bachelor's degree in engineering or architecture is preferred.

Employers: Local government

Salary range: $48,000–$60,000

Housing Development Officer

This position is responsible for assisting clients in securing private and public financing for single-family and/or multi-family projects as well as assembling development sites. This position requires strong analytical and problem-solving ability as well as familiarity with multi-family programs made available to developers by the city, local banks, nonprofit organizations, the U.S. Department of Housing and Urban Development, and other federal, state, and local sources. The housing development officer will provide substantial guidance to developers of city-sponsored housing. Developing contacts with the above-mentioned agencies and with developers is an essential part of this position. Work is performed independently according to guidelines of the city and other agencies involved in financing larger-scale single-family and multi-family developments.

Responsibilities:

- Identify and refine residential development opportunities.
- Meet with developers to discuss their housing development plans as well as the financing alternatives available to meet those plans.
- Review and analyze initial applications for permanent and/or construction financing for larger-scale single-family and/or multi-family developments with regard to eligibility of the development, financial feasibility of each development, and capability of each development team.
- Schedule, coordinate, and analyze examinations made by the city's technical staff of the project site and loan package in an effort to support acceptance or rejection of proposals.
- Assist with understanding of public requirements and approvals.
- Work with lenders, insurers, syndicators, and other funders to develop financing packages for larger residential developments.
- Provide guidance to junior staff in their efforts to evaluate, underwrite, and process loans for residential developments.
- Work with various federal, state, and local agencies and lenders to coordinate project financing.
- Assemble and present completed packages to the city's Real Estate Loan Review Committee.
- Review closing documents for accuracy, completeness, and conformity to city and department policies, guidelines, and regulations.
- Develop working relationships with other housing finance entities.
- Represent the city at banking, finance, development, and other conferences, seminars, etc., as required by city needs or interests.
- Assist in assessing the effectiveness of the housing development programs.

Skills:

- ability to analyze and package loan applications;
- ability to establish and maintain relationships with coworkers and with private and public lending institutions;
- knowledge of housing financing.

Experience: Over five years' experience as a loan officer or in housing development.

Education: This position requires a master's degree in business administration or related field.

Employers: Local government

Salary range: $35,000–$50,000

Housing and Community Development Specialist

With supervision, performs a variety of technical duties to assist senior staff in the development and implementation of housing and community development projects and programs; and to do related work as required.

Responsibilities:

- Collects, researches, and analyzes a wide range of data relative to housing and community development projects and programs.
- Gathers specific land and site data, utilities and conditions information; assists in preparing housing and community development budgets.
- Drafts responses to inquiries for information relating to development activities; responds to numerous citizen inquiries; prepares letters for supervisor's signature.
- Participates in collection of data for preparation of the annual plans; assists in preparation of assistance applications and packaging of requests for mortgage financing.
- May function as project officer for small or routine development and construction projects; conducts research in the area of housing market data.
- Schedules meetings with various development team members.

Skills: Familiarity with housing law and programs, site planning, real estate land-use law, and/or building construction practices; ability to assist in the collection and analysis of data and information as required; ability to speak and write effectively; ability to prepare clear and concise reports.

Experience: Three years of experience working in community development or planning, with direct experience in data collection, analysis, and presentation.

Education: Bachelor's degree in planning, economics, architecture, engineering, or related field.

Employers: Local government

Salary range: $30,000–$50,000 (Note: Positions grow in salary to $70,000 with increased experience, responsibility, and personnel supervision.)

Housing Development Specialist

Develops affordable housing; responsible for entire development process, including land acquisition, financing, construction, marketing, and project lease-up.

Responsibilities:

- Obtain zoning and building permit approvals for project development. Become familiar with licensing requirements and determine sources of funds available to pay for shelter and/or care in an assisted-living facility.
- Prepare applications and financing proposals, negotiate loan terms with lenders, and ensure that all loan and subsidy requirements are satisfied.
- Select and contract with architects, engineers, and general contractors

familiar with the local and state requirements for assisted-living facilities.

- Oversee the design phase of the development process to ensure that all local and state requirements are satisfied.
- Prepare the project development, operating, and leasing budgets.
- Identify and contract with agencies to manage each facility and provide supportive services to the residents.
- Prepare, package, and make presentations for approval of projects to the board of directors.
- Coordinate the preparation and review of loan closing documents with financing institutions, attorneys, engineers, architects, and contractors.
- Work closely with construction manager to monitor construction and compliance with plans and specifications; work with all parties to reach construction completion on budget and on schedule. Attend on-site owner's meetings. Process construction draws and change orders.
- Prepare application for and obtain the license to operate each project.
- Ensure that all permanent loan requirements are met and the appropriate documents are executed.
- Identify and facilitate ties between each project and local community organizations and work with property manager to establish annual budgets.
- Oversee project management to ensure compliance with all local and state requirements and ensure that quality services are being provided to the residents.
- Complete any other tasks necessary to ensure that projects are successfully identified, financed, constructed, leased, and managed.

Skills: Proficiency in use of spreadsheet software; problem-solving skills; excellent oral and written communication skills; and the ability to confidently and effectively work with individuals at all levels of company and client organizations.

Experience: Three years of related experience providing the ability to evaluate the feasibility of a real estate development as well as knowledge of development, construction, and finance of affordable housing.

Education: Bachelor's degree; graduate degree in finance or planning preferred.

Employers: National intermediaries, large nonprofit and for-profit developers, CDCs

Salary range: $40,000–$65,000

Economic/Workforce Development

Employment Specialist

The employment specialist is responsible for a broad range of activities resulting in increased employment for neighborhood residents.

Responsibilities:
- Identify and contact employers in surrounding communities to develop partnerships.
- Establish formal partnerships with employers and create employment opportunities for participants.
- Identify resources such as education, training, and transportation that are directly related to employment opportunities.
- Ensure that referrals of participants to employment opportunities and other related resources are tracked and linked to outcomes, including completion of training programs and employment and job retention data.
- Serve on local committees and task forces, network with resource organizations, perform public speaking assignments as appropriate.

Skills: Excellent written and verbal communication skills, public speaking, and at least some basic computer skills.

Experience: Minimum five years of experience in job development and case management experience with welfare-to-work program.

Education: BA degree in social service or related field; graduate-level experience preferred.

Employers: CDCs, community-based economic development and workforce programs

Salary range: $30,000–$40,000

Community Economic Development Specialist

The community economic development specialist works to improve the business and working opportunities in low-income communities.

Responsibilities:
- Provide direct assistance to community-based organizations.
- Engage in a range of economic development activities.
- Plan and implement a comprehensive community economic development training program.
- Help develop new community economic development programs and initiatives.

Skills: Strong skills in training, program development, project management, and technical assistance.

Experience: At least ten years of demonstrated experience in economic

Experience: At least ten years of demonstrated experience in economic development in community settings, including small business development, commercial real estate development, and/or workforce development; experience providing training and/or technical assistance to community-based organizations.

Education: Master's degree in business, urban planning, or related field

Employers: Local intermediaries

Salary range: $40,000–$60,000

Director, Economic Development

Coordinates and directs new economic development initiatives for community group; represents the organization with all sectors in the development of programs.

Responsibilities:

- Assist the executive director in developing new economic development initiatives that are consistent with the organization's mission.
- Coordinate and direct involvement with new economic development initiatives with a focus on increasing the access of the organization to economic, business, and job-creation opportunities.
- Assist the executive director with coordinating new program development in the metropolitan area.
- Assist in preparation and presentation of technical assistance sessions and other information sessions; provide community-based developers with information to enhance their community development activities.
- Since much of the work will involve on-site provision of technical assistance, applicants must be willing to travel approximately 30% to 50% of the time.
- Coordinate involvement with social venture-capital programs; assist nonprofit community-based organizations with their for-profit business ventures.
- Assist with fundraising activities and supervise contracts for various initiatives.

Skills:

- familiarity with federal, state, and local economic development, with specific skills in development finance, and the management of nonprofit organizations;
- excellent writing and public speaking skills.

Experience: Considerable experience with economic development planning and analysis. Experience providing technical help to community groups desired. Grant writing and contract supervision desired.

Education: A minimum of a bachelor's degree. A post-graduate degree in real estate, finance, business or public administration, or planning is strongly preferred and may substitute for one year of experience.

Employers: National and local intermediaries, CDCs, local government

Salary range: $45,000–$76,000

Storefront Renovation Director

Under the supervision of the executive director, the storefront renovation director is responsible for the implementation of the facade renovation program in the designated program area.

Responsibilities:

- Administer and manage the storefront renovation program.
- Serve as a liaison between developers, architects, contractors, city staff, and the applicant.
- Manage the loan and rebate program.
- Prepare all narrative and fiscal reports pertaining to the program.
- Provide technical assistance to businesses as needed.
- Make project presentations to loan review and design review committees.
- Plan and implement a marketing strategy for the storefront renovation program, in conjunction with the marketing and fundraising specialist.
- Serve as a liaison between applicant and city economic development programs.
- Assist in planning program goals and in writing any proposals necessary to fund the program.
- Other duties as assigned by the executive director.

Skills:

- Ability to work well with small business owners.
- Ability to manage a loan program, including seeing to it that loans are consistent with program regulations.
- Good writing and presentation skills.
- Knowledge of construction preferred.

Experience: Two to three years operating loan and grant programs

Education: Bachelor's degree

Employers: CDCs, local government

Salary range: $25,000–$40,000

Business Consultant

The position of business consultant is a full-time position with responsibilities in the areas of client services to local businesses.

Responsibilities:

- Provide management consulting services and follow-up to active customers.

- Develop an appropriate plan of action for individual consulting and training with each customer.
- Make appropriate referrals to other agencies where required.
- Prepare written documentation on each session with a customer and deliver to the customer's file.
- With the customer, identify appropriate capital needs and potential resources.
- When the fund itself is an appropriate funding resource, represent the customer through the staff-approved and/or board-approved lending process.

Skills:

- demonstrated ability in financial analysis, business planning, and market analysis;
- strong written and oral communication skills, language skills;
- entrepreneurial spirit, team player;
- strong organizational skills;
- nurturing, helping skills;
- ability to work with a diversity of clients, in terms of ethnicity, life experience, income level.

Experience: Two years' successful experience operating or working with small business

Employers: Economic development local intermediaries, some CDFIs

Salary range: $40,000–$70,000

Technical Assistance/Training

Field Service Officer

Responsible for the development, coordination, implementation, and provision of support services and training to community-based neighborhood groups. Work involves a variety of activities related to the establishment and ongoing support of such groups. Work is performed in both field and office settings.

Responsibilities:

- Conduct assessments in designated cities; meet with representatives of financial institutions, businesses, community groups, local government, and philanthropic institutions as part of assessment process.
- Make recommendations on the feasibility of pursuing the development of neighborhood projects in a given locality or with particular sponsoring groups; analyze the relative merits of localities and sites for programs.

- Provide assistance to local boards of directors in the transitional phase between development and program operations; may be assigned to provide continuing liaison with program, including field visits and occasional attendance at board of director's meeting.
- Represent intermediary in meetings; make presentations to groups and at conferences; serve as a staff resource in workshops and training sessions.
- Require and review written reports from the local staff; keep management staff informed of progress and concerns by preparing and submitting regularly scheduled reports; assist other staff in program areas through field visits and reports, as assigned by supervisor.

Skills:

- considerable knowledge of the basic principles of community organization, local government structure, and community power relationships;
- considerable knowledge of the principles and practices of program development and administration;
- considerable knowledge of the principles and practices of small group process facilitation and leadership development;
- knowledge of home financing, neighborhood preservation strategies, and the operation of neighborhood programs;
- knowledge of programs related to the area of assignment;
- ability to identify, obtain the support of, and organize community resources and groups toward the achievement of program goals;
- ability to train local staff, provide ongoing technical assistance to programs, and serve as liaison between local programs and the intermediaries and other community groups and institutions;
- ability to communicate effectively, orally and in writing;
- ability to maintain required records and to prepare regularly scheduled and special reports.

Experience: Considerable experience in a community-based program or group activity involving demonstrated abilities in program development, community organization and interpersonal relations, or related areas; or any equivalent combination of training and experience that provides the above knowledge, abilities, and skills.

Education: Bachelor's degree required. Master's degree in planning, urban studies, or community development preferred.

Employers: National intermediaries

Salary range: $40,000–$90,000

Financial Services Specialist

This position involves responsible and advanced work in coordinating the regional relationship with the financial services industry and ensuring that affiliate organizations are well informed and positioned to capitalize on strategic opportunities and relationships resulting from the rapid consolidation of the financial services industry. Work involves supporting organizations and district field staff in their efforts to enhance and expand relationships with key financial partners whose service areas transcend district boundaries. Work requires advanced knowledge of programs related to the area of assignment, community and political structures, and neighborhood preservation and reinvestment strategies. Work also requires strong organizational abilities and interpersonal skills.

Responsibilities:

- Research and document specific network needs for loan products, fees, and equity.
- Structure deals involving financial institutions that will incur minimum risk and maximum benefit.
- Interact on complex deals, as requested by the district director.
- Identify regional (and national) institutions that represent resources or opportunities for partnerships. Research structure, contact, products, market interests, and current network relationships. Design and implement strategies to elicit resource commitments and active involvement with the network.
- Develop and maintain understanding of current CRA requirements and potential changes and interpretations. Maintain awareness of opportunities including loan participation, capital investments, and other collaborative instruments that fit within the CRA context.
- Maintain an understanding of regulatory changes and directions as they affect the network. Coordinate network comments as appropriate.
- Identify, explore, and document untapped potential for investments and loan products.
- Identify key financial service needs in communities served by the network and assist network organizations to determine how to capitalize strategically on those needs.

Skills: Understanding of financial institution types, structures, powers, and constraints. High degree of organization skill. Excellent oral and written presentation skills. Successful experience in group facilitation. Ability to communicate effectively, orally and in writing. Ability to maintain required records and to prepare regular special reports. Ability to travel as required.

Experience: Substantial experience working in or with the financial services industry. Technical understanding of lending issues and loan products, including loan structuring, investment opportunities, secondary-market offerings. Demonstrated commitment to community reinvestment and to enhancing lending opportunities for lower-income communities and

borrowers.

Education: Bachelor's degree from an accredited college; business major preferred.

Employers: National intermediaries

Salary range: $40,000–$60,000

Director, Native American Programs

The director oversees the organization's Native American housing programs, staff, and fundraising efforts. This individual provides training and technical assistance to Native American tribes throughout the country, in compliance with HUD requirements for community planning, asset and property management, and development of affordable housing.

Responsibilities:

- Manage a small staff in the creation and development of technical assistance, training products, and programs to increase housing and planning resources.
- Monitor development of training curriculum on housing production and management and community planning for Native American tribes.
- Assist tribes in developing affordable housing to include financial packaging, adopting required tribal laws or regulations and overseeing the development process.
- Work with the development group staff in identifying and applying for potential funding sources for the program.
- Keep public policy staff informed of legislation and trends that affect Native American housing and planning issues.
- Establish and manage the Native American program advisory committee.
- Establish annual goals for the Native American program and ensure continual program evaluation and review.
- Work in conjunction with other Native American intermediaries to ensure full delivery of services to tribes and their members.

Skills: Excellent verbal and written skills.

Experience: Minimum of eight years' experience in Native American housing, economic, and/or community development issues.

Education: Bachelor's degree

Employers: National and local intermediaries, Native American assistance providers

Salary range: $45,000–$60,000

Community Services

Case Worker

The case worker works as part of a property management team to build a liveable low-income housing community.

Responsibilities:

- Identify with each tenant the steps that lead to employment and create motivation to design and implement daily activities that promote recovery, employment, and personal growth of residents.
- Work to ensure that building standards are met by residents.
- Communicate with team on building operations.
- Keep property records on case management, petty cash, and other activities.

Skills: Community organization and program development.
Experience: Minimum of two years working with homeless and/or low-income population.
Education: Bachelor's degree in related field.
Employers: CDCs, property management companies
Salary range: $20,000–$35,000

Coordinator of Resident Services

The resident services coordinator establishes, directs, and reports on programs aimed at improving the lives of low-income residents in selected communities through the provision of needed services.

Responsibilities:

- Develop and oversee clear policies and procedures for support services program.
- Work with program staff to develop a vocational training and employment program for tenants.
- Work in conjunction with building management and volunteers to plan social and recreational activities.
- Develop policies and procedures for program operations.
- Assist in fundraising and grant writing.
- Prepare regular program and statistical reports for government and private foundation funding.

Skills: Strong management skills, computers, education training, event planning, and program development.
Experience:

- One to two years' supervisory experience.
- Experience in assessing, developing, and implementing resident groups.

- Experience in mobilizing or organizing resident groups in a community setting, and knowledge of community resources and services with particular emphasis on neighborhoods.
- Knowledge of and sensitivity to the needs of residents in low-income housing.

Education: Bachelor's degree in social work or related field or five years' experience (or combined education and experience) in providing support services to homeless, formerly homeless, or low-income populations.
Employers: CDCs, property management companies, public housing authorities
Salary range: $30,000–$45,000

Resident Services Manager, Community Service Center

The resident services manager develops and supervises the resident services program in properties owned by the CDC for providing housing to formerly homeless people.

Responsibilities:

- Develop and oversee clear policies and procedures for support services program.
- Work with program staff to develop a vocational training and employment program for tenants.
- Oversee development and monitoring of program budgets.
- Assist staff in gathering statistical data and maintaining records.
- Plan and coordinate social and recreation activities; work in conjunction with building management and volunteers.
- Coordinate services within the supportive housing employment collaborative.
- Develop policies and procedures for program operations.
- Implement and coordinate programs to build a community environment among residents; facilitate peer support groups.

Skills: Strong management skills, computers, education training, event planning, and program development.
Experience:

- One to two years' supervisory experience.
- Experience in assessing, developing, and implementing resident groups.
- Experience in mobilizing or organizing resident groups in a community setting, and knowledge of community resources and services with particular emphasis on neighborhoods.
- Working with dually diagnosed populations.
- Knowledge of and sensitivity to the needs of residents in low-income housing.

Education: Bachelor's degree in social work or related field or five years' experience (or combined education and experience) in providing support services to homeless, formerly homeless, or low-income populations.

Employers: CDCs, property management companies, social services organizations

Salary range: $30,000–$45,000

Youth Coordinator

The youth coordinator manages a staff providing in-home crisis intervention for households with youth in designated neighborhoods.

Responsibilities:

- Supervise staff providing in-home crisis intervention services; program staffing, procedures, and goals.
- Work closely with county staff, law enforcement, and other agencies.
- Perform public speaking and networking activities.
- Manage program development and implementation.
- Provide accountability to clients and funders.

Skills: Strong in project management, staff team supervision, operations and service delivery, and program development.

Experience: Five years of experience in youth and family counseling. Two years' experience in supervising a team of staff. Experience in program design and implementation.

Education: Master's degree in social work or psychology.

Employers: Youth services, agencies, some CDCs

Salary range: $30,000–$40,000

Drug Elimination Coordinator

The drug elimination coordinator works with the neighborhood group, youth, and parents to reduce drug use in the community.

Responsibilities:

- Act on behalf of the neighborhood group in monitoring youth and working with parents to eliminate gang and drug activities in the area.
- Act as liaison with community policing activities.
- Prepare monthly reports on activities.
- Transport youth to various activities.
- Produce a newsletter.

Skills: Strong interpersonal and communication skills, and the ability to work with youth and parents.

Experience: Three to five years in establishing and coordinating programs that mentor youth and work with their parents.

Education: Degree in social work, business, or a related field.

Employers: Some larger CDCs, public housing authorities

Salary range: $25,000–$35,000

Homeownership Counselor

The homeownership counselor provides information, training, liaison, and support services to people who purchase their first homes. The counselor educates homeowners so they can maintain their self-sufficiency and live in safer, more stable neighborhoods. A primary focus of this position is to maintain relationships with communities and individual homeowners.

Responsibilities:

- Organize, coordinate, and lead homeowner workshops.
- Organize material for and coordinate publication of homeowner newsletters.
- Assist individual homeowners in accessing needed services.
- Assist with establishing homeowners' associations and other ongoing activities in each development community.
- Meet regularly with homeowners and arrange a first-year anniversary visit with each new homeowner.
- Develop and maintain referral information on available services for homeowners and their families.
- Complete other duties as assigned.

Skills: Excellent communication skills; group presentation and workshop leading skills. Knowledge of real estate development; experience in foreclosure and pre-purchase counseling. Ability to work with diverse populations and a willingness to travel in all types of neighborhoods. Demonstrated ability to take initiative and work independently to develop and carry out programs; at least 50 percent of time will be out of office. Willingness to travel and work extended hours.

Experience: Experience in provision of social services to families and/ or community development.

Education: Bachelor's degree preferred.

Employers: CDCs, private developers of homes for ownership.

Salary range: $25,000–$45,000

Program Planner, Community Service Agency

The program planner has overall responsibility for the service agency's strategic planning activities.

Responsibilities:

- Coordination and implementation of a five-year strategic plan.
- Response to government RFPs (requests for proposals) and foundation

grant requests, and various other functions in areas of advocacy, public
policy, and communications.

Skills:

- Knowledge and expertise in the areas of strategic planning, child welfare,
  RFPs, and state and city funding issues are essential.
- Understanding of private and public child welfare agency issues.

Experience:

- Effective management experience.
- Knowledge of quality control helpful.

Education: Master's degree in social work or related field.
Employers: Community service organizations
Salary range: $40,000–$100,000

Real Estate Management

Assistant Property Manager

The assistant property manager works closely with and assists property
managers with all administrative functions at the sites assigned. This includes responsibility for occupancy, rent collection, rent recertification, and
voucher and subsidy management, as well as budget preparation and financial management. The assistant manager also works closely with residents to
assist them in taking full advantage of the CDC's housing opportunities.

The assistant manager conducts apartment and building and grounds inspections, markets apartments, completes internal and external reports,
and screens applicants. The assistant manager is assigned to a site or sector
that generally includes 150–200 apartments.

Responsibilities:

- Screen prospective applicants, including preparing screening package,
  contacting landlords and other information sources, and reviewing that
  information to determine eligibility for occupancy.
- Work with manager to ensure prompt rent up of apartments, including
  coordination of maintenance.
- Work within established rent collection strategies to meet rent collection
  standards.
- As directed by property manager, inspect apartments to determine routine repair needs.
- Recertify subsidized rents in compliance with state and federal requirements.
- Respond to resident concerns and questions. Refer residents to appropriate social services.

- Assist manager in financial planning and oversight.
- Assist with or coordinate site or company purchasing functions.
- Maintain adequate office and custodial supply inventory for site or com-
  pany needs.

Skills:

- excellent interpersonal and analytic skills;
- knowledge of state and federal subsidized housing programs;
- good written and verbal communication skills;
- computer literacy;
- good problem-solving skills;
- ability to learn about state sanitary code, state and federal regulations,
  and other housing concepts;
- proficiency in two languages (English and Spanish) helpful.

Experience: Subsidized housing and/or property management experi-
ence.
Education: High school diploma or equivalency.
Employers: CDCs, housing authorities, property management companies
Salary range: $18,000–$28,000

Property Manager

The property manager is responsible for all the agency's housing developments in the range of 130 units to 500+ units.

Responsibilities:

- Supervise office staff and work closely with leasing and resident initiative
  departments.
- Oversee occupancy, rent collections, evictions, housekeeping and perform other related responsibilities associated with managing a development.

Skills: Must be computer literate, possess excellent leadership and
communication skills, and be able to establish and maintain effective work-
ing relationships with people from varying socioeconomic backgrounds.
Must possess a valid driver's license.
Experience: Two to three years of subsidized housing and/or property
management experience.
Education: Bachelor's degree in public administration or related field.
Employers: Housing authorities, CDCs, and private management firms
Salary range: $24,000–$37,000

Senior Property Manager

The senior property manager is responsible for several properties and directly supervises property managers, site managers, and assistant managers. The senior property manager must possess strong leadership skills and have a proven track record of problem solving and property management knowledge. The senior property manager must also have a history of dedication to the community and a commitment to low-income housing and residents as well as community development and community-based property management.

Responsibilities:
- Supervise office staff and work closely with leasing and resident initiative departments.
- Oversee occupancy, rent collections, evictions, and housekeeping and perform other related responsibilities associated with managing a development.

Skills:
- professional, business-level, English verbal communication skills with people within and outside of the organization;
- ability to write clear and concise letters using correct English business language;
- ability to write technical documents (such as sections of an operations manual) in clear and understandable terms using correct English business language without error;
- ability to represent the interests of residents, agencies, and the community;
- strong financial management analysis skills related to property management;
- computer literacy.

Experience: Two to three years of subsidized housing and/or property management experience.
Education: Bachelor's degree in public administration or related field. Certification from an accredited property management institution.
Employers: Housing authorities, CDCs, and other property owner and private management firms
Salary range: $30,000–$45,000

Housing and Community Development Property Management Supervisor

Under general supervision, the person in this position directs the operation, management, and maintenance of housing projects and performs related work as required.
Responsibilities:
- Direct maintenance staff in rendering services to residents and maintaining physical plants, buildings, and property.
- Supervise housing and resident managers in ensuring that units are maintained, resident requirements are met, and services are delivered.
- Assist in establishing budgets and controls for agency properties.
- Assist in determining the nature of and implementing a modernization program to improve physical structures and services of the housing complex; and manage programs to completion.
- Develop maintenance contract requirements and assist in procuring contractors.
- Develop and maintain standard operating procedure manuals and policy statements of property management.
- Make recommendations on property management policies to the board of directors.
- Work closely with tenant associations and special interest groups to ensure good tenant-landlord relations.
- Attend board meetings as requested.

Skills:
- knowledge of the procedures, regulations, and policies of public housing projects;
- knowledge of building maintenance principles, tools, and materials;
- knowledge of the various phases of municipal government that affect public housing;
- knowledge of administrative procedures;
- ability to use initiative and resourcefulness to solve complex housing management problems;
- ability to plan and supervise the work of others;
- ability to establish and maintain effective working relationships with tenants and tenant associations;
- ability to communicate effectively, both orally and in writing.

Experience: Three years of progressively responsible housing management experience.
Education: Graduation from an accredited college with a degree in business or public administration, mechanical engineering, or a related field.
Employers: Public housing authorities, private for-profit housing management companies, large-scale nonprofit housing owners
Salary range: $47,000–$72,000

Research and Public Policy

Research Assistant, Public Policy

The research assistant for public policy assists in a broad range of survey and other research to foster the organization's public policy goals.

Responsibilities:

- Collect and review survey data, and conduct data collection interviews with a considerable level of autonomy.
- Conduct library research and literature review on assigned research topics.
- Produce high-quality and complex charts, tables, and data reports using standard word-processing software.
- Assist the research director and participate in all phases of the design and administration of survey research and case studies.

Skills: Ability to communicate effectively orally and in writing. Ability to establish and maintain effective working relationships with a broad range of people in an ethnically diverse setting.

Experience: Work experience in research methodology and statistical analysis and in using Word, Excel, Access, and Powerpoint.

Education: Informal or formal training in using Word, Excel, Access, and Powerpoint.

Employers: National intermediaries, public policy and advocacy groups
Salary range: $20,000–$30,000

Legislative Assistant

The legislative assistant tracks legislation of importance to the national intermediary, makes recommendations on actions, and works with legislative staff to represent the organization.

Responsibilities:

- Draft legislative updates for corporate staff.
- Draft correspondence to members of Congress and their staff.
- Prepare written case studies and track housing legislation and relevant community development issues and news.
- Produce reports and summaries for officers and legislative director as needed.
- Maintain congressional key contact database.
- Monitor and track housing legislation and news and produce reports and summaries to officers as needed.
- Represent the organization at coalition and association meetings; assist board liaison with organizational needs for board relations.
- Prepare materials and logistics for congressional site tours.
- Conduct research for legislative director and officers as needed.
- Serve as liaison to district director assistants and legislative office.
- Travel to events as needed.

Skills:

- strong administration and legal skills;
- demonstrated writing and editing skills;
- proficient in use of computers and word-processing and spreadsheet software;
- ability to manage multiple tasks and to develop and manage database;
- ability to communicate effectively both orally and in writing;
- ability to establish and maintain effective working relationships with others.

Experience: Experience and training that provides the knowledge, abilities, and skills listed above.

Education: College-level course work and any equivalent combination of education, experience, and training.

Employers: National intermediary and public policy organizations
Salary range: $20,000–$40,000

Policy Analyst and Legislative Assistant

This position is responsible for a broad range of activities aimed at producing legislation to benefit community development.

Responsibilities:

- Monitoring and analysis of federal legislation and regulations in policy areas assigned.
- Advocacy for the community development field with the federal administration and the Congress.
- Coalition representation and leadership.
- Grassroots development and networking.
- Basic federal and state policy research and analysis.
- Preparation of informal materials, policy briefs, and newsletter articles.
- Staffing the Public Policy Committee of the board of directors.
- Communication and working relations with affiliate members on federal and state issues.
- Conference and seminar programming and planning, and other tasks as assigned.

Skills:

- ability to analyze complex information and present it in summary form quickly;
- ability to write well and make good presentations;
- ability to work in close coordination with membership and related groups.

Experience: Two or more years of experience in policy analysis, a membership organization, or legislative affairs.

Education: BA in political science or a related field

Employers: National intermediaries, advocacy and policy groups, some local governments

Salary range: $30,000–$35,000.

Director of Policy Development

The director of policy development researches and analyzes public policy issues and develops strategies to advance those issues legislatively through education of the nonprofit sector, government, the business sector, the media, and the general public.

Responsibilities:

- Prepare reports and action alerts for membership.
- Educate the nonprofit sector, the government, the business sector, the media, and the general public on priority issues identified through public policy briefings and face-to-face meetings.
- Develop ongoing communication with local, county, state, and federal officials about the important work being done by community development corporations.
- Attend public hearings.
- Coordinate letter-writing campaigns.
- Sponsor hearings and forums.
- Assist in drafting specific legislation.
- Identify sponsors of legislation beneficial to community development and work with them toward its passage.

Skills:

- ability to analyze complex information and summarize it quickly;
- strong writing and presentation skills;
- staff management and membership management capabilities.

Experience: Five or more years of progressively responsible leadership in policy analysis, membership. or legislative work.

Education: BA in political science or a related field

Employers: Local and national intermediaries, policy and advocacy groups, some local governments

Salary range: $40,000–$65,000

Vice President for Research

The vice president for research is responsible for developing and overseeing all of the organization's research activities.

Responsibilities:

- Identify and obtain the necessary resources to conduct the work.

- Assemble a working team and advisory board.
- Oversee the execution of projects.
- Interact with corporate research partners and work to identify new partners.
- Maintain relationships with funders, corporate partners, and advisory board members.
- Function as an expert on research for trade groups, media, government officials, and other relevant organizations.

Skills:

- demonstrated research skills;
- excellent proven writing, project management, and public presentation skills;
- a commitment to the revitalization of the inner city;
- entrepreneurial approach to research;
- self-starter;
- outgoing personality;
- professional and interpersonal skills necessary to work effectively with a wide range of people.

Experience: Five to seven years of relevant experience, preferably in business. Experience in the inner city preferred.

Education: Master's in business administration, economics, or government

Employers: National intermediaries, advocacy and policy organizations, academic institutes

Salary range: $65,000–$75,000

Capital Provision/Community Lending

Loan Processor

The loan fund provides low-cost interim financing for housing development projects. The loan processor carries out the day-to-day paperwork activities of the loan fund.

Responsibilities:

- Work with the loan fund staff to process loan requests, including preparation of documents for loan origination, closing, and servicing.
- Maintain monthly status reports and provide those reports to the loan fund staff members, loan committee, and loan fund investors.
- Prepare loan application materials and other public information about the loan fund.
- Assist in the preparation of periodic reports as required.

- Assist loan fund staff members to develop, implement, and revise loan fund policies and procedures and administrative systems.
- Develop and maintain housing and real estate expertise and professional contacts within the lending industry and with housing developers.
- Assist with borrower and other loan fund calls.
- Assist and communicate with employees and client agencies.
- Monitor the receipt of project status reports from loan fund borrowers.
- Monitor loan fund disbursements and maintain project spreadsheets.
- Assist in processing periodic draw requests.
- Maintain all loan fund files as needed.

Skills:
- knowledge of real estate finance, development, and law; lending practices and procedures; and the escrow process;
- use of personal computers, including spreadsheet, word-processing, and other business software;
- knowledge of construction and/or development loans and draws;
- planning and managing contract and loan performance;
- participating in work teams;
- working independently and analyzing and solving problems;
- effective writing skills;
- knowing standard office procedures and practices;
- relating well with the public;
- working with minimum supervision and arranging work tasks;
- following detailed instructions;
- familiarity with public and nonprofit agencies and programs;
- working knowledge of relevant software;
- speaking, reading, and writing Spanish;
- knowledge of various investment vehicles.

Experience: Three years of applicable experience in the mortgage lending field.
Education: Bachelor's degree from an accredited four-year college.
Employers: CDFIs, banks, some local governments
Salary range: $30,000–$40,000

Community Lending Specialist

This position is responsible for promoting products and services to housing partners.
Responsibilities:
- Provide high-profile customer service to both traditional mortgage lenders and new nontraditional partners.
- Provide follow-through and implementation of initiatives in conjunction with other housing impact and customer management staff.
- Coordinate weekly and monthly reports to management.

Skills:
- good working knowledge of all aspects of mortgage lending, including underwriting and secondary-market requirements;
- must be well versed in affordable housing issues and have strong familiarity with the lending requirements of the low- and moderate-income community and of rural areas;
- presentation skills, excellent interpersonal skills, and basic knowledge of word-processing and spreadsheet software.

Experience: This position requires at least five years of experience in the mortgage finance industry.
Education: Bachelor's degree in business administration, finance, real estate, public administration or a related field, or related equivalent work experience.
Employers: National intermediaries, Fannie Mae, Freddie Mac, bank regulatory entities, banks
Salary range: $40,000–$50,000

Finance Specialist

The finance specialist is responsible for assisting senior staff in the financial management of city programs. This position requires strong analytical skills and problem-solving abilities, familiarity with computer applications, and a general knowledge of the programs made available by the city. Individuals filling this position will be required to develop contacts with private lending institutions, private and public mortgage insurers, loan servicing and collection agents, and federal and state agencies, as well as borrowers of city funds. Work is performed independently according to sound financial management practices and procedures as established by the city and other agencies involved in the financing of city-sponsored loans.
Responsibilities:
- Review loan packages for completeness, which includes preparing city loan documentation and working with participating lending institutions to obtain required documentation as necessary for loan purchasing.
- Purchase loans and review purchase requisitions for accuracy.
- Analyze monthly servicing reports for the purpose of tracking the performance of the program's loan portfolio.
- Develop monthly management reports that highlight loan production, status of fund balances, loan delinquencies and foreclosures, and lender performance.
- Maintain documentation on loan account funds and contractor payment

requests, which includes the approval of such requests.
- Develop and maintain a computerized database for loan purchasing and servicing.

Skills:
- Knowledge of the basic documentation procedures associated with purchasing and servicing mortgage loans is required.
- The ability to explain loan program guidelines clearly and to maintain effective communication with the parties involved in loan packaging is also necessary.

Experience: This position requires two years' experience in a related field. Another combination of education, experience, knowledge, and abilities demonstrating the qualifications necessary to perform the duties of this position would be considered.

Education: This position requires an associate degree (preferably in real estate or a lending-related field).

Employers: Local governments, CDFIs
Salary range: $33,000–$40,000

Senior Loan Officer
The senior loan officer is responsible for a broad range of activities involving the organization's community development lending activities.

Responsibilities:
- Implement lending pool, guarantee program, home ownership and land banking programs, special needs housing developments, public-sector coordination, training and technical assistance, etc.
- Evaluate proposals for financial services: direct loans, loan packaging on behalf of nonprofit borrowers with conventional lenders, mortgage guarantees, interest subsidies, technical assistance.
- Make recommendations to executive director and board of directors concerning provision of such services. Coordinate agreed-upon services and provide technical assistance services. Maintain documentation of services provided.
- Assist agencies whose staff may have little or no housing experience.
- Research and assist in the development of new programs and policies; various administrative tasks.
- Extensive work with public agencies and governing bodies, nonprofit organizations, private-sector representatives, and the public in general.

Skills:
- knowledge of lending techniques and loan packaging;
- knowledge of housing finance;

- knowledge of basic types of low-income housing and sources of funding for low-income housing;
- knowledge of multi-unit housing development and rehabilitation;
- knowledge of housing policy as it relates to low-income housing;
- familiarity with computers;
- a high degree of self-motivation, initiative, dedication, creativity, and perseverance;
- flexibility with his or her time and willingness to accept an extremely demanding position;
- capacity to work closely with a wide range of organizations and people.

Experience: Minimum of one year of experience with community-based housing or other low-income housing or lending organization; or, minimum of three years' work or academic experience in loan packaging, housing finance, housing development, or business finance.

Education: Master's degree in city planning, business, public administration, or related subject; or bachelor's degree in relevant subject with minimum five years' relevant professional experience.

Employers: National intermediaries, CDFIs, banks
Salary range: High $30s to mid $50s

Senior Finance Specialist
The senior finance specialist is responsible for assisting senior staff in the financial management of city programs. This position requires strong analytical skills and problem-solving abilities, familiarity with computer applications, and a general knowledge of the programs made available by the city. Individuals filling this position will be required to develop contacts with private lending institutions, private and public mortgage insurers, loan servicing and collection agents, and federal and state agencies, as well as borrowers of city funds. Work is performed independently according to sound financial management practices and procedures as established by the city and other agencies involved in the financing of city-sponsored loans.

Responsibilities:
- Review loan packages for completeness, including preparing city loan documentation and working with participating lending institutions to obtain required documentation as necessary for loan purchasing.
- Purchase loans and review purchase requisitions for accuracy.
- Analyze monthly servicing reports for the purpose of tracking the performance of the program's loan portfolio.
- Develop monthly management reports that highlight loan production, status of fund balances, loan delinquencies and foreclosures, and lender performance.
- Maintain documentation of loan account funds and contractor payment

requests, which includes the approval of such requests.

- Develop and maintain a computerized database for loan purchasing and servicing.
- Provide guidance to finance specialists as necessary in performing their duties.
- Assist in assessing the effectiveness of the city's consumer programs.

Skills:

- Knowledge of the basic documentation procedures associated with the purchasing and servicing of mortgage loans is required.
- The ability to explain loan program guidelines clearly and to maintain effective communication with the parties involved in loan packaging is also necessary.

Experience: Five years' experience with real estate lending practices.
Education: An associate degree, preferably in real estate or a lending-related field.
Employers: Local government, CDFIs
Salary range: $35,000–$45,000

Housing and Community Development Loan Officer

Under direction, the person in this position is responsible for underwriting, processing, and servicing numerous home improvement and rehabilitation loans and grant programs; and performing related work as required.

Responsibilities:

- Review formal loan applications and credit documents.
- Analyze all applications for financial ability and risk.
- Prepare recommendations on all applications and submit them to the loan committee for action.
- Appraise real estate used as collateral for proposed loans.
- Coordinate commitment and settlement of all loans and grants.
- Oversee preparation of all loan packages and present them to investors.
- Review and approve all loan disbursement requests.
- Resolve all financial and legal difficulties, such as defaults on loans.
- Ensure that program funds are used to meet predetermined volume goals.
- Establish and administer the program in conformance with approved policies and procedures.
- Market the home improvement loan program to civic associations and to the general public.
- Supervise FHA loan administration and servicing of mortgage loans originated by the organization.

Skills:

- knowledge of principles, laws, and practices related to the appraisal of real estate, loan processing and administration, and credit underwriting;
- knowledge of rules, regulations, policies, and laws governing home improvement loan programs;
- ability to establish and maintain effective working relationships with federal, state, and local government officials, officers of lending institutions, and civic associations;
- ability to communicate effectively both orally and in writing;
- ability to supervise the work of subordinates.

Experience: Four years of experience in appraisal of residential real estate, loan processing, credit underwriting, loan administration, and/or construction lending.
Education: Graduation from an accredited college with a degree in business administration, finance, real estate, or a related field.
Employers: Local government, CDFIs
Salary range: $45,000–$70,000

Senior Asset Manager

The person in this position is responsible for managing a portfolio of community development loans.

Responsibilities:

- Manage a portfolio of assets in a timely and professional manner.
- Monitor all designated accounts including quarterly loan rating.
- Provide an excellent level of customer service to maximize productivity and profitability of relationships.

Experience: Substantial commercial real estate finance experience, including multifamily housing.
Education: Bachelor's degree in business, accounting, finance, or real estate or equivalent work experience. Relevant post-graduate degree and/or training preferred.
Employers: CDFIs, intermediaries, banks
Salary range: $60,000–$85,000

Senior Vice President, Real Estate Lending

The senior vice president in real estate lending is responsible for the strategic and operating activities of the Real Estate Lending Department. This person develops and implements the plan for strategic growth of the portfolio. In conjunction with the subsidiaries and affiliates, this person takes a leadership role in executing the overall strategic mission goals and objectives for real estate development in priority neighborhoods.

Responsibilities:
- Create, develop, implement, and evaluate strategy, products, and services that effectively address customer needs and niches in the real estate market of the priority area.
- Lead and manage the lending and administrative team of the Real Estate Lending Group; educate team and bank staff on real estate development strategy, goals, and objectives.
- Review and establish department lending policies and operating procedures; ensure consistency with overall bank policies and operating objectives.
- Oversee marketing and community as a real estate lender of choice.
- Generate and/or catalyze department's generation of loan volume that meets development and budget objectives.
- Manage loan portfolio for department; direct department participation in loan committee.

Skills:
- Strong communication and relationship-building skills; demonstrated ability to deal with all levels of management, as well as staff, boards of directors, community and corporate leadership, customers at all levels.
- Excellent organizational skills; demonstrated accuracy in handling financial accounts, credit analysis, deal structuring, etc.
- Excellent knowledge of real estate development in underinvested markets and ability to construct a vision for systematic revitalization in priority neighborhoods.
- Ability to generate the enthusiastic participation of colleagues, customers, and the community in bringing vision to reality.

Experience: Eight to ten years of progressive, related experience, including staff management; multiple phases of real estate lending and/or development, particularly underwriting of inner-city single-family and multifamily rehab and new construction; experience in community and economic development helpful.
Education: College degree or equivalent experience
Employers: Banks, large CDFIs
Salary range: $85,000–$90,000

Portfolio Manager and Vice President
The portfolio manager is responsible for managing efforts to achieve the investment objectives of the portfolio once all properties have been acquired; for providing reports and analysis to the corporate investors; and for assisting with development of new products and value-added services

that are profitable and appealing to investors.

Responsibilities:
- Assist with fund management, including setting overall portfolio strategy, contributing to fund and business unit planning process, and participating in resolution of portfolio or property issues.
- Perform annual residual analysis for stabilized portfolios and revise underlying assumptions as needed.
- Contribute to efforts to develop new investment products for corporate investors, including devising reporting options.
- Prepare quarterly and annual investor reports discussing investment performance, fund operations, and property performance.
- Solicit and interpret investor feedback into opportunities for value-added products and service delivery, including innovating new technology and resource solutions, wherever feasible, to leverage existing staff resources.
- Plan and execute investor due-diligence meetings and other client presentations.
- Participate in decision-making process for creation of new funds, ensuring that prospective funds have the necessary diversification and investor protections to achieve investment objectives.

Skills:
- excellent written and verbal communication skills;
- strong organizational skills;
- strong technology and computer skills;
- familiarity with relevant real estate databases.

Experience: Five to seven years' relevant experience.
Education: Bachelor's degree required; master's-level training in a real estate or finance discipline recommended.
Employers: Intermediary equity subsidiaries, banks, Fannie Mae, Freddie Mac
Salary range: $80,000–$100,000

Portfolio Analyst
The portfolio analyst must provide primary analytical support to the sales team for both financial analysis and new product development. To assist the sales team in preparing investment packages and due-diligence reports on new development acquisition projects.

Responsibilities:
- Organize and analyze information used by investors to evaluate properties for potential development or acquisition.
- Perform advanced spreadsheet analysis of investor benefits on existing

and new investment products.
- Assist in the preparation of investor presentations.
- Interface with acquisitions to obtain, model, and distribute information to investors.
- Special projects as necessary.

Skills:
- excellent written and verbal communication skills;
- strong organizational skills;
- strong technology and computer skills;
- familiarity with standard demographic databases.

Experience: Demonstrated interest in real estate, proven facility for financial analysis and financial modeling.
Education: Bachelor's degree (prefer accounting or finance background)
Employers: Intermediary equity subsidiaries, banks, Fannie Mae, Freddie Mac
Salary range: $40,000–$60,000

Director of Community Development Programs, Community Foundation

The director of community development programs is a senior staff officer responsible for overseeing the community development grants program of the foundation. As such, the director is a member of a team of program specialists, administrators, and financial and fund development professionals who make up the staff of the foundation. The director of community development programs is responsible for the development, articulation, and management of a grants program. The position offers an unusual blend of roles in grant making, institutional planning and development, senior management, and donor services. The director reports to the president of the foundation.

Responsibilities:
- Plan and implement a broad range of community development grants programs, including development and management of significant new initiatives.
- Supervise program officers, project consultants, and support staff.
- Work with foundation donors and prospects to develop their giving programs and provide regular reports on programs the donors have supported.
- Help the foundation attract national and local private funds in support of foundation programs.
- Staff board committees, as assigned.

- Articulate the foundation's mission and priorities to trustees, donors, prospects, and the not-for-profit community.

Skills:
- mature judgment;
- previous experience that demonstrates an ability to achieve results and a knowledge of civic and community issues;
- strong evidence of supervisory ability;
- demonstrated ability to meet deadlines and to articulate and work independently toward institutional goals;
- ability to function effectively as an integral part of a small senior management team;
- ability to work effectively with people from a broad range of backgrounds, particularly with trustees, major donors, and the not-for-profit community.

Experience: Considerable experience in the community development field. Excellent communication and personnel management skills.
Education: Bachelor's degree required; graduate degree(s) highly desirable.
Employers: Foundations involved with community development
Salary range: $60,000–$80,000

Director of Regional Office, CDFI

The director of the regional office is responsible for developing and maintaining a lending program to clients serving low-income households and communities.
Responsibilities:
- Direct the marketing of loan programs in the region, supervise staff in the marketing of lending activities in the region, and identify market opportunities in the region.
- Administer a community banking advisory committee.
- Identify needs within the community development field in the region and propose products to meet those needs; work with the president and deputy director for lending and portfolio management to develop and implement those products.
- Prepare business plans and program guidelines for new program areas and administer those programs.
- Provide support to fundraising efforts in the region, including identifying needed sources, making presentations, responding to requests for information, and coordinating development department.
- Work with staff to develop strategies for troubled loans and workouts and to ensure that strategies developed with the deputy director for lend-

ing and portfolio management are effectively implemented.

Skills:
- Demonstrated skills and knowledgeable experience in lending and credit analysis, structuring and deal underwriting in the real estate industry; experience in community development is preferable.
- Management, preferably in nonprofit organizations, including oversight of underwriting and portfolio management staff.
- Experience in and/or knowledge of the community development and affordable housing fields; knowledge of and experience working with public subsidy programs are preferred.
- Strong written and verbal communication skills and an ability to make public presentations.

Experience: Experience working with loan committees and nonprofit boards of directors. Mastery of community development finance and an understanding of market niches in the field that would lead to the creation of innovative products and tools for the entire field.

Education: Degree work in a relevant area of study such as business administration or finance. A bachelor's degree is required, an advanced degree preferred.

Employers: Large CDFIs, some government agencies

Salary range: $65,000–$90,000

Planning

Housing and Community Development Policy Planner

The person in this position is responsible for managing the work of a countywide, interjurisdictional body charged with making recommendations to elected officials in the city on affordable housing resources and related policies. This is a lead-staff position overseeing the planning commission's professional and technical support staff.

Responsibilities:
- Develop policy planning initiatives for the city's Bureau of Housing and Community Development.
- Assist with developing and implementing an annual work plan.
- Assist in the planning, development, and analysis of policy planning initiatives.
- Oversee staff support for planning commission meetings, develop agenda, create agenda packets, and arrange presentations.
- Provide leadership and assistance for staff assigned from affected jurisdictions.

- Advise planning commission regarding policy needs and performance evaluations pertaining to increasing the effectiveness of the publicly funded housing and community development delivery systems.
- Coordinate review of jurisdictions' budgets and action plans related to housing and community development.
- Oversee development of Five-Year HUD Consolidated Plan; monitor local governments' compliance and assist with annual performance updates.
- Develop and maintain formal and informal relationships to other commissions working on housing and community development issues.
- Monitor legislative activities relating to housing and community development.
- Coordinate advocacy efforts.
- Act as primary liaison to HUD.

Skills:
- considerable knowledge of and experience in organization and policy planning theories and practices;
- considerable knowledge of housing and community development–related regulations, policies, and procedures, including citizen involvement and public review processes;
- considerable knowledge of citizen participation processes and practices;
- knowledge of current literature, trends, and developments in the area of housing and community development and related fields;
- knowledge of management principles and practices, including management of people and funds;
- skill in the utilization of performance measurement and evaluation techniques;
- knowledge of the principles and techniques involved in coordinating complex work schedules;
- knowledge of budgeting concepts and principles;
- knowledge of the process and requirements for working with diverse public and private interests;
- considerable skill in the development and presentation of recommended policies and procedures;
- considerable skill in oral and written communication;
- ability to develop program and policy objectives, output measures, and performance indicators;
- skill in managing group planning and decision-making processes;
- skill in the analysis and synthesis of divergent forms of information and their application to complex planning and policy issues;
- ability to establish and maintain effective working relationships with elected officials, bureau managers, other public employees, and groups representing a wide variety of interests and backgrounds.

Experience: Five years' or more experience in housing planning or community development.

Education: BA required; master's degree preferred.
Employers: Local governments, state governments
Salary range: $40,000–$60,000

Community Planner

This position involves entry-level professional work encompassing planning and community development.

Responsibilities:

- Perform planning studies and conduct research, analyze data, and formulate recommendations.
- Prepare documentation for grants.
- Prepare and update other department documents.
- Coordinate programs with other agencies.

Skills: Knowledge of the standard principles, practices, and methods related to professional planning.
Education: Bachelor's degree in urban planning; master's preferred.
Employers: Local government, city planning departments
Salary range: $25,000–$50,000

Fundraising (Development) Jobs

Grant Writer

The primary function of this position is to write proposals for grants to support the mission of the organization. The grant writer works for the director and in collaboration with the vice president and resource development managers.

Responsibilities:

- Help to develop the case for proposals and the framework for requests.
- Collect and organize program and background information and prepare the final proposal document.
- Work with program services staff to develop and write reports on grants.
- Prepare concisely written, strategically focused proposals to foundations, corporations, and individuals utilizing the program and communications resources of the organization to include the Resource, Evaluation and Documentation Department for demographic, market, and anecdotal information.
- Work with local and national program directors to gather program materials from which to develop clear, precise descriptions of goals and objectives.

- Synthesize and articulate concepts, goals, and strategies for both professional and nonprofessional prospects.
- Work with the research and donor systems manager to develop and maintain a tickler on expiring.

Skills:

- excellent writing and editing abilities, including demonstrated sensitivity to the unique style, process, and subtleties of grant and proposal writing;
- ability to synthesize and articulate concepts, goals, and strategies for both professional and nonprofessional prospects;
- ability to work in a high-pressure, creative environment where accountability for results is a priority.

Experience:

- documented success in obtaining grants to support the mission of a nonprofit organization;
- a minimum of four or five years of grant writing experience; background in social services or affordable housing a plus.

Education: Bachelor's degree in a related field
Employers: National intermediaries, large CDCs, local governments
Salary range: $35,000–$50,000

Vice President of Development

The vice president of development reports to the chief operating officer. Primary responsibilities include: (1) managing the fundraising process for the national organization, (2) providing support for the national board of directors, (3) managing national events, and (4) providing fundraising support to the local affiliates.

Responsibilities:

- Manage a tracking system of all current and prospective funders.
- Develop and update funding proposals, standardizing process and customizing when necessary.
- Cultivate and maintain relationships with current and prospective funders.
- Develop and implement a communication process via mailings, phone calls, events, tours, etc.
- Develop and implement a donor recognition plan.
- Keep staff, board, and others informed of fundraising status and opportunities, etc.
- Manage three yearly board meetings.
- Maintain regular communication with the board.

- Support board committees: Executive and Finance (fundraising).
- Help identify new board members.
- Provide support for a new organization for high-net-worth individuals.
- Manage yearly fundraising event.

Skills:
- excellent written and public presentation skills;
- excellent project management skills;
- familiarity with database management and word-processing software;
- private-sector experience strongly preferred;
- ability to work effectively with leaders from the public, private, and community sectors.

Experience: Five to fifteen years' professional experience; some business preferred. Experience with major gifts, sales and marketing, and event management strongly preferred. Experience with board of directors preferred.

Education: Minimum bachelor's; master's preferred.

Employers: National intermediaries

This appendix offers descriptions of universities and colleges with curriculum relevant to community development. Some offer undergraduate minors or majors; others offer graduate degree programs; a few offer both. In addition, some colleges and universities also offer certificate degree programs.

Appendix K, on training and professional development, also has some entries for schools offering certificate programs (schools that do not offer undergraduate or graduate degrees).

The list in this appendix is organized alphabetically by discipline and by colleges within each discipline. It also includes faith-based institutions and Historically Black Colleges and Universities. At the beginning of each discipline is information, if applicable, about the professional association for that discipline, as well as information about any subgroup focusing on community development within the discipline. The disciplines covered are:

Business, Nonprofit Management, and Public Administration
Community Economic Development
Economics
Faith-Based Colleges and Universities
Historically Black Colleges and Universities
Law
Planning
Public Policy
Real Estate Development
Social Work
Sociology
Urban Affairs

Business, Nonprofit Management, and Public Administration

Baruch College, City University of New York
School of Public Affairs
17 Lexington Avenue, Box F-1228
New York, NY 10010
212-802-5912; fax: 212-802-5912
email: SPA_Admissions@baruch.cuny.edu
www.baruch.cuny.edu/spa
Baruch offers an intensive fourteen-month program, combining a nine-month internship with a senior executive and academic course work, conferring a Master of Public Administration, for fellows of the National Urban Fellows Program. Mid-career professionals apply to National Urban Fellows, Inc. See appendix J on internships. The curriculum includes, among other courses, the Political Setting of Public Administration and Public Policy, Public Budget Administration, Computer Methodology and Information Systems, Introduction to Research and Data Analysis, and Managing Organizations in the Public and Nonprofit Sectors.

Other students may earn an MPA with a total of forty-two to forty-five credits, taking many of the same courses; an internship is required for students without professional experience in a public policy setting.

Carnegie Mellon University
H. John Heinz III School of Public Policy and Management
5000 Forbes Avenue
Pittsburgh, PA 15213-3890
412-268-2164; fax: 412-268-5161
www.heinz.cmu.edu
Carnegie Mellon offers a one-year, mid-career Master of Public Management (MPM), as well as a Master of Science in Public Policy and Management, with a concentration in economic development and urban planning, and a PhD in public policy (see the "Public Policy" section of this appendix).

The Master of Public Management is designed for experienced professionals seeking training in management, analytic, and policy skills that will position them for advancement within their organization or industry. Students have the option of either specializing in an existing concentration such as nonprofit organizations, or designing an individual study program in an area like urban economic development. The program may be completed in one year of full-time study, or two to three years of part-time study.

The Center for Economic Development, within the Heinz School, brings academic resources to bear on key issues in regional and national economic development and serves as a resource for student internships.

Cleveland State University
Maxine Goodman Levin College of Urban Affairs
Urban Affairs Building
Cleveland, OH 44115-2105
216-687-2136; fax: 216-687-9291
http://urban.csuohio.edu
Ranked as one the top ten university programs in city management and urban policy, the Levin College of Urban Affairs places a special emphasis on combining classroom teaching with direct public service to seek solutions

to urban problems. Located near inner-city Cleveland, it offers several degree programs: BA in urban studies, Master of Science in urban studies, PhD in urban studies; Master of Urban Planning, Design and Development (MUPDD); and Master of Public Administration. (See the "Urban Affairs" and "Planning" sections of this appendix.)

The Master of Public Administration is a multidisciplinary program designed to accommodate part-time, mid-career students, as well as full-time students. Courses are offered primarily during the evening and on weekends. The curriculum includes seven required core courses that introduce students to the principles of management and public administration and provide a solid foundation in public finance, economics, and research methodology. Students choose one of the following specialization tracks, in which they must take at least four courses: (1) the economic development track, which includes courses in Urban Development Finance, Anti-Poverty Policy, Economic Development Plans and Strategies, Labor Markets, Economic Development and Poverty, and Economic Development Policy; (2) the public works management track, (3) the public management track; and (4) the public health management track. Students are required to complete an internship of at least four hundred hours of fieldwork in a public service agency, supervised by both a practitioner in the field and a faculty member. A waiver for the internship may be granted to individuals with significant professional managerial experience in a government or nonprofit organization. A thesis is not required.

Columbia Business School

Uris Hall, Room 105
3022 Broadway
New York, NY 10027
212-854-1961; fax: 212-662-6754
www.columbia.edu/cu/business

Columbia Business School's four-semester Master of Business Administration (MBA) program offers two curricular areas related to community and economic development: Public and Nonprofit Management, and Real Estate. The Public and Nonprofit Management area is designed to prepare students for careers in the nonprofit and government sectors or for businesses that provide services to these sectors. This curriculum area includes two clusters of study: Future of the Metropolis (an interdisciplinary survey of the forces shaping New York City) and Public Capital Investment.

The Real Estate area prepares students for careers in real estate finance, investment banking, asset management, consulting, entrepreneurial development, and public and nonprofit development. The real estate curriculum at Columbia mainly emphasizes financial analysis, taxation, real estate law, real estate finance, and strategic investment management. Students who ma-

triculate in September typically do an internship after their second term.

The Business School also participates in dual degree programs with other schools at Columbia.

New School University

Robert J. Milano Graduate School of Management and Urban Policy
66 Fifth Avenue, 7th floor
New York, NY 10011
212-229-5950; fax: 212-229-5904
email: mgsinfo@newschool.edu
www.newschool.edu/milano

The Milano Graduate School offers a Master in Nonprofit Management, covering nonprofit governance, general management, fundraising and development, strategic planning, program development, financial and human resources management, policy analysis and problem solving, ethics, earned income, and marketing. Students also gain a comprehensive understanding of the complex political, legal, and competitive environments within which nonprofit organizations must function to accomplish their missions. (See also the "Planning" and "Public Policy" sections of this appendix.)

The forty-two-credit requirement for the degree includes three components: a required schoolwide core of twelve credits, a required program core of fifteen credits, and fifteen credits of elective courses. The schoolwide core includes four required courses: Economic Analysis, Policy Analysis, Quantitative Methods, and Management and Organizational Behavior. The five required program core courses are Theory and Practice of Nonprofit Management, Financial Management in Nonprofit Organizations, Fundraising and Development, a Laboratory in Issue Analysis, and either Managerial Decision Making or an advanced seminar in Nonprofit Management.

Students may organize the five elective courses to suit their individual academic and professional interests and goals. Among the choice of eighteen courses are: Governance of Nonprofit Organizations, The Law of Nonprofit Organizations, Social Movements and Advocacy, The International Nonprofit Sector; and Religious Nonprofit Organizations' Role in Community Building and Social Responsibility. Alternatively, students may organize their elective courses to obtain in-depth knowledge in a particular area of professional interest, e.g., fundraising or urban policy. Students may also select electives from the Cardozo Law School.

Students without previous professional experience in the field of nonprofit management and those planning a mid-career change are strongly urged to undertake a summer internship between the first and second years of study.

Northwestern University

J.L. Kellogg Graduate School of Management
2001 Sheridan Road
Leverone/Andersen Complex
Evanston, IL 60208-2001
847-491-3300
www.kellogg.nwu.edu

Kellogg offers a Master of Management degree, which is comparable to an MBA. Students take a set of required core classes. Those interested in community development are able to major in public and nonprofit management and often combine it with another such as finance. The school supports students interested in public-sector work through summer internship placement and a loan forgiveness program; the students themselves have an active Public and Nonprofit Management Club and alumni network.

Kellogg operates on a quarter system. Full-time students complete the degree in six quarters, or four if they have an undergraduate degree in business. Weekend and evening classes are offered for part-time students as well.

Stanford Graduate School of Business

350 Memorial Way
Stanford, CA 94305-5015
650-723-2766; fax: 650-723-7831
email: mbaapps@gsb.stanford.edu
www.gsb.stanford.edu

The Public Management Program (PMP) at the Stanford Graduate School of Business is a full-time, two-year Master of Business Administration (MBA) program with a focus in public and nonprofit sector management and social entrepreneurship. In addition to taking regular MBA as well as PMP-related courses during the first and second years, all second-year PMP students focus as a class on a public policy topic—the Public Management Initiative (PMI). As part of the PMI, students organize events and activities on a public policy topic chosen their first year. Activities include speakers, site visits, panel discussions, workshops, case studies, partnerships, and public service projects. For example, second-year students during 1993–94 focused on the topic of urban development and created a microbusiness training program and revolving loan fund to support the formation of new businesses in east Palo Alto. In addition to the PMI, the PMP sponsors public service projects, study tours, and visits by leaders in public service. The Stanford Management Internship Fund provides funding for MBA students who work in public service during the summer between their first and second years. The school also offers a loan forgiveness program for graduates working in lower-paying jobs in the nonprofit and public sectors.

The Union Institute

440 E. McMillan Street
Cincinnati, OH 45206
513-861-6400 or 800-486-3116
www.tui.edu

The Union Institute is a private, accredited university that offers BA or BS and PhD programs that are highly individualized and learner centered. It maintains undergraduate centers in Cincinnati, Miami, Los Angeles, and Sacramento as well as the Cincinnati-based Center for Distance Learning, which generally serves anyone seeking a bachelor's degree but living more than one hundred miles from one of the other centers. The doctoral program operates entirely as distance learning. The university actively seeks out underserved populations, nationally and internationally. The average age of the undergraduate learner is thirty-eight; the typical doctoral learner is forty-seven.

The Union Institute calls its students "learners" to emphasize the active role that students play in their education. Academic instruction is delivered primarily through self-paced, independent study so that each learner may maintain professional, family, and community commitments while earning a degree. Faculty members mentor and guide the learner throughout each phase of the program.

On the undergraduate level, a learner may craft a degree plan for a traditional area of concentration, such as social work or justice studies, or may choose to develop a highly personalized area of concentration. A minimum of 128 semester credit hours is required for the bachelor's degree. The average time of completion is sixteen to twenty months. The Union Institute can award credit for college-level, documented learning gained through prior experience.

Doctoral learners may concentrate on areas such as community development, multicultural studies, peace studies, philanthropy and leadership, public policy, sociology, gender studies, and all fields of psychology. There is no "required" curriculum or course work. Instead, learners progress through a series of stages based on their detailed and highly individualized learning agreement for the entire program of study and research. Though self-paced, the university requires doctoral learners to be continuously and actively enrolled for at least twenty-four months. The undergraduate senior project (thesis) or the doctoral dissertation may take one of many forms, including a documented project of social action or change.

University of Alabama

Department of Government and Public Services
901 15th Street South
Birmingham, AL 35294

205-934-9680

www.sbs.uab.edu/gps/mpa.htm

The Master of Public Administration is a fifty-three-hour program. After taking seven required core courses, students may specialize in urban planning and management (sample course: Urban Development), or public policy analysis (sample course: Democracy, Politics, and Power), or public finance and economic development. Students also choose between a subspecialization in either general organizational management or nonprofit management (sample course: Community Planning and Organization). An internship is required.

The Community Planning and Organization course is associated with the University's CENTER OF URBAN AFFAIRS. Through the Center, a neighborhood is selected each year, and students put together a comprehensive plan for that neighborhood, with input from the residents.

University of Louisville

College of Business and Public Administration
School of Economics and Public Affairs
Louisville, KY 40292
502-852-6005; fax: 502-852-4721
email: caford01@ulkyvm.louisville.edu
http://cbpa.louisville.edu

The University of Louisville offers two accredited degrees: a Master of Public Administration with a concentration in urban and regional development, and a PhD in Urban and Public Affairs. (See also the "Urban Affairs" section of this appendix.) The Master of Public Administration is a forty-two-credit multidisciplinary program focused on urban issues and the techniques of modern management, planning, and public policy. The curriculum includes nine required core courses, plus five classes in a specialization track, of which urban and regional development is one. Other areas of concentration include public policy, labor and public management, and health policy and administration. Students have the option, depending on their level of professional expertise, of either completing a fieldwork practicum or writing a thesis.

THE CENTER FOR SUSTAINABLE URBAN NEIGHBORHOODS, affiliated with the University, offers students internship opportunities. The Center provides technical assistance and resources to support community residents in their efforts to physically, economically, and socially revitalize the West End Enterprise Community.

University of San Francisco

Institute for Nonprofit Organization Management
2130 Fulton Street
San Francisco, CA 94117-1080
Main Office: 415-422-6867; education programs: 415-422-6000
email: inom@usfca.edu
www.inom.org

The Institute for Nonprofit Management at the University of San Francisco offers a Master of Nonprofit Administration (MNA), an Executive Certificate in Nonprofit Management, and a Development Director Certificate.

In the MNA program, students meet for one four-hour class each week and take courses sequentially over a period of two and a half years. Courses include marketing, financial management, organizational theory, law, human resource management, and others focusing on general and nonprofit management. To complete the degree, students must also complete a thesis or pass a comprehensive exam. In the twelve-month Executive Certificate in Nonprofit Management program, participants take six classes in total in the areas of governance and planning, marketing, development and fundraising, human resource management, and financial management. The Development Director Certificate, also a twelve-month program, offers an opportunity to fundraising professionals to learn the principles and practices of effective financial and fundraising development for nonprofit agencies. The program can be completed in one of three formats: (1) one four-hour meeting per week; (2) an eight-hour meeting every other Saturday; or (3) a mixed format of online and on-campus meetings.

University of Southern California

School of Policy, Planning, and Development
Von KleinSmid Center, Room 232
Los Angeles, CA 90089-0626
213-740-6842; fax: 213-740-7573
email: sppd@usc.edu
www.usc.edu/schools/sppd

The School of Policy, Planning, and Development combines the School of Public Administration and the School of Urban Planning and Development. (See also the "Planning" section of this appendix.) A dual master's degree combining public administration and planning is available.

The MPA requires sixteen credits in core courses (Public Sector Economics plus twelve credits in areas of competency, such as Organizational Behavior) and an internship. Students also elect courses focusing on an area of study; one such area is local government, another is nonprofit management.

Among the centers where students can pursue research and practice are the Center for Economic Development, which works with distressed communities in a five-county area of Southern California; the Community Development and Design Forum, which provides consultative services to projects involving local community groups and residents, among others; and the Nonprofit Studies Center.

University of Texas at Austin
Graduate School of Business
Austin, TX 78712
512-471-7612; fax: 512-471-4243
email: Cgep@bus.utexas.edu
http://texasmba.bus.utexas.edu

The Graduate School of Business at the University of Texas at Austin offers two full-time, two-year degrees: a Master of Professional Accounting (MPA) and a Master of Business Administration. The MBA program has a concentration in entrepreneurship applicable to business careers in community and economic development. Students in both the MPA and MBA programs are eligible for participation in the Community Growth & Entrepreneurship Program (CGEP). CGEP provides tuition and stipend fellowships, mentorship and internship opportunities, and special courses and field visits to students committed to building businesses in underdeveloped U.S. communities. CGEP participants spend their summer semesters working in underdeveloped communities; students who receive tuition and/or stipend fellowships are in addition obligated to work for one or two years in underdeveloped communities after graduation.

Yale University
School of Management
P.O. Box 208200
135 Prospect Street
New Haven, CT 06520-8200
203-432-5932; fax: 203-432-7004
email: som.admissions@yale.edu
www.yale.edu/som/

The Yale School of Management (SOM) offers a two-year Master of Business Administration. Many SOM students are engaged in joint degree programs, and all students are encouraged to use the resources of the entire university.

The MBA is a full-time program that requires completion of eighteen courses. Because Yale emphasizes leadership training, students are required to complete a year-long Perspectives on Leadership program in their first year. An internship is also required during the summer between the first and second years. Of the concentrations at SOM, Public Management and Nonprofit Management are the two fields most applicable to students interested in community and economic development careers.

The SOM On-the-Road program at Yale takes twenty to thirty students to public and nonprofit organizations for personal interactions with leaders and firsthand observations of real-life situations and challenges. The Local Initiatives Support Corporation (LISC) in New York participates in this program. Although Yale SOM is increasingly focusing on the business side of its curriculum, the school does have resources and a community available for students interested in other career tracks. One SOM graduate states that Yale alumni are a "significant thread throughout economic development in this country" and that the program includes a significant international development component.

Community Economic Development

Cornell University
College of Agriculture and Life Sciences
812 Warren Hall
Ithaca, NY 14853-6701
607-255-1685
www.cals.cornell.edu/dept/cardi

Cornell's College of Agriculture and Life Sciences provides a concentrated, one-year Master of Professional Studies in community and rural development, covering rural and small-town development issues. The program is designed for mid-career rural and community professionals in the community development field, including practitioners, public administrators, economic development specialists, human service providers, and extension agents.

The degree requires thirty credit hours and two units of residence. Required core courses include: Community and Rural Development; Theoretical and Methodological Approaches to Community and Rural Development; and a project seminar course within which students conceptualize and develop their degree projects.

Students take three courses in one of the following four areas of concentration: community development process (e.g., community analysis, participatory action research, group facilitation, principles of intercultural communication, processes of collaborative planning, and promotion of local initiatives); economic development (e.g., analyzing local economic structures and designing approaches to promote community economic development such as through improved small business management and improved community infrastructure); program development and planning (e.g., physical concerns confronting rural areas including land use and techniques for managing land development, the environment, transportation and communication systems); and local government organization and operations (e.g., how decisions are made by local governments and how they affect rural communities).

This program is associated with the COMMUNITY AND RURAL DEVELOPMENT INSTITUTE, which engages in research, education, and policy analysis in community development, economic development, environmental management, human services, and local governance, with a focus on New York State rural communities.

Los Angeles Trade-Technical College

Community Development Technologies Center
2433 South Grand Avenue
Los Angeles, CA 90007
213-763-2520; fax: 213-763-2729
www.cdtech.org

The Community Planning and Economic Development program offers both a one-year certificate program and a two-year Associate of Arts degree. Courses are interdisciplinary, including, among other subjects, an introduction to community economic development, community organizing, business law, economics for workers, spreadsheets, affordable housing, sectoral development and employment strategies, nonprofit management, real estate finance, and law. The College has received work-study funds from HUD for students who need financial aid as they pursue the associate degree.

New Hampshire College

Community Economic Development Program
2500 North River Road
Manchester, NH 03106
603-644-3103; fax: 603-644-3130
http://merlin.nhc.edu

The New Hampshire College Community Economic Development program is the first program in the country with a degree specifically in community economic development. The program offers a Master's and a PhD in Community Economic Development, a Master's in International Economic Development, and various certificate programs. Courses are taught by practitioners who excel in their field of expertise. Programs use a "learner-centered approach" that encourages active participation and values participant experience. The student body is composed of adult practitioners who are already active in their own communities and are committed to returning to their communities with their advanced knowledge.

The Weekend National Master's in Community Economic Development Program meets one long weekend (Friday–Sunday) per month for seven-

teen months beginning in September of each year. It includes courses such as Financing Community Economic Development, Financial Management, Organizational Management of Community Organizations, Development of Cooperatives, Housing Development, Negotiation Strategies and the Training of Trainers, Microenterprise Development, and Indigenous Economics.

As a requirement of their degree, each student designs and puts into action a real development project that meets a market need in their community. Examples of such projects include a new credit union and a new housing development corporation. A computer electronic network is used by students and instructors to communicate with each other when they are not in class. Student networking and sharing of their experiences as practitioners from around the world are considered to be key dynamics of the program.

The Master's in International Community Economic Development is a one-year residential program of course work and field trips tailored to be directly relevant to the participants' work and their management and community development experience.

The PhD in Community Economic Development is intended for leaders in community-based development organizations and those interested in college-level teaching, policy development, and consultation. It entails two years of course work, the pursuit of a research topic with direct community applicability, and a dissertation.

The program also offers training workshops in community economic development. See appendix K.

Economics

University of Massachusetts—Lowell

Department of Regional Economic and Social Development
61 Wilder Street
Lowell, MA 01854
978-934-2900; fax: 978-934-4028
email: resd@uml.edu
www.uml.edu/Dept/RESD/

Launched in September 1998, this program offers two graduate programs in the economic and social development of regions—a Master of Arts as well as a Graduate Certificate. Within the master's program, students may choose from three concentrations: Regional and Community Development (to analyze and plan economic development at the local and regional level); Organization, Technology, and Public Policy (to understand and manage innovation, technical change, and industrial development); and Social and Historical Dynamics (to analyze and act on social aspects of development, including income distribution and environmental issues). The program is highly interdisciplinary; program professors include economists, sociologists, historians, political scientists, community social psychologists, and planners. Students may choose electives from other degree programs, including MBA, Management Science, and Community and Social Psychology.

Since many of the students are employed full-time, the program permits both full- or part-time enrollment; a master's requires thirty credits as well as a project or thesis. The certificate program requires four courses for twelve credits.

Students may work on relevant projects through two University centers: the CENTER FOR FAMILY, COMMUNITY AND WORK, and the CENTER FOR INDUSTRIAL COMPETITIVENESS (which focuses on large and small community businesses).

University of Vermont
Department of Community Development and Applied Economics
106 Morrill Hall
Burlington, VT 05405
802-656-2001; fax: 802-656-1423
email: chalbren@zoo.uvm.edu
www.uvm.edu/~cdae/dept

At the undergraduate level one major is offered—community development and applied economics—with three areas of concentration: Small Business, Consumer Economics, and International Development and Agricultural Economics. Students must fulfill the basic distribution requirements of all students seeking a BS degree in the College of Agriculture and Life Sciences and must complete an internship (usually in the fourth year).

At the graduate level an MS is offered in agricultural economics or international development. Thirty credits are required, including advanced courses in agricultural and resource economics, general economics, or related fields, and thesis research. Areas of concentration include international development, agricultural marketing, and community development. A thesis is required.

Faith-Based Colleges and Universities

Eastern College
Economic Development Program
1300 Eagle Road
Saint Davids, PA 19087-3696
610-341-5800; fax: 610-341-1466
email: econdev@eastern.edu
www.eastern.edu/academic/grad/programs/ecodev/econdev.html

The Economic Development Program at Eastern College offers three full-time degree programs leading to a Master of Science in urban economic development, which is a program for faith-based neighborhood development, or an MS or a Master of Business Administration degree in economic development. Eastern College, a college of Christian faith, prepares students to become development practitioners and agents of change around the world by teaching holistic, community-based development. The College stresses the multiple dimensions of poverty as a problem of economic, social, political, and spiritual nature. Themes emphasized in the program's curriculum are urbanization, servanthood, empowerment, globalization, reconciliation, and multiculturalism. Both of the MS programs and the MBA program teach principles, issues, and models of economic and community development. However, the MBA curriculum also emphasizes the quantitative aspects of business, with courses in finance, accounting, economics, and mathematical decision making. The MS and MBA in economic development require eight core curriculum classes, a course in theology, and an integrative project (internship) or community-based development training (fieldwork experience). The MBA incorporates two additional classes. The faith-based MS in urban economic development is a hands-on practical skills development program for working adults who are practitioners, laypeople, or pastors.

McCormick Theological Seminary
African American Leadership Partnership Program
1313 East 60th Street
University of Chicago Campuses
Chapin Hall
Chicago, IL 60637
773-753-2470; fax: 773-753-2480
email: aalpinfo@aalpinc.org
www.aalpinc.org

McCormick Theological Seminary, jointly with the nonprofit African American Leadership Partnership Program, offers a Master of Arts in theological studies degree program focused on community revitalization for African American religious leaders. The eighteen-course curriculum includes, among others, Theology for Transforming Communities, Community Organizing in a Multicultural Society, and Strategic Processes for Neighborhood/Community Economic Development. Theological courses include the Old and New Testaments. Twelve of the courses are taught in an intensive format with weekend and evening hours to accommodate working participants. The program also requires a ministry project.

See also "Faith-Based Training" in appendix K for a description of a parallel training program.

Mennonite Brethren Biblical Seminary

4824 East Butler Avenue
Fresno, CA 93717
559-251-8628; fax: 559-251-7212
email: MBSeminary@aol.com
www.MBSeminary.com

The Seminary offers a Master of Divinity in Christian community development (CCD). The Christian community development major is designed to prepare leaders to impact at-risk neighborhoods for national and international contexts and establish community-based ministries. Two courses deal with the economic development side of Community Development: Models of Christian Community Development and Community Economic Development.

A major feature of this program is living in "at-risk communities" and working with Christian community development agencies. Currently working in eight at-risk communities in the Fresno area, the program gives interns the opportunity to work in community development agencies.

Seminary Consortium for Urban Pastoral Education

200 North Michigan Avenue, Suite 502
Chicago, IL 60601
312-726-1200; fax: 312-726-0425
and

North Park University

3225 West Foster Avenue
Chicago, IL 60625
773-244-6200; fax: 773-244-4952
email: scupe200@aol.com
www.northpark.edu/acad/macd

SCUPE, in conjunction with North Park University, offers a two-and-a-half-year program for a Master of Arts in church-based community development. The thirty-seven-semester-hour program begins at the end of August; classes are on Friday afternoons and Saturdays to accommodate those with full-time jobs. The program is designed both for those already within the community development field and for those who have no prior experience but would like to make community development a career.

Required courses include, among others: Ethical and Biblical Foundations of Urban Community Development, Principles and Practices of Community Development, Social Change, Community Organizing, and Human Resources Development.

In lieu of a thesis, the program involves a master's project in which students already within the field, with the help of experienced mentors, work to push their organizations to new frontiers of community development, while students who are new to the field are offered supervised internships in community development organizations with the same goals in mind.

Historically Black Colleges and Universities

Howard University

Center for Urban Progress
2006 Georgia Avenue, NW, LL
Washington, DC 20001
202-806-9558; fax: 202-806-9566
email: Rgreen@fac.howard.edu
www.howard.edu

Howard University, a Historically Black University, has launched an interdisciplinary undergraduate minor in community development. An undergraduate major as well as a Master's and a PhD in community development and a professional development certificate program are envisioned in the future.

The community development minor is taken in combination with the student's choice of major. The minor entails two required courses—Community Development from an Interdisciplinary Perspective, offered by the Department of Political Science, and The Economics of Black Community Development, taught through the Department of Economics—two elective courses (two examples: Urban Political Economy, and Environmental and Community Health); and an internship or laboratory course.

Opportunities for student internships are available through the CENTER FOR URBAN PROGRESS, which studies and proposes solutions to the multifaceted urban crises of the nation by: (1) assembling teams of Howard University faculty for applied research on urban issues; (2) providing facilitation support for selected community-based organizations and collaborations engaged in community and urban development; (3) developing collaborations with professional organizations; and (4) providing technical support and expertise to nongovernment organizations and federal and local government agencies.

Mississippi Valley State University

Community Development Leadership Program
1400 Highway 82 West
Ittabena, MS 38941
601-254-3652; fax: 601-254-3302
www.mvsu.edu

Mississippi Valley State offers a Community Leadership Program that can be either a concentration within the undergraduate major in public administration, or a certificate program for students and experienced practitioners from the community who want additional training. For the Bachelor of Science in public administration, students must complete fifty-six hours of general education courses, forty hours of department core courses, and at least thirty-four hours in the community development concentration. Experiential learning, as well as weekly lectures by various community leaders and practitioners, are important components of the program. Courses within the concentration include Principles of Community Development Leadership, Conflict Management and Resolution in Local Communities, and Community Leadership Development Seminar. A minor in community development, which requires six courses, is also available.

The Community Development Certificate Program is multidisciplinary, with eleven required courses taken from public administration, political science, business, and sociology. Required classes include Entrepreneurship and Business Management, Leadership Theory, and Grantsmanship and Business Management.

Morgan State University
Jenkins Building 334
Baltimore, MD 21251
443-885-3225; fax: 410-319-3786
www.morgan.edu

Morgan State offers an accredited, practice-oriented Master in City and Regional Planning, with a concentration in community planning. Initiated in l970, this planning program is one of the oldest and the first to be accredited, among the Historically Black Colleges and Universities. Located in the urban core of Baltimore, the MCRP's priority is the education of African-American planning professionals. A community development emphasis is integrated throughout the curriculum, which is designed to meet the needs of working, experienced professionals. Evening programs are available. Special attention is given to the professional skills and tools needed to understand and to intervene in urban life and phenomena. Requirements include eleven core classes in the history, theory, methods, institutions, and practice of urban planning. A thesis is also required. The remaining six courses may be taken either as electives or clustered as a specialization in one of the following areas: community planning, physical planning, urban design, and international planning. Students have the option of taking these elective units either within the City and Regional Planning Program, or, with faculty approval, from other programs within the University as a whole or from other universities within the metropolitan Baltimore area.

THE INSTITUTE FOR URBAN RESEARCH, under the School of Graduate Studies and Research, undertakes a wide variety of research studies, on areas ranging from community and economic development to family planning and education. Within the Institute, the Community Development Research Center was founded to support community efforts throughout Baltimore; the Center develops in-depth community-by-community data profiles, puts on seminars and workshops, convenes forums, provides practical technical assistance to community groups, and pursues research. Students are typically hired as research assistants in these programs. The Center serves as a clearinghouse for agencies across Baltimore for internship placements. When funding permits, the Center places several students each year in internships.

St. Augustine's College
Certificate Program in Community Economic Development
1315 Oakwood Avenue
Raleigh, NC 27610-2298
919-516-4001; fax: 919-516-5801
www.st-aug.edu

St. Augustine's College, in partnership with the North Carolina Association of Community Development Corporations, sponsors the North Carolina Community Economic Development (NCCED) Studies Program, which awards a Certificate in Community Economic Development. The NCCED Program is an intensive eleven-day training session, divided over three months, which focuses on the fundamentals of community economic development in North Carolina. Specifically developed for staff of nonprofit community organizations, the major course of study involves work performed in small groups around real project opportunities and community situations. The curriculum includes training by renowned professionals from around the country in such essential areas as implementing and managing the development process and team, strategic community-based planning and organizational development, job creation and small business development, residential and commercial real estate development, and grant development.

St. Augustine's is also in the process of developing an undergraduate major in community economic development. The degree will be interdisciplinary and practical, drawing on faculty across the College's academic departments as well as from a broad body of local and national community development practitioners. Practicums will begin in the second year of study, using the state's existing community development corporations as learning laboratories.

Texas Southern University
3100 Cleburne
Houston, TX 77004
713-313-1851; fax: 713-313-1853
www.technology.tsu.edu

The Community Development Leadership Program is offered as a minor by Texas Southern University, in collaboration with the Third Ward Community Development Corporation. The program uses an interdisciplinary approach and covers a range of community and economic development categories that include new business development, housing construction and rehabilitation, leadership development, land assembly, and commercial development. Students take courses in program evaluation; urban community life; construction management, contracts, and specifications; engineering economy; cooperative education; and business and organization management.

Experiential learning is an essential part of the program; students are required to take a six-month internship in a public, private, or nonprofit organization. These field internships, which complement the instructional component of the program, are designed to give students practical skills in both nonprofit management and in the development of programs, partnerships, policies, and methodologies that will improve the quality of life of urban dwellers.

Law

Besides the listings that follow, you will find other leads through the ABA Forum on Affordable Housing and Community Development Law, an interest group maintained by the American Bar Association; contact www.abanet.org and click on the Forums section.

For further information, we suggest two books: *Law Schools: Building a Corps of Lawyers Skilled in CED*, Leslie Newman (1993), National Economic Development and Law Center, 2201 Broadway, Suite 815, Oakland, CA 94612, 510-251-2600; and *Law School Involvement in Community Development: A Study of Current Initiatives and Approaches*, David Beaning (undated), U.S. Department of Housing and Urban Development, Office of University Partnerships; copies available from HUD User, 800-245-2691.

George Washington University
Law School
1819 H Street, NW
Washington, DC 20006
202-994-6592
www.law.gwu.edu
The GWU Law School runs a SMALL BUSINESS CLINIC (telephone 202-994-5795), which provides free legal assistance to small businesses, primarily micro-enterprises, and nonprofit organizations. Each year some two hundred second- and third-year law students engage in hands-on law practice while earning academic credits through the clinic. Law students create legal structures for new businesses and also guide local entrepreneurs through the tax and legal regulatory requirements for starting a new business. The Community Economic Development Project, a project of the clinic, provides legal assistance, counseling, and representation to new and existing non-

profit groups engaged in micro-enterprise development, job training, and community building initiatives on behalf of low-income people.

Law school courses related to the clinic's work include those on corporate law, business planning, and federal tax law; there is also a course on housing rights with a focus on low-income housing.

Georgetown University Law Center
600 New Jersey Avenue NW
Washington, DC 20001-2095
202-662-9000
www.law.georgetown.edu
The Clinical Program at Georgetown University Law Center is one of the largest in the nation. Students receive instruction from fourteen full-time faculty and twenty-four graduate fellows. The twenty-four Clinical Graduate Fellowships provide new lawyers with the opportunity to develop skills as teachers and litigators, receive a considerable stipend plus tuition and fees, and finish with a master of laws degree.

The school runs, among others, an ENVIRONMENTAL JUSTICE CLINIC and the HARRISON INSTITUTE FOR PUBLIC LAW (telephone: 202-662-9600). The Harrison Institute offers two clinics: one on housing and community development practice, and the other on public policy. The housing and community development clinic works in three areas: multifamily housing, facilities development (e.g., day care centers), and economic development. The curriculum covers topics of housing regulation, multifamily ownership, nonprofit governance, and community development financing. The policy clinic covers community health (including programs that create community jobs), land use, and democracy and trade (including international development).

Harvard Law School
Pound Hall
Cambridge, MA 02138
617-495-3100
www.law.harvard.edu
The Harvard Law School's Hale and Dorr Legal Services Center sponsors ninety HLS students each semester through a variety of clinical course placements, including (among others) the Housing Workgroup, which represents tenants in defending against evictions and improving their housing conditions, and the COMMUNITY ENTERPRISE PROJECT. CEP is primarily a transactional legal practice engaged in business law, real estate law, nonprofit law, for-profit law, tax-exemption law, and other related business law, assisting community-based organizations and social enterprises that promote corporate investment in minority and/or low-income communities.

CEP complements an HLS course on community economic development, which studies legal issues that arise in connection with efforts by governments, businesses, and nongovernmental organizations to foster economic development in local communities, especially low-income communities.

Northeastern Law School

P.O. Box 728
Boston, MA 02117-0728
617-373-2395; fax: 617-373-8865
email: lawadmissions@nunet.neu.edu
www.slaw.neu.edu

The three-year, full-time JD degree program offered by Northeastern Law School is the only cooperative legal education program in the United States. Through the co-op program, students have the opportunity to experience firsthand the actual practice of the law. After the first year of classroom-based legal education, all students are required to complete four, full-time legal internships using their summers and academic semesters. Co-op partners include law firms of all sizes, trial and appellate judges in federal and state courts, public defender and legal services organizations, government agencies, corporate and union legal departments, and a variety of advocacy groups. Co-op partners are located in more than forty states and countries. Students generally find their post-graduate prospects substantially enhanced through the co-op program.

Northeastern students may find internship opportunities at the Law School's URBAN LAW AND PUBLIC POLICY INSTITUTE. The Institute provides support to community organizations in Massachusetts through legal assistance, public policy advocacy, information dissemination, networking, and stimulating university-community partnerships.

Saint Louis University School of Law

3700 Lindell Boulevard
St. Louis, MO 63108
314-977-2778; fax: 314-977-3334
http://lawlib.slu.edu

Saint Louis University School of Law offers a solid range of courses relevant to community development: Housing and Community Development, Nonprofit Organizations, Land-Use Control, Landlord and Tenant, and Real Estate Transactions.

Through the HOUSING AND FINANCE LAW CLINIC, students work with local nonprofit housing developers on all aspects of housing development. Students of the clinic spearhead an annual conference on affordable housing for community groups and other nonprofit organizations.

State University of New York at Buffalo Law School

O'Brian Hall, Amherst Campus
Buffalo, NY 14260
716-645-2052; fax: 716-645-5968
www.buffalo.edu/law

The Buffalo School of Law offers a program in Affordable Housing and Community Development Law. The program consists of legal assistance to community development organizations through law school clinics, a curriculum concentration, and research and policy work. The program also houses the quarterly ABA *Journal of Affordable Housing and Community Development Law,* on whose editorial board a number of clinical students serve.

Students are required to complete fifteen credits on the community development law practice, including, for example, courses on corporations, real estate finance, poverty law, and state and local government law. Students are also required to complete nine credits in concentration courses, including work at the AFFORDABLE HOUSING DEVELOPMENT CLINIC, which provides legal counsel to community-based housing groups, and the COMMUNITY ECONOMIC DEVELOPMENT CLINIC, which works with local organizations on job retention and enterprise development strategies. Other concentration courses include those in worker ownership, tax credit finance, and finance transactions. A paper on community development is also required.

University of Maryland School of Law

500 West Baltimore Street
Baltimore, MD 21201-1786
410-706-3492
www.law.umaryland.edu

The COMMUNITY TRANSACTIONS LAW CLINIC at the University of Maryland Law School provides legal assistance to a variety of community-based nonprofit organizations engaged in community economic development across the state of Maryland. Students are closely supervised but work independently to assist clients with legal matters such as corporate formation, tax exemption, and real estate acquisition, financing, and development. In addition to working with clients an average of twenty-eight hours per week, students participate in a weekly seminar on community economic development and a weekly individual tutorial.

The Business Associations course is a prerequisite for the Community Transactions Law Clinic. Students are also encouraged to enroll in two related community economic development seminars: Affordable Housing Development and Economic Development Initiatives. Students with a long-term career interest in community economic development are encouraged to enroll in other related law school courses, including Real Estate Transactions and Tax-Exempt Organizations.

Under the Law School's Recent Graduate Mentoring Program, clinic faculty currently mentor five recent law school graduates who are developing community-based legal practices in a variety of nonprofit settings, including some who are acting as in-house counsel to experienced community development corporations.

University of Michigan Law School

539 Legal Research
Ann Arbor, MI 48109-1210
734-763-9152
and
8109 East Jefferson Avenue
Detroit, MI 48214
www.law.umich.edu

The University of Michigan Law School offers a highly regarded clinical program, the LEGAL ASSISTANCE FOR URBAN COMMUNITIES CLINIC, which combines clinical practice for law students under the supervision of a staff of three attorneys with a bimonthly seminar to discuss various substantive legal issues in community economic development. The clinic provides technical assistance to nonprofit organizations on specific community economic development projects, including housing development and corporate structuring, as well as to environmental law projects. Students earn academic credit for the clinical program.

Yale University Law School

P.O. Box 208329
New Haven, CT 06520-8329
203-432-4995
email: admissions.law@yale.edu
www.law.yale.edu

In addition to the three-year, full-time JD program, Yale Law School offers joint degree programs with most of the other schools and departments at Yale, including but not limited to the following degrees: JD and MPA; JD and MBA; JD and MA. Law students also have access to all of Yale University's schools and departments for cross-disciplinary work. Because Yale sees the study of law as fundamentally interrelated with practical experience, the school encourages special programs (including joint degree programs), intensive semester experiences outside the law school, and experience-based work with the Connecticut state legislature, local and community organizations and student-run programs.

With a strong policy focus, Yale encourages student participation in clin-

ics. Some clinics of interest are the HOUSING AND COMMUNITY DEVELOPMENT CLINIC (HCD), the PROFESSIONAL SCHOOLS NEIGHBORHOOD CLINIC, and the NONPROFIT ORGANIZATIONS CLINIC. HCD, for example, is a two-term, multidisciplinary workshop involving students from the Schools of Law, Management, and Architecture. Under the supervision of faculty and members of the local bar, participants work on behalf of nonprofit organizations, small businesses, and government agencies to promote job creation, neighborhood revitalization, low-income housing, and social service delivery in the New Haven neighborhoods.

Planning

There are two associations for planners. The American Planning Association (122 South Michigan Avenue, Suite 1600, Chicago, IL 60603, 312-431-9100) publishes comprehensive guides to undergraduate and graduate education programs. The Planners Network (c/o Pratt GCPE, 200 Willoughby Avenue, Brooklyn, NY 10205, 718-636-3461) considers itself the association that represents progressive planners; it publishes the bimonthly *Planners Network* newsletter.

California State Polytechnic University, Pomona

Department of Urban and Regional Planning
College of Environmental Design
3801 West Temple Avenue
Pomona, CA 91768
909-869-2688; fax: 909-869-4688
www.csupomona.edu/~urp

The Department of Urban and Regional Planning offers a four-year undergraduate Bachelor of Science degree in urban and regional planning that emphasizes learning by doing, including substantial fieldwork experiences. Students are required to complete a senior project based on a client-based project that they have worked on as part of two community planning studio classes.

The Master of Urban and Regional Planning, offered only as a part-time program, has been designed to meet the needs of working professionals. The multidisciplinary program consists of ten foundation core courses and at least five electives. Students are also required to select a three-course specialization module from among the following four areas: community development, environmental planning and policy, land use and design, and transportation planning and policy. Courses within the community development specialization focus on the residential community as the basic unit of a city, based on the view that by strengthening and reshaping communities, cities in turn will be strengthened.

Cleveland State University

Maxine Goodman Levin College of Urban Affairs
Urban Affairs Building
Cleveland, OH 44115-2105
216-687-2136; fax: 216-687-9291
http://urban.csuohio.edu

The Master of Urban Planning, Design and Development combines planning, design, and development to train individuals to address physical development and economic and social issues in metropolitan areas. The program includes eight required foundation and methods courses. Students may develop their own concentration or may specialize in housing and neighborhood development, economic development, public works, or the environment. Electives may be taken from any department in the University. A capstone seminar, in which students analyze a critical issue in urban problems, is required at the completion of course work. An internship, exit project, or thesis is not required but may be taken as additional electives, which are chosen in consultation with an academic advisor.

The University's Urban Center houses the CENTER FOR NEIGHBORHOOD DEVELOPMENT, which provides leadership training, data and information, and assistance to community businesses for neighborhood-based organizations. Among its programs is Neighborhood Leadership Cleveland, a free, twelve-session program designed to increase the participation and effectiveness of neighborhood residents in deciding the future of their communities.

Cornell University

College of Architecture, Art and Planning
Sibley Hall
Ithaca, NY 14853-6701
607-255-6848; fax: 607-255-6681
www.crp.cornell.edu

The Department of City and Regional Planning offers a Master of Regional Planning (MRP) degree, a two-year, sixty-credit program, of which thirty credits must be taken within the department. Each student creates an individualized academic program, based in the department's core curriculum and reflecting his or her personal goals, and works one-on-one with faculty members. Students can major in economic development. (See the "Community Economic Development Schools" section of this appendix for another Cornell master's program.)

A thesis, research paper, or project, based on summer internship experience, is required, as is one course in each of the six areas: planning institutions, quantitative methods, law, urban development, economic analysis, and project planning. The department offers matching funds to support stu-

dents' summer work. Beyond these requirements, students are encouraged to choose from departments and research centers throughout the University. MRP students may specialize in one of five concentrations—land use and urban environmental planning, international development, planning theory and practice, local and regional economic analysis, and historic preservation planning—or they may choose to develop their analytical, GIS, and research skills more generally. Within these areas of concentration, students may specialize in areas such as housing, community development, participatory planning in communities, and social and environmental policy.

To obtain a PhD, each student designs his or her distinctive program under the supervision of a special faculty committee. Students may concentrate on city and regional planning, planning theory and systems analysis, regional science, urban and regional theory, and urban planning history. Students select minor subjects from throughout the University.

Florida Atlantic University

College of Architecture, Urban and Public Affairs
220 Southeast 2nd Avenue
Fort Lauderdale, FL 33301
954-762-5652; fax: 954-762-5673
www.fau.edu/academic/ccr/urban.htm

The College offers both a Bachelor and a Master of Urban and Regional Planning. The bachelor's program is interdisciplinary and designed to give undergraduates an experiential education in community development and environmental issues, methods, and solutions. Among the requirements are three semester units of community development fieldwork experience.

The master's blends planning knowledge, skills, and values with specialized study in the areas of community and economic development, environmental planning, and/or urban development planning. Requirements include the selection of two electives that complement the area of specialization, as well as practice and synthesis courses that stress the application of academic skills and knowledge to actual planning problems. Also required are the completion of a final report on a planning-related subject, which is studied during a term of work experience in a public or private planning-related organization. Students may enroll on either a full-time or a part-time basis, with evening and summer courses also being offered to accommodate working professionals.

The University's CENTER FOR URBAN DEVELOPMENT AND EMPOWERMENT provides fieldwork opportunities for students based on its philosophy of integrative community development—the interrelatedness of different aspects of the problems and the solutions to a community's quality-of-life issues.

Georgia Institute of Technology

College of Architecture
Atlanta, GA 30332-0155
404-894-2350; fax: 404-894-1628
www.arch.gatech.edu/cp

Both a Master of City Planning (MCP) degree and a PhD in Urban and Regional Planning are offered. A joint master's degree program in city planning and public policy is also available.

The MCP requires two years of full-time course work, plus a salaried internship for one quarter in an approved planning position. A research thesis or option paper is also required. In this flexible program, half of the required curriculum is made up of core courses and the remainder may be chosen from among electives, which may be taken in other departments in the university or in affiliated universities, as well as from within a specialization area. Specializations, which include economic development, land development, transportation, land use, environmental design, and geographic information systems, may be further tailored to more specific focus areas. For example, within economic development, housing or real estate can be a sub-concentration. A field practicum is required; students are placed in jobs in low-income communities through the department's collaboration with the Community Design Center and membership in the Atlantic Outreach Consortium, a HUD university-community partnership of ten southern universities.

The PhD is a small program in which students may individualize their study of community development, taking courses anywhere in the university or from affiliated universities.

Massachusetts Institute of Technology

School of Architecture and Planning
Department of Urban Studies and Planning
77 Massachusetts Avenue
Cambridge, MA 02139
617-253-4409; fax: 617-253-2652
http://web.mit.edu/dusp/www

MIT offers a Bachelor of Science in planning, as well as a Master in City Planning (MCP). A PhD is also offered and is designed to be individualized to meet the needs of people interested in advanced academic pursuits.

The MCP is a two-year full-time program that requires core courses in planning practice, methods, and the economic and social institutions within which planners work. Students design, with faculty guidance, an individual program of study that matches their special interests and select an area of specialization from among the five major types of planning practice: hous-

ing, community and economic development (which focuses on the economic, physical, and social structure of cities and includes a subspecialization in urban schools); city design and development; environmental policy; planning support systems; and international development and regional planning.

Students are encouraged to integrate fieldwork and internships, which are optional, with academic course work. A thesis is required for graduation.

Students may pursue dual degrees in any other departments at MIT including architecture, real estate, transportation, business and management, and operations research. For example, it is possible to obtain a master's in city planning and a master's in real estate. In addition, students may cross-register for classes at Harvard and other area universities as part of their studies.

See also the "Real Estate" section of this appendix.

Michigan State University

College of Social Science
201 UPLA Building
East Lansing, MI 48824-1221
517-353-9054; fax: 517-355-7697
email: urp@msu.edu
www.ssc.msu.edu/~urp

The Bachelor of Science in urban and regional planning requires thirty-five credits in urban and regional planning courses such as graphic design, research methods, law, economics, and planning practice. In addition, students must complete a nine-month senior practicum; remaining credits are earned as guided electives in urban planning related courses or as free electives, which may be multidisciplinary.

The Master in Urban and Regional Planning is organized into three major components: (1) eight classes of core requirements, including a fieldwork practicum; (2) a concentration in economic development, urban policy, or environment and land management, in which students take at least four classes in their concentration (of which a required planning course must be one) and at least three other classes, which may be taken from departments other than planning; and (3), either a master's thesis or a course in a planning-related subject combined with writing a major research paper.

The PhD in Urban and Regional Planning is a research-oriented, multidisciplinary degree that prepares students for academic or professional urban planning practice.

New York University

Robert F. Wagner School of Public Service
4 Washington Square North
New York, NY 10003-6671
212-998-7400; fax: 212-998-7440
www.nyu.edu/wagner/

In addition to a Master of Urban Planning, the Wagner School offers a Master of Public Administration (see the "Public Policy" section of this appendix).

The Master of Urban Planning, with a concentration in urban and social development, explores the economic, social, and technological forces that affect the physical development of a community and the role of planning in such areas as community development and land use. Requirements include core courses that deal with urban planning and its political context, including a class in Measuring Social and Economic Change. A capstone experience, which includes a professional practice workshop or a final paper, is also required, along with a demonstration of computer proficiency. Five elective courses, which may be clustered as a concentration in urban, social, and economic development, are also required. Within this concentration are included courses in such areas as real estate and urban economics, housing and community development, urban and regional policy analysis, and urban economic development. Students have the option of taking their electives from within the Program in Urban Planning, from within other departments of the University, or, selectively, from other academic institutions in the New York area.

The Advanced Professional Certificate Program is an opportunity for professionals to update their knowledge and skills by taking four courses part-time, many of which are offered in the evening. Specialization modules are available in urban planning and policy (including housing and public economics) and public and nonprofit management and policy (including urban public policy and public policy analysis).

Portland State University

School of Urban Studies and Planning
P.O. Box 751
Portland, OR 97207-0751
503-725-4045; fax: 503-725-5199
email: carleen@upa.pdx.edu
www.upa.pdx.edu

Portland State offers one of the most comprehensive academic programs in the United States for studying urban planning, urban studies, and community development. The following degree programs are available: a Bachelor of Arts and a Bachelor of Science in community development, a Master of Urban and Regional Planning (MURP), a Master in Urban Studies, and three interdisciplinary doctoral degrees in which it is possible to focus on community development–related issues: a PhD in Urban Studies, which can be combined with a Master of Urban and Regional Planning; a PhD in Urban Studies and Regional Planning and a PhD in Public Adminstration and Policy.

The undergraduate BA and BS programs evolved from an earlier initiative in training community activists. These two-year interdisciplinary programs build on a required three-term core colloquium in the philosophy, theory, and methods of community development, which students begin in their junior year. Students are also required to select a concentration from one of three tracks: community organization and change; housing and economic development; and communication and community development. Within each concentration they must take an additional six urban planning–related courses, as well as at least four other elective classes, and complete a senior field placement.

The MURP includes eight core foundation courses and the selection of an area of specialization from one of the following: community development, urban and regional analysis, policy planning and administration, land use, environment, and urban transportation. Within the community development specialization, the focus is on the economic, social, and physical needs of neighborhoods and smaller cities. Special attention is given to understanding social and political dynamics and to developing skills in such areas as citizen organizing, participatory planning processes, housing, and economic development.

Pratt Institute

Graduate Center for Planning and the Environment
200 Willoughby Avenue
Brooklyn, NY 11205
718-399-4314; fax: 718-399-4332
email: info@pratt.edu
www.pratt.edu/arch/gcpe

The GCPE is a leading international center for the study of community-based and participatory planning. Courses are offered primarily in the evening and on weekends, giving students flexibility to work full time. GCPE offers two degree programs: a sixty-three-credit Master of Science in city and regional planning, which includes three concentrations, in community development, preservation planning, and environmental planning; and the forty-credit Master of Science in environmental planning. Undergraduates may take a combined program in architecture and planning, and a joint degree program in planning and law is offered in conjunction with Brooklyn

Law School.

The MS in city and regional planning program seeks to develop the student's understanding of community development, land use and environmental issues, and various political and social factors that influence the professional environment for planners. The required capstone course may take the form of work related to a specific course, an extension of the student's employment, or a thesis. Employment and internships in community organizations and associated planning practices are considered an important component of the program.

Within the concentration in community development, nine elective courses are offered, ranging from Uprooting Poverty from a Community Perspective to Planning for Sustainable Communities.

The Pratt Graduate Center is the home of the CONSORTIUM FOR SUSTAINABLE COMMUNITY DEVELOPMENT AND PLANNING, which includes planning schools at Cornell and UCLA, three schools in Canada, and three in Mexico. The affiliated nonprofit PRATT INSTITUTE FOR COMMUNITY AND ENVIRONMENTAL DEVELOPMENT provides student internship opportunities in training, technical assistance, and information services with community-based groups addressing urban and environmental issues in low- and moderate-income neighborhoods.

Rutgers University

Edward J. Bloustein School of Planning and Public Policy
Civic Square 33 Livingston Avenue
New Brunswick, NJ 08901-5475
732-932-3822; fax: 732-932-2253
email: yjohnson@rci.rutgers.edu
www.policy.rutgers.edu/uppd

The Department of Urban Planning and Policy Development, within the School of Planning and Public Policy, offers a Master of City and Regional Planning (MCRP) program. The core curriculum includes Development and Theory of Urban Planning, Urban Economy and Spatial Patterns, Planning Studio, Methods of Planning Analysis, and Survey of Planning Law Principles. Areas of concentration related to economic and community development are housing and real estate, regional development and developing nations, and urban and community development. Urban and community development is a particularly strong concentration, integrating theory, practice, and policy development through studio courses that allow students to work directly with urban communities. The interests of students within the concentration include downtown redevelopment, community revitalization, urban poverty, economic development, and housing.

Associated with the School of Planning and Public Policy are the CENTER FOR URBAN POLICY RESEARCH (CUPR) and the AMERICAN AFFORDABLE HOUSING INSTITUTE, which provide students with possible research and other work or in-

ternship opportunities. In particular, Project Community at CUPR provides links between community-based organizations and nonprofits in New Jersey and the Rutgers community.

Tufts University

Department of Urban and Environmental Policy
97 Talbot Avenue
Medford, MA 02155
617-627-3394; fax: 617-627-3377
email: aurosei@tufts.edu
www.tufts.edu /as/uep/

Tufts offers a full- and part-time Master of Arts in urban and environmental planning, with a special focus on the role of nonprofit and community-based organizations. Combined degree programs are available between urban and environmental planning and biology, child development, civil and environmental engineering, and economics. A dual degree program is also available with the Fletcher School of Law and Diplomacy.

The MA in urban and environmental planning offers two broad areas of concentration that students must choose from: urban and social policy and environmental policy. Students develop a multidisciplinary program of study in one of these areas. Courses include the Foundations of Public Policy, Economics for Planning and Policy Analysis, and Program Evaluation. Courses are also taken in the areas of specialization, which, for the concentration in urban and social policy, are: Community Development and Housing, Social Welfare Policy, and Child and Family Policy. A field project and an internship related to the specialization are required, as is an advanced seminar that integrates course work and fieldwork. A thesis must also be completed.

Three certificate programs for the working professional are offered, in Management of Community Organizations, Program Evaluation, and Community Environmental Studies, which emphasize participatory strategies for community self-determination and sustainability. To earn a certificate, students must successfully complete four courses from the master's program for credit or three courses and a certificate field project.

The LINCOLN FILENE CENTER FOR CITIZENSHIP AND PUBLIC AFFAIRS promotes active citizenship and community leadership. Its activities include a yearly training conference on community development and providing educational and employment opportunities for graduate and undergraduate students.

University of Arizona

College of Architecture
Tucson, AZ 85721
520-621-6751; fax: 520-521-8700
http://CAPLA.arizona.edu/planning

The College of Architecture offers a multidisciplinary two-year Master of Science in Planning degree. The University's desert location and close proximity to Mexico and Latin America have enabled it to develop a special expertise on border issues, the Southwest, and planning with desert communities. The curriculum includes core classes such as Ecosystemology, Planning the Built Environment, Financing Public Services, Analytic Methods in Planning and Management, and Land-Use Planning Law.

Students select one concentration from three areas, one of which is land use and community development. This concentration focuses on how land use affects the social and built environments. Required core classes include Urban Social Issues, Land Use and Growth Controls, Public and Policy Economics, and Financing Public Services. An internship is required. Joint programs with Mexico and Panama offer the opportunity for international internship experiences.

THE ROY B. DRACHMAN INSTITUTE for Land and Regional Development, as part of the University of Arizona's Planning Program, offers opportunities to students for internships working on planning issues with community groups. The Institute provides technical assistance and support to urban communities around both design and development issues.

University of California at Berkeley

College of Environmental Design
316 Wurster Hall
Berkeley, CA 94720-1850
510-642-3256; fax: 510-642-1641
www.dcrp.ced.berkeley.edu

The University offers both an accredited Master of City and Regional Planning (MCRP) and a research-oriented PhD in City and Regional Planning, in which community development is an area of specialization. The MCRP is a two-year program with a curriculum that includes at least six core requirements in the history, institutions, economics, research, and analytic methods of planning. Also required is a studio workshop on a client-based planning project. Students select one concentration area to specialize in from among the following: community development and services, regional planning and development, housing and project development, urban design, transportation policy and planning, and environmental planning and policy. Within the community development and services concentration, emphasis is placed on "community building," a comprehensive view of neighborhood revitalization that stresses the need for links among strategies to improve the physical, social, cultural, and economic components of neighborhood life. Students learn the basic skills necessary to empower local communities, urban or rural; provide planning and design services to the disenfranchised; assist the poor in getting an equitable share of built en-

vironment and public social services; and understand how to form innovative public-private ventures and partnerships. An internship or a professional report is also required.

Through the UNIVERSITY-OAKLAND METROPOLITAN FORUM, a college-community partnership that is administered by the University's Institute of Urban and Regional Development, students can find internship opportunities with nonprofit community development organizations. Students who are selected for the Community Development Work Study Program receive financial support for two years while they work in Forum-supervised community internships.

University of California at Irvine

School of Social Ecology
Irvine, CA 92697-7075
949-824-3480; fax: 949-824-2056
email: ksadler@uci.edu
www.seweb.uci.edu

The Department of Urban and Regional Planning, located within the School of Social Ecology, follows an interdisciplinary approach to community development that combines social, ecological, economic, behavioral, health, and policy perspectives. It offers both a Master and a PhD in Urban and Regional Planning.

The MURP curriculum includes twelve required core courses and thirteen elective classes as the foundation to address the complexities of community development and urban change. Seven specializations are offered: community development and social policy, land-use policy, environmental policy and planning, economics and public policy, transportation policy, community health planning, and urban design and behavior. The community development and social policy specialization provides concepts for analyzing various dimensions of community development and strategies for developing and implementing effective social policy. In this specialization, course work focuses on the connection between community development and (1) urban/suburban development; (2) housing and human services; and (3) capacity building and leadership development. Internships are not required but are strongly encouraged.

The requirements for the PhD include eight required courses, a comprehensive examination, completion of both a pre-dissertation research project and a dissertation, and four elective classes. Students have the option of using their electives to acquire a generalist background or to focus on particular planning issues. Areas of specialization are the same as for the master's, with the addition of one: urban design and well-being.

University of California at Los Angeles

School of Public Policy and Social Research
3250 Public Policy Building
Los Angeles, CA 90095-1656
310-825-4025; fax: 310-206-5566
email: upinfo@sppsr.ucla.edu
www.sppsr.ucla.edu/dup/home_dup.htm

Housed within the School of Public Policy and Social Research, the Department of Urban Planning offers a Master of Arts in Urban Planning and a research-oriented PhD in Urban Planning, using a multidisciplinary approach to community development and urban planning that gives special attention to the issues of multiculturalism, race, and gender. Joint degree programs are also offered between urban planning and law, architecture, Latin American studies, and business.

The MA is a two-year, full-time program of eighteen courses, seven of which are required core foundation classes in the history, theory, and methods of planning. Students select one area of concentration, within which they take at least five courses, from among the following: community development and the built environment; environmental analysis and policy; regional and international development; and social policy and analysis. The community development and built environment concentration includes specializations in either community planning and development or physical development and public policy. For students who choose to emphasize community planning and development, classes and projects focus on community economic development, real estate development, community planning, and housing policy. Students may also design their own areas of concentration. Fieldwork is required, depending on the extent of a student's professional experience. Students must complete a thesis, take a comprehensive exam, or do a client project.

The PhD curriculum includes Community Development: Social, Economic, and Physical, as one of the major field areas in which students can specialize in. THE CENTER FOR COMMUNITY PARTNERSHIPS provides students at the master's and doctoral levels with challenging opportunities for internships working with local communities doing technical assistance and training on such issues as housing and community planning and community economic development.

University of Cincinnati

School of Planning
P.O. Box 210016
Cincinnati, OH 45221-0016
513-556-4943; fax: 513-556-1274

email: connie.dean@uc.edu
www.daap.uc.edu/planning

The School of Planning offers five degrees: Bachelor of Arts in urban planning, Bachelor of Science in urban studies, Undergraduate Certificate in planning, Master of Community Planning (MCP), and mid-career Master of Urban Management.

The five-year Bachelor of Arts in urban planning is the only cooperative education program of its kind in the United States. By alternating three-month professional practice assignments with three-month academic periods, students graduate with six quarters of professional experience. The four-year Bachelor of Science in urban studies is a broad-based, liberal arts program.

The MCP, which requires seven full-time quarters of study, prepares students to enter the professional practice of planning in government and within the private sector (e.g., land developers, law firms, economic development agencies). Urban design and economic and/or international development are two possible paths of specialization. A graduate project and an internship with a planning organization are required for completion of the MCP. Joint degrees are offered with the College of Law, the Department of Geography, and the Public Administration Program.

University of Florida

College of Architecture
P.O. Box 115706
Gainesville, FL 32611-5706
352-392- 0997; fax: 352-392-3308
www.arch.ufl.edu

The College offers a Master of Arts in Urban and Regional Planning (MAURP) as well as a PhD degree. The MAURP curriculum includes ten core foundation courses in the history, theory, ethics, and methods of planning. Students also select an area of specialization, within which they take at least three classes. They have the option of designing their own or choosing one from among the following: housing, community, and economic development; growth management; real estate development; urban design; transportation; preservation of historic districts; and planning and geographic information systems. Also required are a professional internship, usually completed in the summer, and two studio courses, which give students experience working, sometimes in multidisciplinary teams, on a real-world planning project. A thesis or a terminal project is also required.

The research-oriented PhD offers three areas of specialization: urban and regional planning, architectural sciences, and building construction sciences. Specializations in community development and housing are also available.

The Department of Urban and Regional Planning has two research centers: THE COMMUNITY REDEVELOPMENT CENTER, which conducts urban revitalization projects, a source of student community development research and field internships; and the GEOFACILITIES PLANNING AND INFORMATION CENTER, which specializes in the use of computer technology to solve planning problems.

University of Illinois at Chicago

College of Urban Planning and Public Affairs
412 South Peoria
Chicago, IL 60607-7065
312-413-8088; fax: 312-413-2314
www.uic.edu/cuppa/upp.

The School of Urban Planning and Public Affairs offers a Master's in Urban Planning and Policy (MURP) and a PhD in policy analysis, as well as a professional certificate through the Urban Developers Program and fieldwork opportunities through the University's many research centers.

The master's includes core and methods courses in planning, specializations, electives, an internship, and a major thesis or project. A specialization from one of the following five areas must be completed: community development, in which students review current theories about local organizing, asset management, citizen participation, ethnic and racial relations, and government development policy, and learn the arts of political communication, neighborhood planning, equity planning, and consensus building at the grassroots level; economic development; international development; physical planning; and urban transportation. There is room in the curriculum to pursue two specializations, if desired.

The PhD in policy analysis offers a core program in advanced theory and methods focusing on the impact of public policy on urban areas.

The Urban Developers Certificate program is a collaboration between the Chicago Rehab Network, a group of experienced community development practitioners, and the College of Urban Planning and Public Affairs. This year-long program combines training in nonprofit community-based development with management and leadership skills. Credits may be applied toward a master's degree in urban planning and policy. Topics include: Nonprofit Financial Management Skills, Development Finance, Development Process, Asset Management, Resident Issues, Essentials for Community-Based Development.

The College of Urban Planning and Public Affairs has six research centers in which urban planning students can gain internship and research experience: the CENTER FOR URBAN ECONOMIC DEVELOPMENT, which addresses the economic needs of Chicago and other urban areas; the VORHEES CENTER FOR NEIGHBORHOOD AND COMMUNITY IMPROVEMENT, which is within the Center for Urban Economic Development and provides technical assistance and research for

community organizations and coalitions in the Chicago area; the GREAT CITIES INSTITUTE, which focuses on the integration of disciplines relevant to urban issues; the INSTITUTE FOR RESEARCH ON RACE AND PUBLIC POLICY; the SURVEY RESEARCH LABORATORY; and the URBAN TRANSPORTATION CENTER.

University of Illinois at Urbana-Champaign

College of Fine and Applied Arts
111 Temple Buell Hall, 611 Taft Drive
Champaign, IL 61820
217-333-3890; fax: 217-244-1717
email: j-terry@uiuc.edu
www.urban.uiuc.edu

The College offers three degrees: a bachelor's, a master's, and a research-oriented PhD in urban planning, all of which are interdisciplinary and include a concentration and fieldwork experience in community development.

The MUP curriculum includes seven foundation courses in planning theory and methods, the choice of up to five units of electives, and the option to take at least three classes in one concentration or to design a specialization to fit specific interests. Concentration areas include community and economic planning, environmental science and sustainable development, international planning, land use and infrastructure, planning analysis and information systems, and preservation planning. A capstone class, in which students work on an actual client's planning project, is also required.

An internship in planning is not required but is encouraged. Students may either complete a thesis or a master's project and workshop, which may be in a student's area of concentration. Fieldwork placements are available for both undergraduate and graduate students through the EAST ST. LOUIS ACTION RESEARCH PROJECT, a collaborative effort between the faculty and students of the University of Illinois at Urbana-Champaign and East St. Louis neighborhood groups. The Project enables students to participate in service learning and research by working collaboratively with community residents who are mobilizing to address the immediate and long-term social, economic, and environmental needs of some of the city's poorest communities.

University of Maryland at College Park

Urban Studies and Planning Program
0100 C Caroline Hall
College Park, MD 20742-9150
301-405-6790; fax: 301-314-9897
email: urspweb@bss2.umnd.edu
www.bsos.umd.edu/ursp/
The Urban Studies and Planning (URSP) program offers a Master of

Community Planning (MCP) with an emphasis on community involvement in the entire planning process. The two-year program combines required courses in the concepts, process, context, and practice of planning, with the flexibility to specialize in one of the following areas: economic development, housing, land use and environmental policy, social planning, organization, and administration, and urban design.

The fifty-one-credit program includes both a workshop and a one-semester fieldwork placement in which students participate in a wide variety of supervised planning projects in the community and in weekly seminars to discuss the practical, theoretical, professional, and ethical issues that arise during field work.

A part-time evening program is also available and may be completed in up to five years.

University of Massachusetts at Amherst

Department of Landscape Architecture and Regional Planning
109 Hills North
Amherst, MA 01003
413-545-6635; fax: 413-545-1772
www.umass.edu/larp

The department confers both a Master of Regional Planning (MRP) and a PhD. A dual degree program with a Master of Landscape Architecture is also offered. Core requirements include Planning History and Theory, Quantitative Methods, Law, and Economic and Social Planning Analysis. In planning studios, students gain experience working on a real-life planning problem with a client. Students also elect an area of concentration: (1) economic and community planning, in which they study industrial locations, regional analysis, social planning and social impact assessment, public and private finance, land-use planning, and spatial analysis; (2) built environment; (3) environmental policy and planning; (4) information technology in planning; or (5) landscape planning.

A doctoral program in regional planning leads to a research degree for students interested in careers in the academic world or in research in public agencies or private corporations.

THE CENTER FOR ECONOMIC DEVELOPMENT provides students with opportunities for applied research on such issues as brownfield sites and rural development. The Urban Places Project provides urban design and neighborhood planning services to low-income neighborhoods in mid-sized cities that have not traditionally had access to design and physical planning assistance.

University of Massachusetts at Boston

College of Public and Community Service
100 Morrissey Boulevard
Boston, MA 02125-3393
617-287-6000; fax: 617-287-7274
www.cpcs.umb.edu

The College actively recruits members of communities that have traditionally experienced limited access to higher education and inadequate services, with the goal of enabling these individuals to become both service providers and active participants in the development of their communities. Through the Center for Community Planning, one of seven academic centers housed within the College of Public and Community Service (CPCS), students may earn a Bachelor of Arts and/or a Certificate in Community Planning. To earn the BA, students must complete a self-assessment program, as well as general education, applied language, and mathematics certificates, and must also earn the certificate. Requirements for the certificate include the completion of core competencies (the basic units of academic credit at CPCS that are used in the place of grades). These competencies emphasize direct, hands-on experience in community and agency situations and recognize a specified set of skills and knowledge that demonstrate the ability to put knowledge into practice. Among the required courses are: Analyzing Organizing Strategies, Community Needs Analysis, Strategies for Economic Development, and Planning Models and Theories.

Opportunities for student internships are available through the COLLABORATIVE FOR COMMUNITY SERVICE AND DEVELOPMENT, which offers learning and research projects that bring together teams of students and faculty to work with community groups in the Greater Boston area.

University of Michigan

College of Architecture and Urban Planning
2000 Bonisteel Boulevard
Ann Arbor, MI 48109-2069
734-764-1298; fax: 734-764-1298
email: pdunlap@umich.edu
www.caup.umich.edu

The College offers a two-year, full-time Master in Urban Planning (MUP) degree as well as a PhD in Urban, Technological, and Environmental Planning. Both programs focus on the interrelatedness of the physical, social, economic, and political concerns of urban regions.

Areas of concentration within the MUP degree include: development planning, which includes subconcentrations in economic development, community development and housing, and international development and housing; and physical planning and urban design.

A major emphasis is placed on learning through community service. While there is no required internship, students are encouraged to work on community development projects, which are available full-time during the summer and as both part-time and integrated fieldwork courses during the year. Service learning opportunities are offered to students through both the University's Center for Learning Through Community Service and the DETROIT COMMUNITY OUTREACH PARTNERSHIP CENTER, a partnership between the University of Michigan and thirteen community-based organizations in Detroit.

Dual degree programs are available between the School of Urban Planning and the Schools of Social Work (Community Planning) and Architecture.

University of Minnesota

Hubert H. Humphrey Institute of Public Affairs
301 19th Avenue South
Minneapolis, MN 55455
612-624-5003; fax: 612-625-3513
email: egoetz@hhh.umn.edu
www.hhh.umn.edu/gpo/degrees/murp

The topic of housing, community development, and social policy is one of the strong focuses of the University's Master of Urban and Regional Planning. It is one of the required core courses as well as an elective concentration. The forty-eight-credit master's includes 12.5 credits in core planning courses (including the housing, community development, and social policy course); in addition, the program calls for twelve credits from Humphrey Institute core courses, such as Urban Economic Theory and Urban Structure. A three-credit capstone workshop is another requirement. Each student must also complete an internship, normally in the summer after the first year of the program, by working for a minimum of four hundred hours in a private or public agency. The Institute has strong connections with a large number of CDCs and other agencies in the Twin Cities area.

Students are encouraged to explore specialized concentrations and can design their own. A concentration in housing, community development, and social policy might include courses, for example, in American cities, private-sector development, community organization and advocacy, and race relations theory. Economic development is another concentration offered, with courses ranging from community economic development to strategies for sustainable development.

See also the "Public Policy" section of this appendix.

University of New Mexico

School of Architecture and Planning
2414 Central Avenue, Southeast.
Albuquerque, NM 87131-0002
505-277-5050; fax: 505-277-0076
email: crp@unm.edu
http:saap.unm.edu

The Community and Regional Planning Program, considered a national leader in progressive, community-driven planning education, offers a full-time and a part-time Master in Community and Regional Planning (MCRP) degree program. Joint degrees are also available: Master of Arts in Latin American studies, Master of Public Administration, Master of Architecture. .

The MCRP emphasizes building community at a local scale and within regions. The program uses small classes and team-based, case-based learning. The full-time two-year program requires fifty credit hours of study, twenty-seven of which must be taken within the Community and Regional Planning Program. Students select one emphasis consistent with their professional objectives in either rural and urban community development or natural resources and environmental planning. A methods course and a foundations course specific to each emphasis, as well as a capstone studio, where students apply planning skills to real community problems, are all required. Students must also complete a thesis or professional project.

The rural and urban community development courses cover: the economic, political, and social aspects of planning in urban and rural settings; the foundations for the preparation, development, and implementation of community and regional plans; understanding sustainable community development within a regional framework; and the concepts and skills necessary to assist and advocate on behalf of communities and to build their capacities to be able to advocate for themselves.

The natural resources and environmental planning emphasis analyzes community development issues in terms of their implications for the natural systems that support them, focusing on sustainable issues of land and water, access to and control of natural resources, and their implications for growth management.

The Master of Public Administration combined with the MCRP is for students who desire public-sector careers requiring the skills of both a planner and an administrator.

The Master of Architecture combined with the MCRP is designed for students who wish to pursue a design-oriented practice along with the policy orientation of community planning. Students entering this dual degree program must have a BA or BS in architecture or a Bachelor of Architecture degree.

University of North Carolina at Chapel Hill

Department of City and Regional Planning
CB 3140
Chapel Hill, NC 27599
919-962-3983; fax: 919-962-2518
www.unc.edu/depts/dcrp

One of the oldest and largest programs of graduate planning study and research in the United States, the Department of City and Regional Planning offers a two-year Master of Regional Planning program and a doctoral program. Dual degrees may be pursued, in planning and law, planning and business, planning and public administration.

The Master of Regional Planning program's core curriculum covers planning theory, urban spatial structure, quantitative methods, economics, a specialization-related course in law, and a capstone workshop in which students apply professional skills to the real-life problems of a client. Students must also complete a master's project. Four courses are taken in an area of specialization; electives may be taken anywhere in the University or in affiliated universities. Specializations are chosen from one of three focus areas, all of which allow for further specialization:

1. Economic development, which develops the student's understanding of policy approaches and professional tools needed for implementing local and regional economies. Three areas of further specialization are local economic development planning, development finance, and infrastructure planning for economic development.
2. Housing, real estate, and community development, which covers the development of subsidized and market-rate residential and commercial properties, and the revitalization of urban neighborhoods. Further specialization can be pursued in housing and community development, covering the theory, method, and practice of increasing the supply of affordable housing and revitalizing urban neighborhoods; or real estate development, which in collaboration with the Kenan-Flaglar Business School, focuses on feasibility and cost-benefit analysis of development projects and examines the regulatory process and public values associated with regulation.
3. Land use, transportation, and environmental planning, which includes specializations in land use and growth management, environmental planning, coastal management, and transportation planning.

In the doctoral program, a mentoring model is used: each student is paired with a faculty member. The program is individualized for each student, and students may take courses in any department in the University or in other affiliated universities.

University of Oregon

School of Architecture and Allied Arts
1209 University of Oregon
Eugene, OR 97403-1209
541-346-3635; fax: 541-346-2040
http://laz.uoregon.edu

The School offers an undergraduate BA or BS degree in planning, public policy, and management, as well as a Master in Community and Regional

Planning (MCRP). A PhD in environmental studies is also available.

The undergraduate program in planning, public policy, and management is action oriented, multidisciplinary, and problem focused. It is open to upper-division students. Program requirements include five core foundation courses and the choice of an individualized area of concentration. Students are also required to complete an internship in their area of concentration. An undergraduate minor in planning, public policy, and management is available to students majoring in social science and the humanities.

The Master in Community and Regional Planning curriculum combines applied, problem-solving, field-based graduate course work with research opportunities. Special emphasis is placed on issues of significance to the communities and regions of the Pacific Northwest and the Pacific Basin. Requirements include seven core planning courses and the selection of one concentration, which may be either individualized or chosen from the following three: community and regional planning (addresses the interrelated economic, social, and environmental issues facing communities and regions); environmental planning (concerned with protecting, enhancing, and sustaining natural environments); social planning (focuses on the interface between social relations and the spatial-physical environment).

Students must also complete a two-term, experiential learning Community Planning Workshop. This workshop is one of the many technical and planning assistance programs offered to communities, agencies, and organizations across Oregon by the University's COMMUNITY SERVICE CENTER. A thesis or the combination of a student research colloquium and terminal project is also required.

University of Pennsylvania

Graduate School of Fine Arts
Room 127, Meyerson Hall
210 South 34th Street
Philadelphia, PA 19104-6311
215-898-8329; fax: 215-898-5731
www.upenn.edu/gsfa

The Department of City and Regional Planning offers a two-year, sixteen-course Master of City Planning program. Themes covered in the program include the physical city, global urbanization, new planning tools, the impact of technology, and community planning. Required courses, taken chiefly during the first year, range from urban economic analysis to the law of planning and urban development. Four practicums (for example, familiarity with GIS systems) to strengthen practical skills are also required. Students select an area of concentration, of which economic and community development is one; among the courses for this concentration are sustainable development, citizen participation, and entrepreneurial inner-city housing.

A summer internship in a planning office is also required. The entire spring semester of the final year is given to a planning studio, in which students work on a planning project within an actual community. Students may also pursue a dual degree.

A PhD degree, under an interdisciplinary faculty, is also offered, requiring twenty courses.

The Department also offers a post-master's certificate program in planning that permits students to concentrate in one area of interest. Housing and community development is one of the concentrations available; international planning, enabling a student to spend the third or fifth semester studying abroad, is another. A certificate is also available in urban design or real estate design and development.

University of Southern California

School of Policy, Planning, and Development
Von KleinSmid Center, Room 232
Los Angeles, CA 90089-0626
213-740-6842; fax: 213-740-7573
email: sppd@usc.edu
www.usc.edu/schools/sppd

The School of Policy, Planning, and Development combines the School of Public Administration and the School of Urban Planning and Development. (See the "Business, Nonprofit Management, and Public Administration" section of this appendix.)

The Master of Planning program requires forty-eight units and usually takes two years, full-time. The curriculum supports four broad concentrations; these include community and local economic development, and social and environmental planning. The community and local economic development concentration includes courses such as Finance of Real Estate Development, Planning and Economic Development, and Urban Economic Analysis.

There are seven required core courses, including Planning Theory, the Social Context of Planning, and the Urban Economy. Two laboratory-workshop courses are required, linking academic education with the world of practice, typically with real clients. Each summer the school offers one or more international workshops. A comprehensive examination is required, as well as an internship of at least four hundred hours in a planning or planning-related position.

University of Virginia

Graduate School of Architecture
Campbell Hall
Charlottesville, VA 22903

804-924-6442
http://minerva.acc.virginia.edu/~arch/

The Graduate School of Architecture offers a two-year, full-time Master of Planning (MP) program. Concentration areas in the MP program include housing and community development, environmental protection and conservation, land use and growth management, and policy analysis and management. Students are free to develop their own concentrations, however.

In their concentration areas, students are expected to develop a specialty through course work and a mandatory summer internship. Joint degree programs are available with the Graduate School of Arts and Sciences, the School of Law, the Darden Graduate School of Business Administration, and the School of Graduate Engineering and Applied Science.

The research and educational resources of the INSTITUTE FOR ENVIRONMENTAL NEGOTIATION AND INSTITUTE FOR SUSTAINABLE DESIGN are available to University of Virginia students. Through graduate research assistantships, the Institute for Environmental Negotiation trains students in negotiation, mediation, and other interactive approaches to planning and policy making. Students are welcome to the projects, seminars, and conferences of the Institute for Sustainable Design, which sponsors research on sustainable designs for products, buildings, neighborhoods, communities, regions, and the global environment.

University of Washington

College of Architecture and Urban Planning
410 Gould Hall
Box 355740
Seattle, WA 98195-5740
206-543-4190; fax: 206-685-9597
www.caup.washington.edu/html.urbdp

The Department of Urban Design and Planning offers a self-directed undergraduate major in community and environmental planning, and a Master of Urban Planning (MUP). Begun in 1995 by a group of students who decided they wanted to start their own major, the undergraduate program was launched as a collaborative effort between faculty and students. Open to students in their junior year, the program emphasizes the education of the whole person, with each student assuming responsibility for his or her learning experience. Within the interdisciplinary major, students can take any course they want, but they must bring back what they learn to core-required community and environmental planning (CEP) seminars; these seminars focus on the topics of The Idea of Community, Environmental Response, Social Structures and Process, Planning in Context, Ethics and Identity, and Community and Environment. Leadership of the CEP program is interactive and by committee, where everyone shares responsibility for the direction of the learning community. The requirements for the program, besides the core seminars, include the student's individualized study plan and at least five methods courses selected from such areas as quantitative and qualitative reasoning and group dynamics and facilitation. Senior year internships are required. Comprehensive evaluations are given in the place of grades.

The Master of Urban Planning program is a two-year, accredited program. Students are each assigned both a faculty advisor and a mentor from Seattle's professional community to help them develop an education program and to establish connections with the professional community. Requirements include seven core foundation courses and a master's thesis. Students may select a specialization, including urban development and housing, which includes issues of community economic development, policy, and physical development and finance. Internships are optional but strongly encouraged.

University of Wisconsin at Milwaukee

School of Architecture and Urban Planning
P.O. Box 413
Milwaukee, WI 53201
414-229-4014; fax: 414-229-6976
email: sbs@uwm.edu
www.sarup.uwm.edu

The School offers a two-year, forty-eight-credit Master of Urban Planning (MUP) program dedicated primarily to policy and economic development planning. In addition to studying planning theories and methods, students acquire knowledge in economics, finance, housing, land use, the environment, energy, and transportation. Land-use planning and economic planning are two possible areas of focus. The department runs an internship program that awards credit for work in urban planning–related organizations, as well as a mentorship program that matches students with area professionals. The program also offers joint degrees in architecture, transportation engineering, and public administration and certificates in urban planning (for undergraduates) and geographic information systems (for graduate students).

Wayne State University

College of Urban, Labor and Metropolitan Affairs
Faulty Administration Building, Room 3198
225 State Hall
Detroit, MI 48202
313-577-2701; fax: 313-577-8800
www.culma.wayne.edu

The Department of Geography and Urban Planning offers a two-year, full-time or evening part-time Master in Urban Planning with a principal orientation toward community planning. Using its location in aging central-city Detroit as its primary laboratory, this program offers a broad-based, multidisciplinary, and flexible curriculum. Of the forty-eight required credits, twenty are electives that allow students to explore various approaches to community development. The curriculum involves taking seven core foundation courses in the nature and process of planning, completing at least two electives that may be in other disciplines, participating in a two-course capstone in which students work with a neighborhood client on a planning issue, and the completion of a final paper, which may be a three-credit master's essay or an eight-credit master's thesis. Students must also select one area of concentration in which at least three classes must be taken. Choices are: 1. Housing and urban development, including classes on urban poverty and racial segregation. Students in this concentration may also participate in the Urban Linkages Internship Program of the CENTER FOR URBAN STUDIES. 2. Urban economic development with classes in state and local public finance; regional, state, urban economic development; labor markets; location theory; and regional economics. 3. Planning and public policy, with classes such as State and Local Finance and Comparative Public Administration.

Public Policy

The professional association for the discipline of public policy is the Association of Public Policy Analysis, P.O. Box 18766, Washington, DC 20036-8766, 202-261-5788.

Carnegie Mellon University

H. John Heinz III School of Public Policy and Management
5000 Forbes Avenue
Pittsburgh, PA 15213-3890
412-268-2164; fax: 412-268-5161
www.heinz.cmu.edu

The School offers an innovative Master of Science in Public Policy and Management (MSPPM), with a concentration in economic development and urban planning, and a one-year, mid-career Master of Public Management (MPM) (see the "Business, Nonprofit Management, and Public Administration" section of this appendix.) A PhD in public policy and management, which focuses on advancing the quality of the research and analysis of issues such as community and economic development, is also available.

The core curriculum of the MSPPM requires courses in policy analysis, management, political science, management information systems, and finance. Also required is a year-long project application course in which team of students, guided by advisory boards of professionals in the field, develop an interdisciplinary approach to understanding a public interest problem and structuring systematic solutions. Students may either specialize in a pre-structured concentration area or design their own interest area. Pre-structured concentration areas include economic development and planning, management information systems, and policy analysis. Within the economic development and planning concentration, among the courses students can select from are: Financing Economic Development and Urban and Regional Economic Development A summer internship in a professional organization is required after students complete their first year of study.

Harvard University

John F. Kennedy School of Government
79 JFK Street
Cambridge, MA 02138
617-495-1155
email: ksg_admissions@harvard.edu
www.ksg.harvard.edu

The Kennedy School of Government (KSG) offers programs in Master of Public Policy (MPP), Master of Public Administration (MPA), and Mid-Career Master of Public Administration (MMPA). Most students are admitted to the two-year, full-time MPP, a program that requires eighteen academic credit hours and a policy analysis exercise (PAE). In the PAE, taken in the second year, students examine existing public- or nonprofit-sector problems and produce real-world solutions. A small number of students are admitted to the two-year, full-time, sixteen-credit MPA program. The MPA program's curriculum is tailored to the needs of individual students with greater experience or education in specific career fields. The full-time, one-year MMPA program is intended for professionals with at least seven years of experience in a related field.

Fields of concentration most relevant to community and economic development careers are housing and economic development and urban economic development. Concentrations such as international development and nonprofit management may also be of interest. Although not specifically known for its community development curriculum, the KSG allows students the resources and flexibility to tailor programs for community and economic development careers. KSG students are allowed to cross-register at all other Harvard schools, MIT, and Tufts. For example, KSG students may take courses at the Harvard Graduate School of Design (GSD), which offers a Master in Urban Planning that focuses on the technical aspects of the built environment.

Associated with the Kennedy School are the HAUSER CENTER FOR NONPROFIT ORGANIZATIONS, the JOINT CENTER FOR HOUSING STUDIES, and the UNIT FOR HOUSING AND URBANIZATION. The Unit in particular is internationally recognized for its work promoting sustainable urban development and helping municipalities and community-based organizations structure public-private partnerships for urban development projects and community-based development initiatives. The three centers offer students opportunities for seminars, conferences, research assistantships, and mentorships.

New School University

Robert J. Milano Graduate School of Management and Urban Policy
66 Fifth Avenue, 7th Floor
New York, NY 10011
212-229-5462
www.newschool.edu/milano

The New School offers a Master of Science degree in urban policy analysis and management and a PhD in public and urban policy. Additional Master of Science degree programs are available in nonprofit management (see the section on "Business, Nonprofit Management, and Public Administration" in this appendix), health services management and policy, and human resources management.

The Master of Science in urban policy curriculum includes required courses in policy analysis, economic analysis, quantitative methods, management and organizational behavior, public economics, and other courses. Students have the option of either taking a generalist approach to their education or taking a clustering of elective courses in concentration areas. Concentration areas are: (1) housing and community development, which focuses on innovative approaches for improving the quality of life within low-income urban communities, including affordable housing and workforce; (2) urban development policy, which examines, among other things, urban labor markets, public policies and institutions designed to improve outcomes for the less advantaged, discrimination, and the changing character and effects of immigration; (3) social policy, which focuses on the policy, management, and programmatic issues professionals encounter in working to improve the economic well-being and social functioning of diverse groups—an advanced seminar is also required, as is the submission of a professional decision report in which the student analyzes an important professional issue and recommends a specific response; and (4) public management.

The PhD in public and urban policy, a multidisciplinary doctoral program, is designed to train specialists particularly from those groups (for example, African-Americans and Latinos) that are still largely underrepresented in academic institutions and in the higher levels of public policy professional fields. Particular attention is given to public policy issues of the New York metropolitan area, including housing and community development, economic development, employment and training, social policy, and health care. Concentrations include health policy, economic development, housing, community development, and social policy.

The COMMUNITY DEVELOPMENT RESEARCH CENTER, housed within the Milano School, offers students opportunities for research and internships working on various aspects of community development and policy issues. The Center provides publications, engages in research, and promotes cutting-edge strategies for community-based development.

New York University

Robert F. Wagner School of Public Service
4 Washington Square North
New York, NY 10003
212-998-7400; fax: 212-998-7440
www.nyu.edu/wagner/

The Wagner School offers a Master of Public Administration, with concentrations in either public policy analysis or public and nonprofit management and policy. A subspecialty cluster in urban and economic development is available for both concentrations. (Also offered is a Master of Urban Planning, with a concentration in urban, economic, and social development. See the "Planning" section of this appendix.) All programs may be taken on either a full- or a part-time basis. Advanced Professional Certificates, with modules in urban planning and policy and public and nonprofit management and policy, are also available. Many joint degree programs are offered.

The curriculum for the Master of Public Administration, with a concentration in public policy analysis, includes five required core courses in management, public policy, and quantitative methods, together with six foundational management classes and a client-based capstone course. The specialization in management for public and nonprofit organizations focuses on the role of managers in formulating organizational goals, translating them into specific programs and objectives, and directing resources to assure their achievement. Students in management for public and nonprofit organizations take the same five required core courses as in the specialization in public policy analysis, together with a minimum of six management-related classes; they must also compete a client-based capstone course. Specializations also share a specialty cluster in urban economic and community development, which emphasizes policy making, planning, and financing of urban development, including institutional, political, intergovernmental, and economic aspects of project development.

The Advanced Professional Certificate Program is an opportunity for professionals to update their knowledge and skills by taking four part-time courses, many of which are offered in the evening. Specialization modules are available in urban planning and policy (including housing and public economics) and public and nonprofit management and policy (including urban public policy and public policy analysis).

Syracuse University

Maxwell School of Citizenship and Public Affairs
215 Eggers Hall
Syracuse, NY 13244-1020
315-443-4000; fax: 315-443-9721
email: maxwell.info@maxwell.syr.edu
www.maxwell.syr.edu

The Maxwell School offers three distinct programs in public administration and policy: Master of Public Administration (MPA), Master of Arts for mid-career executives, and the research-oriented Doctor of Philosophy. All three programs are multidisciplinary and interweave community development issues throughout their curriculums.

The MPA degree is an intensive, twelve-month program with a three-pronged curriculum: the political, economic, and social context of public administration; organizational analysis and management techniques; and the application of qualitative and quantitative analysis to public policy issues. Within the forty-credit curriculum are twenty-five required core credits, which include a public affairs colloquium, a team consulting project, and course work in leadership, policy, management, and finance. The remaining fifteen elective credits may be satisfied by course work in one established and specialized program of study, which is selected from: public and nonprofit management, social policy, state and local financial analysis and management, international development, environmental policy, and technology and information management. Students also have the option of designing their own program of study in a specific interest area such as community economic development.

The Master of Arts in public administration is designed for mid-career executives and recognizes the unique skills brought to the classroom by experienced professionals. The thirty-credit program may be completed in one year of full-time study or at a slower pace in a combination of study and professional employment. There are no requirements, and students are free to design a multidisciplinary program of course work that meets their specific professional needs.

University of Minnesota

Hubert H. Humphrey Institute of Public Affairs
301 19th Avenue South
Minneapolis, MN 55455
612-626-8910
email: jmehr@www.hhh.umn.edu
www.hhh.umn.edu

The Institute offers a Master of Public Policy (MPP) degree and a new Executive Master of Public Affairs (EMPA), which is intended to help working professionals with at least ten years of relevant experience to advance in existing careers or to switch careers.

In the MPP program, students take a primary concentration to strengthen their skills in public management, policy analysis, or planning; and a secondary concentration in substantive issues areas, which include economic and community development, public and nonprofit leadership, and management. Dual degrees may be taken with the Law School, School of Social Work, and the Department of Political Science. A doctoral program is in the process of development.

The MPP requires twenty-one credits of core courses that cover fundamental knowledge and skills in politics, organizations, policy analysis, microeconomics, planning, and quantitative methods (skills required in nearly all public affairs careers); a three-credit capstone workshop or seminar; eighteen more credits in a primary concentration; twelve more in a secondary concentration; and an internship.

The EMPA offers concentrations including economic and community development; housing, social planning, and community development; and public and nonprofit management and leadership; or a concentration tailored to a participant's specific policy interests. Classes are offered evenings. Part-time study is permitted. The degree requires thirty semester credits, including eight credits in a synthesis seminar and workshop; four credits in leadership; nine credits in the concentration; six credits in skills courses; and three credits in electives.

The HHH Institute operates a number of centers, including the CENTER FOR DEMOCRACY AND CITIZENSHIP, which runs, among other programs, the Project Public Life initiative, which promotes citizenship and community development.

University of North Carolina at Chapel Hill

118 Abernathy Hall
Campus Box #3435
Chapel Hill, NC 27599-3435
919-962-1600
email: asta_crowe@unc.edu
www.unc.edu/depts/pubpol

The University offers both a Bachelor of Arts and a PhD in public policy analysis. In addition, a Master's Certificate (a formal minor) is offered for students enrolled in policy-related programs at UNC-CH, including City and Regional Planning, Public Administration, and Social Work.

The undergraduate BA program is interdisciplinary and requires the completion of six core courses, one from each of the following areas: intro-

duction to public policy, ethics and policy analysis, economic analysis, political and administrative feasibility of policies, quantitative analysis, and advanced individual projects in policy analysis. Students must also choose an area of specialization, in which they complete five additional classes. They have the option of defining their own area of specialization in either a substantive area or in an aspect of analysis. Among possible substantive areas are economic development and public policy, and urban development and housing policy. Internships are encouraged but not required. Students may also declare public policy analysis as their minor concentration.

The Master's Certificate may be useful for students seeking jobs that require formal training in public policy analysis. Students are required to complete a course of study that includes a minimum of sixteen hours of approved course work in public policy analysis as well as prerequisite courses in intermediate microeconomics and probability and statistics. All students are required to take the course in public policy analysis and the seminar in public policy analysis.

The PhD in Public Policy Analysis is designed to train scholars and analytic thinkers to solve real-world problems that spill over traditional lines. The exploration of multiple disciplines is encouraged, and core foundation courses, which include state-of-the-art policy analysis methods, are used to teach how to solve problems in public policy areas such as housing and urban development. Along with public policy theory, students also learn the political and administrative aspects of implementation in their chosen policy specialization.

See also the "Planning" section of this appendix.

University of Washington

Daniel J. Evans School of Public Affairs
P.O. Box 353033
Seattle, WA 98195-3055
206-616-1607; fax: 206-543-1096
email: elainec@u.washington.edu
www.evans.washington.edu

The Evans School of Public Affairs offers a traditional program of study leading to the Master of Public Administration (MPA) for full-time day students and an evening degree program for mid-career professionals. The eighteen-credit core curriculum provides students with broad-based public policy analysis and management knowledge, including basic analytical and quantitative skills. Students then enter into specialized study in a chosen policy "gateway." Students with community development interests usually specialize in urban and regional affairs or nonprofit management. They work together with the Evans School's teaching and research faculty who specialize in these issues and take courses in other areas of the University as well, including in the Department of Urban Planning and the Schools of Business Administration, Social Work, and Public Health.

Internships and projects are conducted in the Seattle metropolitan area; degree projects on community development subjects have been carried out in the past with the assistance of such agencies as the Seattle Housing Authority; the Regional Transit Authority; the Washington Department of Community, Trade and Economic Development; the Seattle Department of Neighborhoods; and the Seattle Office of Economic Development.

The evening degree program for professionals with seven to ten years of work experience enables students to work full-time. Degree requirements include fifty-four credits, divided among core courses, electives, and four leadership seminars to develop managerial leadership in relation to their workplace roles and challenges. Internships are not required.

Students are also aided in their community development studies by five policy research centers headquartered at the Evans School. These include the INSTITUTE FOR PUBLIC POLICY AND MANAGEMENT, which hosts a monthly criminal justice policy forum; and the NORTHWEST POLICY CENTER, which assists the city of Seattle in carrying out the federal government's Enterprise Community program and the Seattle Jobs Initiative. The INTERDEPARTMENTAL CONSORTIUM ON NON-PROFIT MANAGEMENT (which includes the Evans School of Public Affairs) offers graduate and professional students interested in nonprofit management the opportunity to take an array of courses to complement existing degree programs.

Real Estate Development

Massachusetts Institute of Technology
Center for Real Estate
77 Massachusetts Avenue, W31-310
Cambridge, MA 02139-4307
617-253-4373; fax: 617-258-6991
email: mit-cre@mit.edu
http://web.mit.edu/afs/athena.mit.edu/org/c/cre/www

The Center for Real Estate (CRE) at MIT offers a Master of Science degree in real estate development. The full-time, eleven-month program combines lectures with case analyses and assignments. Demanding and rapidly paced, the program requires a minimum of eight courses and a thesis. Guest speakers from the real estate industry and field experience gained during thesis research provide students with a real-life perspective. CRE graduates enter diverse fields, from commercial real estate companies to investment and consulting groups to public housing agencies. The CRE offers joint de-

grees with other departments at MIT, including the Department of Architecture, Urban Studies and Planning, and the Sloan School of Management.

Social Work

The Association for Community Organization & Social Administration (ACOSA) is the association for social workers involved in community practice, including community organizing. Its Web page address is www.wvu.edu/~socialwk/acosa/acosa-l.html. The National Association of Social Workers is also developing a section for community practice.

Boston College
Graduate School of Social Work
McGuinn Hall
Chestnut Hill, MA 02467
617-552-4024; fax: 617-552-3199
email: swadmit@bc.edu
www.bc.edu/bc_org/avp/gssw/gssw.htm

Boston College offers a two-year, full-time Master of Social Work (MSW) program called Community Organization, Policy, Planning, and Administration (COPPA), which emphasizes social and economic justice; it prepares students for careers in social planning, policy practice, community practice, and administration. There is also a PhD program.

The six required COPPA courses are Macro Practice Skills, Communities and Organizations, Evaluation Research for Macro Practice, Social Planning and Policy in the Community, Urban Development Planning, and Administration of Human Services Programs. There are a variety of elective classes including a study class that travels to Cuba every year. Two field placements, usually in planning agencies, legislative offices, and neighborhood organizations, are mandatory. COPPA students, faculty, and alumni are active in the development of Boston, and the city offers many educational and internship opportunities. Joint degrees are also available with BC's schools of Law, Business, and Pastoral Ministry.

Boston University
School of Social Work
264 Bay State Road
Boston, MA 02215
617-353-3750; fax: 617-353-5612
www.bu.edu/ssw

The School of Social Work emphasizes social and economic justice in the urban environment. The Master of Social Work program requires eight core courses including those on social welfare policy, research methods, racism, and ethics. Students also must select one specialization, either in clinical social work with individuals, families, and groups, or macro social work practice. Within the macro social work practice method, students are required to take a minimum of four courses and may focus on courses from one of three areas: community organization (the analytic, strategic, tactical, and interactional skills needed for community empowerment and change); social planning (the analytic tools and political skills needed to remedy the multiple social problems affecting a community); and human services management (basic administrative and management competencies). Students must also complete two fieldwork placements, one in each year of study.

The interdisciplinary PhD program in social work and sociology recognizes that today's pressing urban problems require the creative integration of social science theories and research with social work practice. Students take six core courses and six classes in the specialized study of two subfields, one in social work and one in sociology. The program is flexible, and students are encouraged to pursue their own interests, which may be in such areas as community development and urban sociology.

Case Western Reserve University

Mandel School of Applied Social Sciences
10900 Euclid Avenue
Cleveland, OH 44106-7164
216-368-5883 or 800-863-6772; fax: 216-368-5065
email: aet2@po.cwru.edu
http://msass.cwru.edu/

The Mandel School of Applied Social Science offers a Master of Science in Social Administration (MSSA) with concentrations in management and community development. The MSSA curriculum consists of general classes in social work methods, human development theory, social policy, research methods, two semesters of field education, and thirty-five credit hours of classes in a specialization including community development. The community development concentration stresses the rebuilding of local communities through methods of grassroots organizing and economic development.

From 1,080 to 1,208 clock hours of field education (including work in an internship position) are necessary for graduation. Although the program is normally two years in length, students with a Bachelor of Social Work can obtain the degree in one year; part-time students may finish the program in three years. The Intensive Weekend Program allows employed social workers to pursue an MSSA in three years while maintaining full-time employment. Dual degrees are available with programs in Social Welfare, Law,

Management of Nonprofit Organizations, and Business Administration.

Centers of interest affiliated with the Mandel School are the CENTER FOR COMMUNITY DEVELOPMENT, the CENTER ON URBAN POVERTY AND SOCIAL CHANGE, and the Mandel CENTER FOR NON-PROFIT ORGANIZATION. All three centers offer educational opportunities for MSSA students in the form of seminars, conferences, and possible mentorship, internship, and research opportunities. For graduates and professionals, the centers offer continuing education and training opportunities.

Catholic University of America

National Catholic School of Social Service
Cardinal Station
Washington, DC 20064
202-319-5458; fax: 202-319-5093
email: cua-ncss@cua.edu
http://ncsss.cua.edu/

One of the oldest schools of social work in the United States, Catholic University offers a Master of Social Work (MSW) and a research-oriented PhD in social work. An undergraduate program in social work, which is designed to prepare students for both graduate work and direct entry into supervised social work practice, is also available.

The MSW curriculum includes six required courses in theory, practice skills, and professional values. Students also select a concentration and take advanced theory and practice classes in either clinical social work or social policy, planning, and administration. With faculty approval, students may combine these two concentrations. The concentration in social policy, planning, and administration is designed to equip students with the skills needed to support communities' efforts for self-empowerment and capacity-building. For this concentration, students take courses in community organizing, social policy, and planning. Two year-long field placements are also required of students in both concentrations. The University's location in Washington, D.C., gives students unique opportunities for fieldwork in congressional offices, with lobbying groups, and in other professional organizations.

The PhD in social work program offers concentrations in theory and research in policy and administration in social work as well as theory and research in clinical social work. Doctoral students also have the option of an Individualized Educational Contract, which allows them to develop a course of study that is tailored to their specific interests.

Georgia State University

College of Health and Human Services
Department of Social Work

for student field placements. Temple's location in Philadelphia's urban core facilitates student placements in a variety of community development projects with community-based organizations with which the Center contracts to do capacity-building and technical assistance work.

University of California at Los Angeles

School of Public Policy and Social Research
3250 Public Policy Building
Box 951656
Los Angeles, CA 90095-1656
310-825-7737; fax: 310-206-7564

email: msw@sppsr.ucla.edu (for the MSW Program);
swphd@sppsr.ucla.edu (for the PhD program)
www.sppsr.ucla.edu/

Located within the School of Public Policy and Social Research, UCLA's Department of Social Welfare offers a Master of Social Work and a research-oriented PhD, both with an emphasis on policy analysis and community development. A dual degree is also available in law and social work.

The Master of Social Work is a full-time, two-year professional program that combines formal academic course work with practical fieldwork experience. Requirements include core foundation courses in the history, theory, research, and practice methods of social work. Students select one concentration in either social work with individuals, families and groups (which examines the problems of the individual within the larger context of the family and small-group relationships), or social work in organizations, communities, and policy settings (which focuses on the theory and skills needed for community development and organizational management). Courses within the latter concentration include organizational and community innovation and change skills. Along with a concentration, students also select an area of specialization from among four: children and youth services, mental health, health services, and gerontology. Two fieldwork placements are required, one in each year of study.

University of Connecticut

School of Social Work
Greater Hartford Campus
Asylum Avenue and Trout Brook Drive
West Hartford, CT 06117-2698
860-570-9141; fax: 860-570-9139
http://socialwork.uconn.edu

In the University's Master of Social Work program, community organization is one of the concentrations offered. The concentration, with its focus on community empowerment, requires four method courses that explore such areas as social action, social planning, and strategies for community intervention. Students also select a minor method by taking one beginning course in a second method. They have the option of grouping their elective classes into substantive areas that focus in greater depth on such areas as urban issues, the black experience and social work, Puerto Rican studies, and women's issues. Requirements of all master's candidates include a core examining human development, behavior, social policy, and research.

Fieldwork is required and involves two different 560-hour supervised placements over a two-year period. Students may also fulfill their fieldwork requirement in block, full-time placements if they have completed most of their course work.

Dual degrees are available between the School of Social Work and the University of Connecticut schools of Law, Business Administration, and Medicine, and with the Yale Divinity School.

University of Maryland

School of Social Work
Louis L. Kaplan Hall
525 West Redwood Street
Baltimore, MD 21201-1777
410-730-9019
email: ssw.umaryland.edu
http://ssw.umaryland.edu/

The School of Social Work offers both a master's and a research-oriented PhD in social work. Students have the option of either full- or part-time enrollment.

For the Master of Social Work, students complete a core foundation curriculum and must select a primary concentration from among the following: management and community organization, which seeks to develop new and better community services, social programs and policies, and legislative advocacy; clinical social work, which focuses on clinical work helping individuals and families; or a third concentration that combines essential skills from within the clinical social work and community organization concentrations.

In addition, students select an area of specialization in one of seven fields of practice: social and community development, aging, employee assistance programs, families and children, health, mental health, and substance abuse. In the first year, two-day-a-week generalist fieldwork practice placements are required, and in the second year, three-day-a-week field placements that complement and strengthen the student's choice of concentration and specialization may be selected.

Dual degrees are available between the schools of Social Work and Business Administration, Law, and Judaic Studies.

University of Michigan

School of Social Work
1080 South University Avenue
Ann Arbor, MI 48109-1106
734-764-3309; fax: 734-936-1961
email: ssw.msw.info@umich.edu

www.ssw.umich.edu

The School offers both a Master of Social Work and an interdepartmental dual PhD in social work and social science (chosen from anthropology, economics, political science, psychology, or sociology).

The Master of Social Work degree requirements include the selection of a dual concentration in which students take nine credit hours of course work in each of the following: (1) a practice method concentration, which focuses on the theories and interventions related to practice with individuals, families, organizations, groups, and communities; and (2) a practice area concentration, which targets types of social work practice and their specific policies and procedures. Within the practice method concentration, students select one method from among the following four: community organization (social action, capacity building, and change at the community level); interpersonal practice (the transaction between people and their social environment); management of human service (the management and direction of human service organizations); and social policy and evaluation (the analysis, development, and implementation of social policy into social goals). For the practice area concentration requirement, students select one from the following five: community and social systems, adults and elderly in families and society, children and youth in families and society, health, and mental health.

In addition, 912 hours of supervised fieldwork in social work agencies are required for graduation. Dual degree programs are available between the School of Social Work and the schools of Urban Planning (Community Planning), Business Administration, Public Health, and Public Policy.

University of Pittsburgh

Graduate School of Social Work
2104 Cathedral of Learning
Pittsburgh, PA 15260
412-624-6302
www.pitt.edu/~pittssw

The University offers both an accredited Bachelor of Arts in social work and a Master of Social Work, with a concentration in community organization and social administration. A PhD in social work is offered, which may be pursued jointly with the MSW. Additional joint degree programs are available, combining the MSW and a Master of Urban and Regional Development, a Master of Public Administration, a Master of Public and International Affairs, among others. Students enrolled in these programs may cross-register at Carnegie Mellon University.

The BA in social work is a multidisciplinary degree, which, offered during the third and fourth years of college, prepares graduates to work in a variety of social work areas such as community organizing and advocacy

within a multicultural society. The curriculum includes twelve required foundation courses and two semesters of supervised fieldwork.

The MSW degree, with a concentration in community organization and social administration, trains professionals for leadership as organizers, managers, and policy makers in the human services. Within community organization, students may choose a specialty in either human services organizing (ways to organize human service providers on a range of issues from service delivery coordination to coalition building) or community economic development. Within the community economic development specialization, knowledge about community-based economic development is combined with the planning, organizing, management, accounting, and policy core skills. For students who elect to major in social administration, some of the study areas include organizational behavior and management, group process, financial management, and supervision. Along with a concentration, students also select an area of specialization from three: children, youth, and families; health care; and mental health. Two fieldwork placements, one for each year of study, are also required.

Washington University

George Warren Brown School of Social Work
Campus Box 1198
One Brookings Drive
St. Louis, MO 63130-4899
314-935-6676; fax: 919-962-2518
www. gwbweb.wustl.edu

The School confers both a Master (MSW) and a PhD in social work. Dual degree programs are available in law and social work, business and social work, architecture and social work, and Jewish communal service and social work.

The curriculum's required foundation courses, which must be completed in the first year, include Human Behavior in the Social Environment, Social Welfare Policy, History of Social Welfare, and Social Work Values and Ethics. In a first-year fieldwork practicum, students gain experience with social work practice as they apply the knowledge, skills, and values they have learned from their foundation courses. In their second year, students pick a concentration and complete course work while doing a second fieldwork practicum.

The community development concentration, called the social and economic development concentration, centers on problems of underdevelopment in U.S. communities and in other countries, exploring poverty, institutions, and organizations; the engagement of diverse social, economic, political, and cultural groups in the development of communities and strengthening the institutional capacity of communities to bring about sustained development. Requirements are two theory courses, two practice methods courses, a policy, and an evaluation course. The fieldwork practicum must take place in a site where the primary focus is on community, social, and economic development and where the mission is to serve poverty groups and/or groups that are at risk because of race, gender, geography, or other variables. An individualized concentration may be structured, combining, for example, the social and economic development concentration with the study of women in poverty.

The CENTER FOR SOCIAL DEVELOPMENT is a practicum site for many social work students. The Center is a leader in innovative approaches to the concept of "development" that can replace the traditional welfare state emphasis on maintenance, such as asset-building policy initiatives and individual asset accounts.

Wayne State University
School of Social Work
4756 Cass Avenue
Detroit, MI 48202
313-577-4400; fax: 313-577-8770
www.socialwork.wayne.edu

Located in Detroit's urban core, Wayne State's School of Social Work has a commitment in its teaching, research, and service activities to address the problems of people living in this environment. The School offers both an accredited Bachelor and a Master of Social Work, with a concentration in community practice and social action. Professional development certificate programs are also available.

The bachelor program, which has been ranked the number one undergraduate program in social work in the country for the last four years, prepares students for entry-level practice in social work. Course work in this program includes University-wide general education requirements as well as the core knowledge, values, and skills for social work practice. An emphasis is placed on fieldwork education, with students spending two days a week at a community-based placement during three of their four semesters in the program.

The Master of Social Work curriculum includes core requirements in the areas of human behavior and the social environment, social welfare organization and policy services, and practice methods. Students must also choose one of five concentrations, which include community practice and social action. The community practice and social action concentration focuses on interventions in social agencies, institutions, and neighborhoods of the community and society to enhance the quality of life. Students in this concentration will spend three days each week in practicums that relate to such areas as urban social planning, community development, comprehensive

health planning and development, and social change. Students may attend the MSW program on either a full- or a part-time basis.

Sociology

The professional association for sociologists is the American Sociological Association, 1307 New York Avenue, Suite 700, Washington, DC 20005, 202-383-9005; Web site: www.asanet.org. The members also maintain a section on community and urban sociology.

Baylor University
Division of Sociology
Burleson Hall, Suite 300
Waco, TX 76798
254-710-1165
email: Tillman_Rodabough@baylor.edu
www.baylor.edu/~Sociology/soc.htm

Baylor University offers a Master of Arts in sociology and a PhD in applied sociology focusing on the application of sociological knowledge to real-world settings. The master's program offers courses including a seminar on Community, Economics of Poverty and Discrimination, and Social Change and the Industrial Society. The program emphasizes a background in academic sociology, training in applied methodology, and experience in real social, community, and organizational settings. Students gain real-world experience by working at the University's CENTER FOR COMMUNITY RESEARCH AND DEVELOPMENT. The Center works primarily in the Waco community and other Texas cities, including Dallas, San Antonio, and Houston. It encourages faculty and student research projects that positively change organizations and communities.

Loyola University at Chicago
Department of Sociology and Anthropology
6525 North Sheridan Road
Chicago, IL 60626
773-508-3445; fax: 773-508-7099
email: jwittne@luc.edu
www.luc.edu/depts/sociology

The Loyola Sociology Department is an important center for research and teaching on the urban community, with much of the department's research and teaching focused on measuring the quality of life in urban communities and examining possibilities for social change. The department offers a Bachelor of Arts and a Master of Arts in sociology and applied sociology. The MA in applied sociology requires thirty hours of course work and a preliminary qualifying exam, or twenty-four hours of course work, the exam, and the completion of a master's thesis. An internship is required. The program has both part-time and full-time students, and Loyola is supportive of working students who are seeking new or additional skills.

The CENTER FOR URBAN RESEARCH AND LEARNING (CURL) and the POLICY RESEARCH AND ACTION GROUP (PRAG) are valuable resources to Loyola students. Community leaders, faculty, and students come together in CURL to find practical solutions for Chicago's communities. Examples of CURL projects include assessing the impact of welfare reform and evaluating public and private agency programs. PRAG, like CURL, works to connect academia and communities by developing research apprenticeships within community-based organizations and encouraging undergraduate and graduate students to consider career options in community-based research.

Ohio State University
College of Food, Agriculture and Environmental Sciences
Department of Human and Community Resource Development
208 Agricultural Administration Building
2120 Fyffe Road
Columbus, OH 43210-1067
614-292-4624; fax: 614-292-7007
email: thomas.27@osu.edu
www.ag.ohio-state.edu/~ag_educ/index.html
The University's Rural Sociology Program offers a new concentration in community development as part of the Master of Science in rural sociology, with a flexible program to accommodate diverse career plans, workplace settings, and experiences. Students select from a broad range of courses that provide insight into domestic and international community development. The concentration requires that electives be taken in other departments whose courses complement the rural sociology focus and provide more specialized training in aspects of community development.

Students may choose a thesis or an exam option. The thesis option requires a total of forty-five credit hours plus thesis to graduate. The exam option requires a total of fifty credit hours and successful completion of an MS exam to graduate. The program is flexible but requires ten credits of sociological theory; ten credits of research methods related to community development, such as fundamentals of GIS; a five-credit core community development course; ten credits of rural sociology course work related to community development; and ten to fifteen credits of electives organized around a particular theme of community development. Courses offered cover such wide-ranging issues as Rural Poverty, Women in Society, Black Community Politics, and Programming Environments for Human Use.

University of Wisconsin at Madison
Departments of Urban and Regional Planning, and Rural Sociology
1450 Linden Drive, Room 350
Madison, WI 53706
608-262-1510; fax: 608-262-6022
email: ggreen@ssc.wisc.edu
www.wisc.edu/urpl; www.ssc.wisc.edu/ruralsoc/
The program in community development at the University of Wisconsin is an interdisciplinary program offered by faculty from the departments of Urban and Regional Planning, Agricultural and Applied Economics, Rural Sociology, and Human Ecology. It is designed for students who wish to work as community development planners, policy analysts, organizers, and program administrators. The Master of Science degree is actually granted by the Department of Urban and Regional Planning. Students are required to take about one-third of their credits in core planning courses, one-third in community development courses, and one-third electives. There are two required community development courses: Community Economic Analysis and Community Development. Students are required to take at least one course dealing with methods, one dealing with community development processes, and one dealing with the context of community development. The program is a two-year degree; there is no certificate; a thesis is encouraged but not required.

Urban Affairs

The Urban Affairs Association, the professional association for this discipline, is headquartered at the University of Delaware, Newark, Delaware 19716. The Web site is www.udel.edu/uaa; the association produces the *Journal of Urban Affairs*.

Cleveland State University
Maxine Goodman Levin College of Urban Affairs
Urban Affairs Building
Cleveland, OH 44115-2105

216-687-2136; fax: 216-687-9291

http://urban.csuohio.edu

Ranked in 1998 by both *U.S. News & World Report* and *Business Week* among the top ten university programs in city management and urban policy, the Levin College of Urban Affairs combines classroom teaching with direct public service. Degree programs are a BS in urban studies, Master in Urban Studies, PhD in urban studies; Master of Public Administration, and a Master of Urban Planning, Design, and Development.

The BS in urban studies is an interdisciplinary academic program with a wide range of choices both in the classroom and in internships in urban agencies. Degree requirements include fourteen required distribution classes that cover general and introductory topics in urban studies, two core courses in Urban Systems and Contemporary Urban Issues, and two methods courses selected from such areas as proposal writing, program evaluation, and neighborhood analysis. Students must choose a concentration from among five: urban planning, urban management, urban environmental policy and management, historic preservation, and general urban studies. They have the flexibility to design a three-course sequence of their choosing in consultation with their advisor.

The MS in urban studies requires five core courses and three electives; students develop their own concentration or specialize in: economic development, organizational leadership, urban policy analysis, environmental law, or public policy. A thesis or exit project is required. Up to three electives may be taken. Internships are encouraged but not required and may count as an elective.

A PhD in urban studies is offered with curriculum designed in consultation with faculty.

San Francisco State University

Urban Studies Program

1600 Holloway Avenue

San Francisco, CA 94132

415-338-1178; fax 415-338-2391

email: raquelrp@sfsu.edu

www.sfsu.edu/~urbstu

The Bachelor of Arts in the Urban Studies Program at San Francisco State University offers a major and minor in urban studies. In addition to eight required classes and three electives, students must complete an internship entailing twelve to fifteen hours of fieldwork for fifteen weeks, a two-hour seminar every other week, and the submission of regular journals and brief discussion papers. The program welcomes part-time students with classes

in the late afternoon and early evenings. The department also publishes *Urban Action*, an annual student journal devoted to urban studies and development.

Temple University

Department of Geography and Urban Studies

1115 Berks Street

Philadelphia, PA 19122-7833

215-204-1248; fax 215-204-7833

email: oldchief@vm.temple.edu

www.temple.edu/gus/graduate.html

The Department of Geography and Urban Studies offers a two-year, full-time Master of Arts in urban studies. The program focuses on urban problems and public policy in U.S. cities, covering issues such as housing, social and economic justice, the environment, neighborhood quality, urban economic development, and the provision of public services. Completion of the degree requires thirty-three or thirty-six credit hours of classes, a master's thesis, and passage of an oral and written master's comprehensive examination. Internships, although not mandatory, are available to graduate students who wish to gain work experience outside the University, usually between the first and second years. An internship paper that focuses on the operational and policy aspects of the internship organizational may be substituted for the master's thesis. The program also offers interdisciplinary PhD programs with the departments of Anthropology, Economics, Political Science, and Sociology.

University of Delaware

Graduate School of Urban Affairs and Public Policy

184 Graham Hall

Newark, DE 19716-7301

302-831-8289; fax: 302-831-3587

email: suapp@udel.edu

www.udel.edu/suapp/ctcommunitydevel.htm

The School offers a Master of Public Administration degree as well as both a Master of Arts and a PhD degree in urban affairs and public policy. Students may take a concentration in community development and nonprofit leadership.

Known as the Delaware model, the program incorporates three elements of equal importance: academic instruction, public and community service, and applied and practical research. Each year almost all the students are engaged in paid internships, where, supervised by a faculty or professional staff person, they work with a public or community agency, usually thirty-five hours per week for three summer months.

The MA in urban affairs is a thirty-six-credit, two-year program. The CENTER FOR COMMUNITY DEVELOPMENT AND FAMILY POLICY provides support to the community development concentration. Within that concentration, students take courses on issues such as urban housing policy, poverty, neighborhood and community development, and local economic development policy and practice.

The MPA is a forty-two-credit, two-year program in which students specialize in one of five areas, including international development policy, state and local management, and environmental and energy management.

In addition, the Center for Community Development operates professional development training courses (see appendix K); students may take some of the workshops for credit.

University of Louisville

College of Business and Public Administration
Louisville, KY 40292
502-852-6005; fax: 502-852-4721
www.louisville.edu/urbanaff.htm

The University of Louisville offers two accredited degrees: a Master of Public Administration with a concentration in urban and regional development, and a PhD in urban and public affairs. (The Master of Public Administration is discussed in the "Business, Nonprofit Management, and Public Administration" section of this appendix.) The PhD, which is practice oriented, prepares students for careers in university teaching, public and non-profit administration, public policy research, urban development and planning, and program evaluation. The curriculum's core foundation courses give a broad orientation, while field areas enable students to develop expertise in a particular specialty. Field areas include urban policy analysis and infrastructure and environment.

University of Wisconsin at Milwaukee

Urban Studies Program
P.O. Box 413
Milwaukee, WI 53201-0413
414-229-4751; fax: 414-229-4266
email: usp@csd.uwm.edu
www.uwm.edu/Dept/Urban_Studies/

The University of Wisconsin at Milwaukee offers Master of Science and PhD degrees in urban studies. The program offers a liberal arts approach to the examination of urban social processes and seeks to provide students with solid research and policy-making tools. Students have the option of pursuing a generalist program of urban study or a specialist track in economic development, health care delivery systems, or social policy and social structure. The MS program welcomes returning students with established careers by offering most classes in the late afternoon or early evening. The MS program requires thirty credits of classes and a master's thesis.

The cost of an education can be particularly worrisome to students who want to work in the community and economic development field, where job salaries are usually lower than salaries in the for-profit sector. With tuition and living expenses often above $20,000 a year, it is a good idea to apply for as much financial aid as possible when applying to and attending college or going for an advanced degree.

The first step is to contact the financial aid office at your college or university. That office will provide you with financial aid applications and accompanying forms and let you know the deadlines for the forms. You should work closely with your financial aid officer(s) regardless of what you think your financial aid eligibility may be. The more you work with your school's financial aid office, the better chance you have of receiving a package that meets your needs.

As part of your school financial aid package, you might receive federal grants and loans. Grants are the best kind of financial aid, as they need not be repaid. Of all loans, federal loans are the best because they have lower interest rates, are more lenient in repayment than private loans, and may even be waived if you work in certain federal programs (such as the Peace Corps) after college.

Many universities offer special financial aid for students in specific disciplines, and you may find that community development is one of those areas. For example, Cleveland State University's Urban Affairs Department offers the Dively Undergraduate Neighborhood Internship, providing an undergraduate with a paid twenty-hour a week internship at a community-based organization. The University of Southern California offers an impressive list of merit- and need-based scholarships to master's and doctoral students in the School of Policy, Planning and Development.

The last piece of advice is something you will hear again and again: Find out about grants and scholarships and apply for as many as you can. To find out about grants and scholarships, browse your local library, bookstores, and the resources of the college nearest you, and browse the Web. See the recommendations below.

The Federal Work-Study Program

One good way for college students to fund term-time and summer-time internship opportunities in the community and economic development field is to work under the Federal Work-Study Program, a federally funded program designed to create jobs for financially needy students who need to work during college. Through the work-study program, the federal government pays 65 percent of an eligible student's wages if the student works for a federal, state, or local public agency or a private, nonprofit organization, and the sponsoring organization supplies the remaining amount. For some community service organizations, the program pays 75 percent of the student's wages. Because many community and economic development organizations are nonprofits, most of them are eligible to hire Federal Work-Study students.

To work under the Federal Work-Study Program, you should take the following steps. First, contact your school's financial aid office for an application form or find out if Federal Work-Study is already a part of your financial aid package. Second, find out about community and economic development organizations you would like to work for and contact them to see if they have internship opportunities and funding to help pay for their part of the Work-Study Program wages. If you find an organization that is willing to hire you under the Work-Study Program, work with your school's Work-Study Program department to take the next necessary steps.

HUD's Community Development Work-Study Program

Every year colleges and universities across the country compete for grants from the U.S. Department of Housing and Urban Development to fund work-study programs for selected students enrolled in a full-time graduate program in community development or a closely related field such as urban planning, public policy, or public administration. With the HUD grants, the college or university can offer selected students two-year packages of financial aid, including support for tuition, travel, and books, as well as stipends for work. The program is intended to help minority and economically disadvantaged students who lack the financial resources to pursue a professional education and career in community development.

The institution of higher education selects the students, secures work assignments for them, disburses the funds, and monitors student performance.

The institutions with these grants are places where you might be able to find financial aid combined with valuable work experience. To get a list of the year's colleges and institutions with Community Development Work-Study Program awards, use the following Web, mail, or phone information:

University Partnerships Clearinghouse
P.O. Box 6091
Rockville, MD 20849-6091
800-245-2691; fax: 301-519-5767
email: oup@oup.org
www.oup.org

HUD's Hispanic-Serving Institutions' Work-Study Program

Every year community colleges with substantial Hispanic populations compete for grants from the U.S. Department of Housing and Urban Development for funds from which they can offer work-study programs to assist selected students who are enrolled full-time in academic programs that promote community building. With the HUD grants, the college can offer selected students two-year packages of financial aid, including support for tuition, fees, and books, as well as stipends for work. The program is intended to help minority (not only Hispanic) and economically disadvantaged students who lack the financial resources to pursue a professional education and career in community building. Acceptable fields of study include public administration and public policy; urban economics, urban management, and urban planning; and other fields that promote community building.

The institutions are responsible for arranging for local work placement with state or local governments, area-wide planning organizations, Indian tribes, or nonprofit organizations and for monitoring student participation and performance. The institutions eligible for this program are Hispanic-Serving Institutions (as certified by the U.S. Department of Education), whose students are at least 25 percent Hispanic, as well as low-income and first-generation college attendees.

The institution selects the students, secures work assignments for them, disburses the funds, and monitors student performance.

To get a list of the year's Hispanic-Serving Institutions with HUD Work-Study Program awards, use the following Web, mail, or phone information:

University Partnerships Clearinghouse
P.O. Box 6091
Rockville, MD 20849-6091
800-245-2691; fax: 301-519-5767
email: oup@oup.org
www.oup.org

HUD's Doctoral Dissertation Grant Program

HUD also offers about fifteen one-year grants of up to $15,000 each year to students for applied research to complete their dissertation for a PhD degree. The research and thesis must be relevant to HUD's interests. For further information, contact the University Partnerships Clearinghouse as listed above.

Cooperative Education

Cooperative education is a college program that integrates classroom studies with paid work experience in a related field. Students can work full-time and then go to school full-time, or work part-time while going to school. About nine hundred colleges and universities now offer co-op education programs. Businesses of all sizes and types enroll as willing to employ co-op students.

If you are going to a school with a co-op program and find a community development organization you want to work for that is not enrolled with the school's program, you can ask your college to enroll the organization. Most colleges will accommodate such a request.

For a list of colleges and universities participating in this program, contact:

National Commission for Cooperative Education
360 Huntington Avenue, 384CP
Boston, MA 02115-5096
617-373-3770; fax: 617-373-3463
email: ncce@lynx.neu.edu
www.co-op.edu

For More Information about Student Financial Aid:

The most comprehensive Web site is www.finaid.org.

Useful publications include:
College Scholarships and Financial Aid: With ARCO's Scholarship Search Software, John Schwartz et al. (Macmillan Publishing Company, 1997).

Financial Aid for Dummies, Herm Davis and Joyce Lain Kennedy (IDG Books Worldwide, 1999).

Peterson's Scholarships, Grants & Prizes 2000, Jon Latimer, ed. (Peterson's, 1999).

You Can Afford College. Alice Murphey and Kaplan Educational Center Ltd Staff (Simon and Schuster Trade, 1999).

Appendix D: National and Regional Organizations

This appendix contains information about many national organizations active in the community development field. It is by no means a complete listing, but it does offer a look at the field's remarkable breadth and scope of activity. The organizations described in Part I are at the core of the community development movement, while the institutions listed in Part II do related work. They are listed alphabetically by organization name in each part.

Part I: Core Organizations

Association of Community Organizations for Reform Now

739 8th Street SE
Washington, DC 20003
202-547-2500, fax: 202-546-2483
email: resgeneral@acorn.org
www.acorn.org/

The Association of Community Organizations for Reform Now (ACORN) is a grassroots organization composed of community groups made up of low- and moderate-income families. Founded in Little Rock, Arkansas, in 1970, ACORN has grown to a membership of over ninety thousand families in over twenty-six states and the District of Columbia. ACORN groups work on a variety of community issues, such as affordable housing, bank disinvestment, living-wage jobs, welfare rights, the environment of poor neighborhoods, and voter registration. (See appendix K.)

Association for Enterprise Opportunity

1601 North Kent, Suite 1101
Alexandria, VA 22209
703-841-7760; fax: 703-841-7748
email: aeo@assoceo.org
www.microenterpriseworks.org-

AEO is the trade association for organizations that support the development and expansion of very small, "micro," enterprises as an approach to help disadvantaged individuals and communities build their assets. AEO provides networking, information exchange, training, conferences, regional and state meetings, policy, and advocacy.

Center for Community Change

1000 Wisconsin Avenue NW
Washington, DC 20007
202-342-0567 or 0519; fax: 202-342-1132
email: info@communitychange.org
www.communitychange.org

CCC was founded in 1965 to help build the capacity and power of low-income groups and disadvantaged minorities to conceive and launch self-help programs to bring about positive community change. The Center provides on-site assistance to local groups and conducts research on housing, community development, and reinvestment strategies; publishes guidebooks and research studies; and serves as a strong voice on behalf of communities in the public and private sectors. Its newsletters include *Community Change* and *Organizing*.

Center for Neighborhood Technology

2125 West North Avenue
Chicago, IL 60647
773-278-4800; fax: 773-278-3840
email: info@cnt.org, www.cnt.org

The Center for Neighborhood Technology (CNT) is a nonprofit research and education organization, founded in 1978, focusing on sustainable, environmentally sound community economic development. The Center provides technical assistance and develops pilot projects to demonstrate sustainable development. CNT's newsletter, *Place Matters*, is published periodically (and past issues of its newsletter, *The Neighborhood Works*, are available on its Web site).

Christian Community Development Association

3827 West Ogden Avenue
Chicago, IL 60623
773-762-0994; fax: 773-762-5772
email: chiccda@aol.com
www.ccda.org

The Christian Community Development Association (CCDA) is a membership organization for churches, church-related organizations, and individuals who are working for community development with a Christian faith base. The Association publishes a quarterly newsletter and holds regional and national conferences.

Community Development Society

1123 North Water Street
Milwaukee, WI 53202
414-276-7106; fax: 414-276-7704
email: cole@svinicki.com

www.comm-dev.org

The Community Development Society (CDS) is a professional society for public- and private-sector community development practitioners; members represent fields ranging from education, health care, and social services, to government and utility companies. There are branches in many U.S. states as well as overseas. CDS supports information exchange among members through conferences and online groups and provides research and reference information to them. It publishes the *Journal of the Community Development Society* and newsletters and maintains an online discussion group.

Corporation for Enterprise Development

777 North Capitol Street NE, Suite 410
Washington, DC 20002
202-408-9788; fax: 202-408-9793
email: cfed@cfed.org
www.cfed.org

This national organization (CfED) does research and operates pilot initiatives to try to develop new ways to increase small businesses owned and run by low-income people and to provide benefits to low-income communities and people. CfED works with public, private, and nonprofit organizations at the national, state, and community levels and provides a range of services: policy design, analysis, and advocacy, demonstration and project management; consulting; training and technical assistance. It publishes a quarterly newsletter, *Assets*.

Corporation for Supportive Housing

50 Broadway, 17th floor
New York, NY 10006
212-986-2966; fax: 212-986-6552
email: information@csh.org
www.csh.org

The Corporation for Supportive Housing (CSH) is a national nonprofit intermediary organization with the mission of expanding the quantity and quality of service-supported permanent housing for individuals with special needs who are homeless or at risk of becoming homeless. CSH provides technical and financial assistance to supportive housing providers in New York City; Chicago; New Haven, CT; San Francisco; Minneapolis–St. Paul; the State of Connecticut; the State of Georgia; and Phoenix, AZ. CSH's technical assistance services include project development, financial packaging, and service planning. It provides direct project financing through up-front capital, predevelopment loans, bridge loans, and guarantees to stimulate project development and encourage public- and private-sector funders to invest substantial financing in supportive housing.

Development Leadership Network

c/o Warren Conner
11148 Harper
Detroit, MI 48213
313-571-2800
www.picced.org/resource/dln.htm

The Development Leadership Network (DLN) is an association of community development practitioners, most of whom are graduates of the Development Training Institute.

Development Training Institute

2510 St. Paul Street
Baltimore, MD 21218
410-338-2512; fax: 410-338-2751
email: info@dtinational.org
www.dtinational.org

The Development Training Institute is a national nonprofit community development training organization. See appendix K.

The Enterprise Foundation

10227 Wincopin Circle, Suite 500
Columbia, MD 21044-3400
410-964-1230; fax: 410-964-1918
email: mail@enterprisefoundation.org
www.enterprisefoundation.org

The Enterprise Foundation is a national nonprofit intermediary that provides financing, technical assistance, and networking to its network of local community-based development organizations. It provides grants and predevelopment and construction loans. An affiliate, the Enterprise Mortgage Investments, Inc., provides long-term permanent financing for multifamily housing, while the Enterprise Social Investment Corporation provides equity financing through Low Income Housing Tax Credits, and the Corporate Housing Initiative provides equity financing for special needs housing. The organization discusses its activities in its *Enterprise Quarterly* magazine. See also appendix K.

First Nations Development Institute

11917 Main Street
Fredericksburg, VA 22408
540-371-5615; fax: 540-371-3505
email: info@firstnations.org
www.firstnations.org

The First Nations Development Institute is a nonprofit economic development organization dedicated to improving Native American reservations in a culturally viable manner. Through the Eagle Staff Fund and the Oweesta Program, the Institute supports reservation-based, Native-devised economic development initiatives with seed capital, training, and technical assistance in micro-enterprise loan funds and affordable housing.

Habitat for Humanity International

121 Habitat Street
Americus, GA 31709
912-924-6935
email: public_info@habitat.org
www.habitat.org

Habitat for Humanity International is a nonprofit, ecumenical, Christian housing organization dedicated to eliminating substandard housing worldwide and making decent shelter a matter of conscience and action. Habitat works in partnership with people in need throughout the world building simple, decent shelters that are sold to them at no profit through no-interest loans. Funds, building materials, and labor are donated by individuals, churches, corporations, and other organizations that share Habitat's goal of eliminating substandard housing in the world. Habitat is currently building homes in over eight hundred cities in the United States.

Handsnet

1990 M Street NW, Suite 550
Washington, DC 200036
202-872-1111; fax: 202-872-1245
email: hninfo@handsnet.org
www.handsnet.org

Founded in 1987, Handsnet links some five thousand public interest and human service organizations across the United States. Network members include national clearinghouses and research centers, community-based service providers, foundations, government agencies, public policy advocates, legal services programs, and grassroots coalitions. Handsnet on the Web offers daily news from Handsnet on CONNECT (a private network information database); information about Handsnet services, forums, and members; Handsnet's latest *Action Alerts* and its *Weekly Digest;* and a sample from the hundreds of policy, program, and resources articles posted each week by Handsnet members. The information on Handsnet on the Web is currently available at no charge.

Housing Assistance Council & Rural Housing Services, Inc.

1025 Vermont Avenue NW, Suite 606
Washington, DC 20005
202-842-8600; fax: 202-347-3441
email: hac@ruralhome.org
www.ruralhome.org

Housing Assistance Council (HAC) is a national nonprofit corporation created to increase the availability of decent housing for rural low-income people; Rural Housing Services is its technical assistance arm. It works, first, by creating and sustaining interest and action from all levels of government concerning rural housing for low-income people, and second, by helping rural housing associations become more productive and professional. Specifically, HAC provides technical assistance, revolving loan funds, housing program assistance, training, research, and information services. HAC's newsletter, *HAC News,* comes out twice monthly.

The ICA Group

20 Park Plaza, Suite 1127
Boston, MA 02116
617-338-0010; fax: 617-338-2788
email: icaica@aol.com
www.ica-group.org

The ICA Group helps to create and strengthen democratic employee-owned and -controlled businesses in minority, low-income, and blue-collar communities. It specializes in providing business, legal, and educational technical assistance and financial packaging to CDCs, worker-owned companies, cooperatives, employees faced with plant closings, community groups wishing to start their own enterprises, and community-based organizations who wish to create jobs through the development of their own enterprises. The organization's newsletter is the *ICA Bulletin*.

Institute for Community Economics

57 School Street
Springfield, MA 01105-1331
413-746-8660; fax: 413-746-8862
email: iceconomic@aol.com
Web site under development

Working primarily with community land trusts, the Institute for Community Economics (ICE) provides information, research, and referral; training; off-site consultations; and on-site technical assistance on all aspects of community land trust development and management. Its newsletter is called *ICE Update*.

Local Initiatives Support Corporation

733 Third Avenue, 8th floor
New York, NY 10017-3204
212-455-9800
www.liscnet.org

The Local Initiatives Support Corporation (LISC) is a national nonprofit intermediary that provides assistance to its network of local community-based development organizations in five ways: making short-term, below-market loans; guaranteeing loans; providing grants; channeling corporate equity investments through its National Equity Fund; and providing technical assistance in all phases of organizational and project development. It carries news about its own and its network members' activities in its newsletter, *Link*. (Also see appendices J and K.)

McAuley Institute

8300 Colesville Road, #310
Silver Spring, MD 20910
301-588-8110; fax: 301-588-8154
email: info@mcauley.org
www.mcauley.org

The McAuley Institute is a national nonprofit intermediary organization founded by the Sisters of Mercy in 1982. The Institute provides funding and technical assistance to assist individuals, local organizations, and faith communities to produce local, affordable housing. McAuley's priority is to work with community-based groups and groups that serve women.

National Association of Community Action Agencies

1100 17th Street NW, Suite 500
Washington, DC 20036
202-265-7546; fax 202-265-8850
email: info@www.nacaa.org
www.nacaa.org

As a private, nonprofit membership organization, the National Association of Community Action Agencies (NACAA) represents the interests of community action agencies (local agencies formed under the Johnson-era War on Poverty programs) and other public and private groups organized to fight poverty and deliver social services to their communities. Any community services block grant-funded agency is eligible for membership. The NACAA provides national representation, coalition building, a training and educational program, and technical assistance and conducts research and publishes a newsletter, *NACAA Network*, with reports on community action activities.

National Association of Neighborhoods

1651 Fuller Street NW
Washington, DC 20009
202-332-7766; fax: 202-332-2314

NAN is a national organization serving grassroots neighborhood groups across the country with policy advocacy, training, and conferences.

National Coalition for the Homeless

1012 14th Street NW, Suite 600
Washington, DC 20005-3406
202-737-6444; fax: 202-775-1316
email: nch@ari.net
http://nch.ari.net

The Coalition's mission is to end homelessness in the United States through public education, advocacy, and community organizing and by providing technical assistance to community groups.

National Community Building Network

672 13th Street, Suite 200
Oakland, CA 94612
510-893-2404; fax: 510-893-6657
email: network@ncbn.org
www.ncbn.org

The National Community Building Network (NCBN) is an alliance of local initiatives working to reduce poverty and create economic opportunity through comprehensive community-building strategies. It holds periodic meetings and engages in policy development and networking.

National Community Reinvestment Coalition

733 15th Street NW, Suite 540
Washington, DC 20005
202-628-8866; fax: 202-628-9800
email: ncrcmemb@gte.net
www.ncrc.org

The National Community Reinvestment Coalition (NCRC) is a nonprofit membership organization of almost seven hundred national, regional, and local members. In fulfilling its mission to increase the availability of credit for traditionally underserved people and communities, NCRC provides technical assistance to community groups through publications and workshops.

National Congress for Community Economic Development

1030 15th Street NW, Suite 325

Washington, DC 20005
877-44NCCED or 202-289-9020; fax: 202-289-7051
email: memberinfo@ncced.org
www.ncced.org

The National Congress for Community Economic Development (NCCED) is the trade association for community-based development organizations. Founded in 1970, NCCED supports its members and the field through advocacy, research, publications, training, mentorships, and conferences that offer networking and technical presentations. It publishes the *Development Times* and *Resources* newsletters. See also appendix J.

National Council of La Raza

1111 19th Street NW, Suite 1000
Washington, DC 20002
202-785-1670; fax: 202-785-0851
email: info@nclr.org
www.nclr.org

The National Council of La Raza (NCLR) is a constituency-based national organization that serves as a resource and information clearinghouse, coordinating body, training and technical assistance provider, and funding source for local Hispanic organizations. NCLR operates special programs on topics such as literacy, community health, and economic development.

National Economic Development and Law Center

2201 Broadway, Suite 815
Oakland, CA 94612
510-251-2600; fax: 510-251-0600
email: pubinfo@nedlc.org
www.nedlc.org

The National Economic Development and Law Center (NEDLC) is a national public interest law and planning center, established in 1969. NEDLC provides legal counsel and representation to community housing organizations; business and economic development initiatives; comprehensive strategic planning assistance; training in basic community economic development organizational and project issues; and publications.

National Housing Institute

439 Main Street
Orange, NJ 07050
973-678-9060; fax: 973-678-8437
email: hs@nhi.org
www.nhi.org

The National Housing Institute publishes the outstanding magazine *Shelterforce*, which analyzes housing problems from the people's point of view. The organization also serves as an organizing resource for tenants and housing activists around the country.

National Low Income Housing Coalition

1012 14th Street NW, Suite 610
Washington, DC 20005
202-662-1530; fax: 202-393-1973
email: info@nlihc.org
www.nlihc.org

Since 1975, the National Low Income Housing Coalition (NLIHC) has educated, advocated, and organized for decent affordable housing and for freedom of housing choices for all low-income people. A membership organization representing housing advocates, organizers, tenants, and professionals in the housing field, NLIHC works with Congress and the executive branch to obtain adequate federal support for low-income housing and related programs. Areas of concern to the Coalition include rent increases for low-income tenants in federally assisted housing, tax incentives for low-income housing investment, and tenant displacement. It publishes a weekly *Memo to Members*.

National Neighborhood Coalition

1875 Connecticut Avenue NW, Suite 410
Washington, DC 20009
202-986-2096; fax: 202-986-1941
email: nncnnc@erols.com
www.neighborhoodcoalition.org

The National Neighborhood Coalition (NNC) is the nonprofit "coalition of coalitions" whose national, regional, and local member organizations work on neighborhood issues. The Coalition sponsors regular information forums in DC with high-level government speakers, serves as a national voice for neighborhoods, and publishes an outstanding networking newsletter, *The Voice*.

National People's Action

810 North Milwaukee Avenue
Chicago, IL 60622-4103
312-243-3038; fax: 312-243-7044
email: npa@npa-us.org
www.npa-us.org

National People's Action (NPA) is an activist congress of over three hundred neighborhood and community organizations. It is the sister organization to the National Training and Information Center (see below). NPA does organizing and direct action on behalf of issues affecting low-income communities, including bank and insurance company disinvestment.

National Puerto Rican Coalition

1700 K Street NW, Suite 500
Washington, DC 20006
202-223-3915; fax: 202-429-2223
email: nprc@aol.com
www.bateylink.org

Composed of one hundred associations, the National Puerto Rican Coalition (NPRC) works to further the social, economic, and political well-being of Puerto Ricans throughout the United States and in Puerto Rico. NPRC conducts needs assessments and provides technical assistance to organizations based in Puerto Rican communities and/or organizations that provide services to Puerto Ricans.

National Rural Development and Finance Corporation

711 Navarro Street, Suite 350
San Antonio, TX 78205
210-212-4552; fax: 210-212-9159
email: rdfc@dcci.org
www.rdfc.org

The National Rural Development and Finance Corporation (NRDFC) is a national nonprofit intermediary, formed in 1977, to provide financial and technical assistance to economic development activities benefiting low-income rural people. NRDFC provides project-specific technical assistance and manages regional and national revolving loan funds to finance local projects.

National Training and Information Center

810 North Milwaukee Avenue
Chicago, IL 60622
312-243-3035; fax: 312-243-7044
email: ntic@ntic-us.org

The National Training and Information Center (NTIC) was established in 1972 as a resource center for grassroots groups in low-income and minority communities working on housing, community development, and revitalization. It provides training on neighborhood organizing and on confrontation, negotiation, and public relations strategies. Its research and publications focus on analyzing financial lending data for redlining and reinvestment campaigns. NTIC is affiliated with National People's Action, a nationwide network of grassroots groups, for which it sponsors national and regional conferences. (See also appendix K.)

National Trust for Historic Preservation

1785 Massachusetts Avenue NW
Washington, DC 20036
202-588-6064; fax: 202-588-6038
email: john_leith-tetrault@nthp.org
www.nthp.org

The National Trust for Historic Preservation (NTHP) started the Community Partners Program (CPP) to help preserve historic buildings and homes and to support livable and sustainable urban communities. The Community Partners Program targets communities with landmark projects of scale that will stimulate future neighborhood investment once they are developed. CPP's financing and real estate development experience helps local governments, nonprofit sponsors, and private developers rehabilitate historic properties and create stronger market dynamics for the surrounding community.

Neighborhood Funders Group

6862 Elm Street, Suite 320
McLean, VA 22101
703-448-1777; fax: 703-448-1780
email: nfg@nfg.org
www.nfg.org

The Neighborhood Funders Group (NFG) is a membership association of grant makers working to strengthen the capacity of organized philanthropy to understand and support community-based efforts to organize and improve the economic and social fabric of low-income urban neighborhoods and rural communities. The NFG provides its members with information, learning opportunities, and other professional development activities and encourages the support of policies and practices that advance economic and social justice.

Neighborhood Reinvestment Corporation

1325 G Street NW, Suite 800
Washington, DC 20005
202-220-3200; fax: 202-220-2160
email: webmaster@nw.org
www.nw.org

The Neighborhood Reinvestment Corporation (NRC) is a national nonprofit intermediary created by federal legislation in 1978 and funded primarily by congressional appropriations to revitalize neighborhoods by mobilizing local partnerships of public, private, and community actors. The Corporation provides start-up funding and ongoing technical assistance and networking to over 180 local NeighborWorks (tm) organizations, primarily

Neighborhood Housing Services groups, and holds regular training institutes (see appendix K). Its publications are the *Neighborhood Works Journal* and *Bright Ideas*.

Neighborhood Reinvestment also operates the Community Information Exchange, which is the "institutional memory" and information service for the community-based development field. The Exchange offers databases of case studies, funding and financing sources, and technical assistance providers, as well as an annotated bibliography.

Policy Link

101 Broadway
Oakland, CA 94607
510-663-2333; fax: 510-663-9684
email: info@policylink.org
www.policylink.org

This recently established national nonprofit organization focuses on promoting community building. It seeks to link transportation, housing, health, and community development policies and practices to eliminate poverty. It promotes community building through research and policy development, advocacy, communications (telling the stories of successful initiatives), and capacity building through publications, Web site conversations, and convenings.

Pratt Institute Center for Community and Environmental Development

379 Dekalb Avenue, 2nd floor
Brooklyn, NY 11205
718-636-3486; fax: 718-636-3709
email: picced@picced.org
www.picced.org

The oldest university-based advocacy planning organization, dating back to 1963, the Pratt Institute for Community and Environmental Development (PICCED) provides technical assistance and training to neighborhood groups in New York City and the tri-state region in organizational development, participatory planning, real estate financing, affordable and special housing needs, commercial and industrial development, and public policy analysis (see appendix K). Its affiliated Planning and Architectural Collaborative provides architectural, design, and construction services to neighborhood-based housing groups in the New York area. It houses the Planners Network, a national membership association of progressive planners, which publishes the *Planners Network* newsletter.

Rural Community Assistance Corporation

3120 Freeboard Drive, Suite 201
West Sacramento, CA 95691
916-447-9832; fax: 916-447-2878
wfrench@rcac.org
www.rcac.org

The Rural Community Assistance Corporation (RCAC) provides training, technical assistance, and resources to rural disadvantaged neighborhoods through partnerships with communities and local organizations. RCAC has a predevelopment loan fund to assist in housing development; a housing division that assists sponsors of farm labor, rural self-help, and multifamily and rehab housing development; and an environmental service division that assists communities with water and wastewater projects.

Structured Employment/Economic Development Corporation (Seedco)

915 Broadway, Suite 1703
New York, NY 10010
212-473-0255; fax: 212-473-0357
email: ggreen@seedco.org
Web site under development

Structured Employment/Economic Development Corporation (Seedco) is a national nonprofit intermediary that creates local partnerships among community-based groups and institutions, such as universities and medical centers, to launch pilot initiatives to demonstrate new ways to help revitalize impoverished neighborhoods. Projects have involved job training and employment and community health.

Sustainable America

42 Broadway, Suite 1740
New York, NY 10004-1617
212-269-9550; fax: 212-269-9557
www.sanetwork.org

A national nonprofit membership organization recently formed as a national forum to promote "new economies" of sustainable economic development (covering worker's rights, environmental sustainability, "high-road" business development, and community design). Members encompass environmental justice and conservation action, labor, religious, and community development groups. The organization's agenda includes work-study groups, a cooperative bank of technical assistance providers, newsletter, resource materials, and annual general assembly.

Local Organization

Hundreds of resource organizations, within the public and nonprofit sectors, exist at the local level. The Community Information Exchange (now housed in the Neighborhood Reinvestment Corporation; see above) maintains the most extensive database of these organizations. One example is:

Community Economic Development Assistance Corporation

18 Tremont Street
Boston, MA 02108
617-727-5994; fax: 617-727-5990
email: cedac@cedac.org

Established by the Massachusetts legislature in 1978, the Community Economic Development Assistance Corporation provides technical assistance and predevelopment loans to nonprofit developers of affordable and special needs housing; comanages a fund for the development of child care facilities; and engages in public policy work, especially to preserve the state's stock of rental housing units.

Part II: Related Organizations

American Communities

P.O. Box 7189
Gaithersburg, MD 20898-7189
800-998-9999; fax: 310-519-5027
email: comcon@aspensys.com
www.comcon.org

American Communities is an information service operating under contract with HUD that provides information about HUD's housing and community development programs as well as information and materials on housing and community development activities and resources.

American Community Gardening Association

100 North 20th Street, 5th floor
Philadelphia, PA 19103-1495
215-988-8785; fax: 215-988-8810
email: smccabe@pennhort.org
www.communitygarden.org

ACGA is a national nonprofit membership organization of professionals, volunteers, and supporters of community gardening in urban and rural communities, including community food gardening, urban forestry, and the preservation of open space. It works to form state and regional community gardening networks, develop resources, encourage research, and conduct educational programs.

American Institute of Architects

R/UDAT
1735 New York Avenue NW
Washington, DC 20006
202-626-7532
www.aia.org

The Rural/Urban Design Assistance Team (R/UDAT), a public service of the American Institute of Architects (AIA), offers multidisciplinary teams to solve community problems. Responding to a local request, R/UDAT members will participate in a phase of community discussion to define the issues; R/UDAT next picks a multidisciplinary team from outside the locality, including, for example, an architect, a land-use lawyer, and community development specialist, to address the issues and participate in intensive town meetings; then it assists the locality to develop and implement a plan. R/UDAT's work is pro bono; the community pays only the team's travel and living expenses.

American Planning Association

122 South Michigan Avenue, Suite 1600
Chicago, IL 60603
312-431-9100; fax: 312-431-9985
www.planning.org

The APA is a national nonprofit membership organization providing research, information, and advice to planners and others interested in planning issues. It has local chapters that develop educational programs.

American Public Human Services Association

810 First Street NE, Suite 500
Washington, DC 20002-4267
202-682-0100; fax: 202-289-6555
email: info@aphsa.org
www.aphsa.org

The American Public Human Services Association (APHSA), formerly known as the Public Welfare Association, is a membership organization concerned with the improvement of human services policies and programs, including welfare, child welfare, health care, and other issues involving families and the elderly. The Association publishes a directory of public welfare programs and agencies and numerous publications on public policies and program initiatives involving welfare.

Association for Community Based Education

1805 Florida Avenue NW
Washington, DC 20009
202-462-6333

The Association for Community Based Education (ACBE) was founded in 1976 to assist community-based educational institutions serving the needs of people who are not served adequately by traditional public and private institutions. ACBE seeks to promote indigenous community development and local control by providing financial and technical support to the field of community-based education through direct technical assistance, research, program evaluation, mini-grants, scholarships, and advocacy to its member organizations. It is currently focusing its energies in literacy, economic development, community-based colleges, and rural development.

Association for Community Design

c/o Portland Community Design
2014 Northeast Martin Luther King
Portland, OR 97212
503-281-8011; fax: 503-281-8012
email: rcurry7@ix.netcom.com
www.communitydesign.org

The Association for Community Design is the trade association for dozens of local nonprofit community design centers across the country, which through their own staffs or volunteers from architectural firms provide planning and architectural services to distressed communities.

Brookings Institution

Center on Urban and Metropolitan Policy
1775 Massachusetts Avenue NW
Washington, DC 20036
202-797-6139; fax: 202-797-2965
email: mlambert@brook.edu
www.brook.edu

The Brookings Institution Center on Urban and Metropolitan Policy seeks to shape a new generation of urban policies that will help build strong cities and metropolitan regions. In partnership with academics, private- and public-sector leaders, and locally elected officials, the Center informs the national debate on the impact of government policies.

Center for Budget and Policy Priorities

820 First Street, NE, Suite 510
Washington, DC 20002
202-408-1080, fax: 202-408-1056
www.cbpp.org

The Center for Budget and Policy Priorities (CBPA) analyzes national data and prepares reports on current policy and budget issues on poverty, income, labor, hunger and welfare, and state fiscal policy. CBPP also reports on income distribution trends as they affect certain minority sectors, special analysis of federal budget issues including the president's budget request, and other analyses of poverty issues including one based on the most current census data.

Center for Cooperative Housing

1614 King Street
Alexandria, VA 20002
703-684-3185; fax: 703-549-5204
email: coophousing@usa.net
www.coophousing.org

The Center for Cooperative Housing (CCH) was formed in 1991 as a subsidiary of the National Association of Housing Cooperatives to focus on the creation of homeownership opportunities through the development of housing cooperatives. The CCH mission is to assist community-based groups, resident associations, nonprofit organizations, and government agencies in developing and organizing housing cooperatives for low- and moderate-income individuals and families by linking potential local sponsors with the cooperative housing network and with the emerging set of federal programs designed to facilitate the creation of cooperative housing.

Center for Labor and Community Research

3411 West Diversey Avenue, Suite 10
Chicago, IL 60647
773-278-5418; fax: 773-278-5918
www.clrc.org

The Center (formerly known as the Midwest Center for Labor Research) is a nonprofit consulting and research organization that works for unions, community and development organizations, the religious community, state and local governments, and others who share a commitment to jobs, justice, and economic stability.

Center for Living Democracy

289 Fox Farm Road, P.O. Box 8187
Brattleboro, VT 05304-8187
802-254-1234; fax: 802-254-1227
email: info@livingdemocracy.org
www.livingdemocracy.org

The Center inspires people and helps them enhance their skills to solve local problems and as citizens to "do democracy" to create a healthier community and society. The Center's American News Service (www.americannews.com) researches and reports on the problem-solving initiatives taking place in organizations and communities across the country and provides stories of those innovations to the media.

Center for Policy Alternatives

1875 Connecticut Avenue NW, Suite 710
Washington, DC 20009
202-387-6030; fax: 202-986-2539
email: info@cfpa.org
www.cfpa.org

The Center for Policy Alternatives, founded in 1975, is a progressive nonpartisan, nonprofit, public policy center that promotes innovation at the state and local levels in economic development, the environment, women's economic justice, governance, and leadership. CPA's activities include the regular publication of its *Ways & Means* newsletter, the publication and dissemination of books and state reports, the provision of technical assistance, the dissemination of model legislation, and the convention of state and regional conferences.

Center for Youth Development

1875 Connecticut Avenue NW
Washington, DC 20009
202-884-8267; fax: 202-884-8404
email: cyd@aed.org

The Center for Youth Development and Policy Research, established in 1990 by the Academy for Educational Development, is devoted to researching, publicizing, and advocating on youth issues. Projects have included Community YouthMapping, the mobilization of young people to research and document what resources and support systems are available to them; and the development of a youth-worker training curriculum in collaboration with the National Network for Youth, Inc.

Consortium for Housing and Asset Management

10227 Wincopin Circle, Suite 500
Columbia, MD 21044
410-772-2724; fax: 410-964-1918
www.cham.org

The Consortium for Housing and Asset Management (CHAM) provides asset and property management training for providers of low-income nonprofit housing. See appendix K.

Institute for Local Self Reliance

2425 18th Street NW
Washington, DC 20009-2096
202-232-4108; fax: 202-332-0463
email: ilsr@agc.apc.org
www.ilsr.org

The Institute for Local Self Reliance is a nonprofit technical assistance provider and research organization established in 1974 to foster community development through local resource utilization. The Institute specializes in assisting businesses, municipalities, community organizations, and state government agencies in the areas of economic development, municipal waste utilization, utility economics, community planning, and public policy change.

Ms. Foundation for Women

120 Wall Street
New York, NY 10005
212-742-2300; fax: 212-742-1653
email: info@ms.foundation.org
www.ms.foundation.org

Ms. Foundation for Women is a multi-issue women's foundation with a special Economic Development Program that provides grassroots women's organizations with education and training in economic development; on-site technical assistance; internship opportunities; information and referral; and short-term problem-solving assistance. The Foundation sponsors an annual national training institute on women and economic development.

National Affordable Housing Network

P.O. Box 3706
Butte, MT 59702
406-782-8145; fax 406-782-5168
email: nahn@nahn.com
www.nahn.com

The Network, whose members are practitioners and researchers working to change the way low-cost housing is built for disadvantaged households, provides research design and evaluation assistance, education, information, and policy and program design assistance.

National Association of Community Health Centers, Inc.

1330 New Hampshire Avenue NW, Suite 122
Washington, DC 20036
202-659-8008; fax: 202-659-8519
email: nachc@erols.com

www.nachc.com

Founded in 1970, the National Association of Community Health Centers, Inc. (NACHC) advocates on behalf of community-oriented primary health care programs and the medically underserved and uninsured people they serve. The mission of the NACHC is to achieve access to quality health care for all Americans by ensuring and supporting the continued existence and expansion of community-oriented primary health care programs. NACHC provides legislative advocacy, education, training, and resource opportunities, as well as information and technical assistance on issues and trends affecting health care programs and medically underserved and uninsured populations.

National Association of Housing Cooperatives

1614 King Street
Alexandria, VA 22314
703-549-5201; fax: 703-549-5204
email: coophousing@usa.net
www.coophousing.org

The National Association of Housing Cooperatives is a federation of housing cooperatives, professionals, organizations, and individuals promoting the interests of cooperative housing communities. NAHC provides technical assistance on the development, organization, and operation of housing cooperatives, and also provides extensive training programs.

National Association of Housing and Redevelopment Officials

630 Eye Street NW
Washington, DC 20001-3736
202-289-3500; fax 202-289-8181
email: nahro@nahro.org
www.nahro.org

The National Association of Housing and Redevelopment Officials (NAHRO) is a professional membership organization representing local housing authorities, community development agencies, and individual professionals in the housing and community development fields. Founded in 1933, NAHRO works to provide safe, decent, and affordable housing for low- and moderate-income people, providing its nine thousand members with information on federal policy, legislation, regulations, and funding. NAHRO also provides professional development and training programs in all phases of agency operations, including management, maintenance, and procurement. NAHRO's *Monitor* newsletter is published semimonthly.

National Council for Urban Economic Development

1730 K Street NW, Suite 700
Washington, DC 20006
202-223-4735; fax: 202-223-4745
email: mail@urbandevelopment.com
www.cued.org

The National Council for Urban Economic Development (CUED) is a membership organization serving public and private participants in economic development across the United States and around the world. CUED members include public economic development directors, chamber of commerce staff, utility executives, and academicians. CUED provides professional development opportunities, publications (including the *Economic Development* newsletter), and networking conferences to its members.

National Development Council

211 East 4th Street
Covington, KY 41011
606-291-0220; fax: 606-291-3774
email: NDCCVG@aol.com
www.ndc-online.org

The National Development Council is a private, nonprofit corporation that focuses on job creation, small-business growth, and the economic revitalization of our nation's small towns.

National Housing & Rehabilitation Association

1726 18th Street NW
Washington, DC 20009
202-328-9171; fax: 202-265-4435
email: gpeth@aol.com
www.housingonline.com

The National Housing & Rehabilitation Association (NH&RA) is a network of technically sophisticated and highly experienced multifamily developers, lenders, mortgage and investment bankers, attorneys, general contractors, and state and local housing officials. It showcases innovative and replicable development projects; updates members on available financing and federal and state legislative and regulatory issues; and facilitates business development opportunities.

Partners for Livable Communities

1429 21st Street NW
Washington, DC 20036
202-887-5990; fax: 202-466-4845

email: partners@livable.com
www.livable.com
Partners for Livable Communities seeks to strengthen local communities economically and socially through the arts and improved public spaces. It produces publications, convenes conferences, and provides technical assistance on a fee-for-service basis.

Urban Institute

2100 M Street NW
Washington, DC 20037
202-261-5709
email: paffairs@ui.urban.org
www.urban.org
The Urban Institute is a nonprofit research organization established in 1968. The Institute investigates social and economic problems confronting the nation and the government policies and public and private programs designed to alleviate them. A multidisciplinary staff of about 230 works in seven policy centers: Health Policy, Human Resources Policy, Income and Benefits Policy, International Activities, Population Studies, Public Finance and Housing, and State Policy.

This appendix describes a sampling of the organizations that do community development work overseas. Entries are listed alphabetically.

Accion International

120 Beacon Street
Somerville, MA 02143
617-492-4930; fax: 617-876-9509
email: info@accion.org,
www.accion.org

Accion International is a nonprofit organization started in 1961 to reduce poverty by providing loans and other financial services to poor and low-income people who start their own tiny businesses. An international lender in the micro-lending field, Accion is an umbrella organization for a network of microfinance institutions in thirteen Latin American countries and eight U.S. cities. Accion provides loans and training to local microfinance organizations and then helps replenish loan funds by repackaging and selling the microloans made by the local organizations to U.S. investors. Accion believes that bringing financial services to the smallest of small business people at fair rates of interest can help people get off welfare, rebuild inner-city neighborhoods, and provide a valuable alternative for those left behind by factory closings and corporate downsizing.

American Friends Service Committee

1501 Cherry Street
Philadelphia, PA 19102
215-241-7000; fax: 215-241-7275
email: idgeneral@afsc.org,
www.afsc.org

The American Friends Service Committee (AFSC) is a Quaker organization that includes people of various faiths who are committed to social justice, peace, and humanitarian service. Today, AFSC conducts social and technical assistance programs designed to enable people to develop their own power and resources. For example, AFSC has worked with rural women to improve economic and living conditions in Mozambique; worked in Asia (Laos, Vietnam, and Cambodia) to help villages with irrigation, drinking water, and food production; and worked on sustainable development and the promotion of alternative markets with farmers in Honduras.

Foundation for International Community Assistance

1101 14th Street NW, 11th floor
Washington, DC 20005
202-682-1510; fax: 202-682-1535
email: finca@villagebanking.org
www.villagebanking.org

The Foundation for International Community Assistance (FINCA International), started in 1984, now serves more than 121,000 poor families in fourteen countries. FINCA's mission is to support the economic and human development of families trapped in severe poverty by providing small loans (microlending) of working capital to low-income families. FINCA provides microlending services through "village banks," groups of people (usually women) who guarantee each other's loans. FINCA currently has projects in Central and South America, the Caribbean, Asia, Africa, and North America.

Grassroots International

48 Grove Street, #103
Somerville, MA 02144
617-524-1400; fax: 617-524-5525
email: grassroots@igc.org
www.grassrootsonline.org

Grassroots International is an independent development and information agency that channels humanitarian aid to democratic social change movements in the third world. Grassroots International builds grassroots-level partnerships that provide a unique basis to inform the media and the public about Third World conflicts and crises.

Specifically, this organization focuses on strategically important areas where national movements are challenging economic and social inequalities and repressive political systems. Grassroots International currently funds relief and development projects in Africa, the Middle East, and the Philippines. To safeguard its independence, they do not accept funds from any government agency, and all programs are wholly supported by private donations.

Habitat for Humanity

See appendix D.

Institute for International Cooperation and Development

P.O. Box 103
Williamstown, MA 01267
413-458-9828; fax: 413-458-3323
email: iicd@berkshire.net
www.iicd-volunteer.org

The Institute for International Cooperation and Development (IICD) is a nonprofit educational organization founded in 1986 to promote global understanding and international solidarity. IICD trains volunteers, eighteen years old and over, for work in community development projects in Africa, Latin and South America, and Asia. Volunteer programs are six to twenty months long and include preparation and follow-up periods at the Institute in Massachusetts. After returning to the United States participants spend time creating educational materials and giving presentations to schools, organizations, and the general public.

InterAction (American Council for Voluntary International Action)

1717 Massachusetts Avenue NW, Suite 801
Washington, DC 20036
202-667-8227; fax: 202-667-8236
email: webmaster@interaction.org
www.interaction.org

InterAction is the nation's largest coalition of international development, disaster relief, and refugee assistance agencies, with more than 150 member organizations. InterAction member organizations share the same fundamental goals of easing human suffering and strengthening people's ability to help themselves. InterAction seeks to foster partnerships, collaboration, and leadership among its member organizations, especially in the areas of disaster relief, sustainable development, and American foreign policy. For people interested in international community development, InterAction's Web site is a great resource on organizations and volunteer and job opportunities.

InterAmerican Foundation

901 North Stuart Street, 10th floor
Arlington, VA 22203
703-841-3800; fax: 703-841-0973
email: correa@iaf.gov
www.iaf.gov

Funded by the American Congress under the Foreign Assistance Act of 1969, the InterAmerican Foundation is an independent agency of the U.S. government, created in 1969 as an experimental U.S. foreign assistance program. Since then, IAF has worked in Latin America and the Caribbean to promote equitable, responsive, and participatory self-help development by awarding grants directly to local organizations throughout the region. The IAF also enters into partnerships with public- and private-sector entities to scale-up support and mobilize local, national, and international resources for grassroots development.

International Development Exchange

827 Valencia Street, Suite 101
San Francisco, CA 94110-1736
415-824-8384; fax: 415- 824-8387
email: idex@igc.org
www.idex.org

The International Development Exchange (IDEX) was founded in 1985 to support community-based development in Africa, Asia, and Latin America and to offer U.S. citizens an avenue for international understanding and action. IDEX helps link sponsors in the United States (usually student groups, religious organizations, corporations, foundations, and individuals) with community-based projects overseas. Typical projects supported by IDEX are organized around women's groups, agricultural development, social services, cooperatives, and micro-enterprise development.

Oxfam International

Oxfam America Office
26 West Street
Boston, MA 02111-1206
617-482 1211; fax: 617- 728-2595
email: info@oxfamamerica.org
www.oxfamamerica.org

Oxfam International, an international development, advocacy, and relief agency, is a partnership of ten different Oxfams in three continents dedicated to ending poverty worldwide. Oxfam works in seventy countries in Africa, Asia, the Middle East, Latin America, the Caribbean, Eastern Europe, and the former Soviet Union countries, with the basic belief that poor people have to have a say in the decisions that shape their lives. In partnership with local organizations, Oxfam helps develop programs in emergency relief and sustainable development.

Peace Corps

1111 20th Street NW
Washington, DC 20526
800-424-8580
www.peacecorps.gov

Since the organization's creation in 1961, more than 150,000 Peace Corps volunteers have fought hunger, disease, illiteracy, poverty, and lack of opportunity around the world. Currently, 6,500 volunteers are serving in eighty-four countries in Africa, Asia, the Pacific, Central and South America, the Caribbean, Central and Eastern Europe, and the former Soviet Union. Peace Corps volunteers work in the fields of agriculture, business, education, the environment, forestry, health, engineering, the skilled trades, the teaching of English, urban development, and youth development. All expenses of Peace Corps service are paid by the program. Volunteers also receive foreign language training; opportunities for graduate school scholarships and fellowships; partial cancellation of eligible Perkins loans and deferment of most federally guaranteed college loans; complete medical and dental care during Peace Corps service; easier access to federal jobs after service through noncompetitive eligibility; and a readjustment allowance of $5,400 after twenty-seven months of service. Peace Corps volunteers must be U.S. citizens, be at least eighteen, possess skills and experience requested by Peace Corps host countries, and meet Peace Corps medical and legal requirements. Contact the national Peace Corps office for a listing of regional offices.

U.S. Agency for International Development

Information Center
Ronald Reagan Building
Washington, DC 20523-0016
202-712-4810; fax: 202-216-3524
email: info@usaid.gov
www.info.usaid.gov

The United States Agency for International Development (USAID) is the independent government agency that provides economic and humanitarian assistance to advance U.S. economic and political interests overseas. USAID addresses the needs of developing countries and emerging democracies by funding the efforts of private voluntary organizations (PVO) and nongovernmental organizations (NGO) working on sustainable programs in such countries' organizations overseas.

Visions in Action

2710 Ontario Road NW
Washington, DC 20009
202-625-7403
email: visions@igc.org

Visions in Action is an international nonprofit organization that sends volunteers to work for six months or one year with nonprofit organizations, health centers, and the media in Tanzania, Uganda, Zimbabwe, South Africa, Burkina Faso, and Mexico. Volunteers work in human rights, journalism, micro-enterprise, public health education, community development, housing, research, youth coordination, social work, the environment, refugee relief, theater, and other sectors. Vision in Action also publishes the newletter *Action Notes* four times a year.

The institutions described here are those that have financing as their central function. In addition, many national community development organizations have a financing role, as do federal, state, and local agencies. The entries that follow are organized alphabetically, within sections that include: community development financial institutions, mainstream banking institutions, regulatory agencies, secondary market institutions, and syndicators.

Community Development Financial Institutions

A community development finance institution (CDFI) is a nongovernmental financing entity with the primary mission of promoting community development by investing in and providing financial services to low-income communities or other underserved markets. Examples of CDFIs include community development credit unions, development banks, community banks, community development venture capital funds, micro-enterprise funds, community development corporation–based lenders, and revolving loan funds.

Coalition for Community Development Financial Institutions

924 Cherry Street, 2nd floor
Philadelphia, PA 19107-2411
215-923-5363; fax: 215-923-4755
email: cdfi@cdfi.org
www.cdfi.org

The Coalition is the public policy and advocacy organization for the CDFI sector. It also maintains a directory of CDFIs across the country and hosts a national conference yearly. Its sister organization is the National Community Capital Association.

Community Development Financial Institutions Fund

U.S. Department of the Treasury
601 13th Street NW, Suite 200-South
Washington, DC 20005
202-622-8662; fax: 202-622-7754
Email for outreach department head: szablyah@cdfi.treas.gov
www.treas.gov/cdfi

The CDFI Fund, created by Congress in 1994 and organized as a wholly owned government corporation within the U.S. Department of the Treasury, seeks to make credit, capital, and financial services available to underprivileged communities in two ways. It offers grants and technical assistance to enhance the effectiveness of local community development financial institutions (CDFIs); and it offers grants to traditional banks as incentives to expand their services to meet distressed communities' credit needs.

It also publishes a newsletter, *CDFI Fund Quarterly*, and supports a special training program to enhance the skills of CDFIs.

Community Development Venture Capital Alliance

9 East 47th Street, 5th floor
New York, NY 10017
212-980-6790; fax: 212-980-6791
email: cdvca@cdvca.org
www.cdvca.org

CDVCA is an association of venture capital funds, community development corporations, and other organizations that provide equity investments in fledgling companies that have the potential for creating sustainable development, corporate citizenship, and social entrepreneurship. CDVCA members integrate community and environmental concerns into their professionally managed venture capital portfolios. CDVCA provides its members with information access, training, and opportunities for peer learning; it organizes teams of consultants to provide technical assistance to members' projects and to help new funds get organized.

Housing Trust Fund Project

1113 Cougar Court
Frazier Park, CA 93225
805-245-0318; fax: 805-245-2518
email: Brooksm@commchange.org

The Housing Trust Fund Project (HTFP), initiated by the Center for Community Change, tracks housing trust funds across the country, and provides information and technical assistance to those working to create and implement housing trust funds. Typically established by units of government, housing trust funds are a dedicated source of revenue used for the production and preservation of housing affordable to lower-income households. HTFP publishes a directory, *A Status Report on Housing Trust Funds in the United States*, and a quarterly report, *News from the Housing Trust Fund Project*.

National Community Capital Association

924 Cherry Street, 2nd floor
Philadelphia, PA 19107
215-923-4754; fax: 215-923-4755
email: ncca@communitycapital.org
www.communitycapital.org

The National Community Capital Association is a national membership organization for nonprofit community development financial institutions (CDFIs). NCCA serves its members with training, technical assistance, and performance-based financing. Its newsletter is the *Community Investment Monitor*. (Until December 1997, the Association was known as the National Association of Community Development Loan Funds, or NACDLF.) NCCA's sister organization is the Coalition for Community Development Financial Institutions, which does public policy and advocacy for the field.

National Federation of Community Development Credit Unions

120 Wall Street, 10th floor
New York, NY 10005
212-809-1850; fax: 212-809-3274
email: email@natfed.org
www.natfed.org

NFCDCU, founded in 1974, is the membership association for credit unions that serve predominantly low-income and minority communities. It has a growing membership of over 165 credit unions nationwide, providing advocacy, capital, training, and program support and provides assistance to groups interested in forming credit unions.

NCB Development Corporation

1401 Eye Street NW, Suite 700
Washington, DC 20005
800-955-9622 or 202-336-7680; fax: 202-336-7804
email: webmaster@ncb.com
www.ncb.com

NCBDC is the development finance affiliate of the National Cooperative Bank, an institution founded to provide financial services to the nation's cooperative business sector. NCBDC provides risk capital and other types of development financing to community-based nonprofit organizations as well as to start-up and newly established cooperatives, with emphasis on low- to moderate-income facilities (from limited equity housing cooperatives to community health centers).

Shorebank Advisory Services

1950 East 71st Street
Chicago, IL 60649
773-753-5694; fax: 773-753-5880
email: helen_dunlap@sbk.com
www.sbk.com/advisory.htm

This consulting group, established in 1988, is a subsidiary of the Shorebank Corporation, the bank holding company that includes the South Shore Bank. The South Shore Bank was the first bank deliberately organized to use the tools of a private financial institution to stimulate market forces and create a systematic cycle of economic development in its South Shore neighborhood. It became the model for other development banks across the country and the inspiration for the creation of the federal CDFI Fund.

Shorebank Advisory Services provides technical assistance to communities in setting up their own development financial institutions and other community economic development initiatives.

Mainstream Banking Institutions

American Bankers Association

1120 Connecticut Avenue NW
Washington, DC 20036
800-338-0626
email: custserv@aba.com
www.aba.com
Center for Community Development
202-663-5103; fax: 202-663-5359
email: choward@aba.com

The ABA is the national trade association for banks, including smaller community banks, mid-sized and money center banks, and some thrifts. The national ABA holds an annual conference on community and economic development banking. The ABA's Center for Community Development, founded in 1993, promotes nondiscriminatory lending practices and encourages investments in minority and low- and moderate-income communities. Each state also has a state bankers association.

American League of Financial Institutions

900 19th Street NW Suite 400
Washington, DC 20006
202-628-5624; fax: 202-296-8716
email: info@alfi.org
www.alfi.org

Established in 1948, the league represents the interests of African-American-, Asian-American- and Hispanic-American-controlled community banks. Besides its involvement in federal legislation and policy making, ALFI offers technical assistance to financial institution start-up groups.

America's Community Banks

900 19th Street NW, Suite 400
Washington, DC 20006
202-857-3100; fax: 202-296-8716
email: info@acbankers.org
www.acbankers.org

Representing some five thousand progressive, community-oriented savings banks and thrift associations across the country, ACB lobbies on the federal level and provides services to its members and the public, including publications, conferences, and training programs.

Independent Bankers of America Association

One Thomas Circle NW, Suite 400
Washington, DC 20005-5802
800-422-8439 or 202-659-9216; fax: 202-659-9216
email: info@ibaa.org
www.ibaa.org

IBAA represents some 5,500 smaller, independently owned and operated community banks nationwide. The IBAA's Community Banking Network subsidiary provides education seminars and regular newsletters to members and makes services available to member banks in common, enabling them to issue credit cards, offer competitive mortgage rates, and provide other financial services. Each state also has an IBAA network.

Mortgage Bankers Association of America

1125 15th Street NW
Washington, DC 20005-2766
202-861-6500
www.mbaa.org

This is a trade organization that serves as the voice of the real estate industry. Members include mortgage companies, commercial banks, private mortgage insurers, and others engaging in mortgage banking. The MBAA lobbies at the federal level, produces publications, and provides business insurance. It runs education programs, including a School of Mortgage Banking, correspondence and distance-learning courses, and conferences and seminars.

National Association of Affordable Housing Lenders

2121 K Street NW, Suite 700
Washington, DC 20037
202-293-9850; fax: 202-293-9852
email: naahl@naahl.org

NAAHL is a trade association for lending institutions particularly interested in community development lending. It holds training conferences, advocates on behalf of affordable housing, and publishes a monthly newsletter that tracks affordable housing finance innovations and relevant national legislation. It publishes a periodic newsletter about affordable housing finance innovations and relevant national legislation. It has published a booklet, *Best Practices in Community Development Lending: Building Sustainable Communities*, with twelve case studies of successful affordable housing and development projects.

National Association of Urban Bankers

1801 K Street NW, Suite 200A
Washington, DC 20006
202-861-0000
email: naub340@aol.com
www.naub.org

This is the national nonprofit membership organization for minority professionals working in the banking and financial services industries. The Association has fifty-four chapters across the nation. It offers members networking, information, and education opportunities.

National Bankers Association

1513 P Street NW
Washington, DC 20005
202-588-5432
www.nationalbankers.org

This is a trade association for 103 minority- and women-owned banks (MWOBs) that primarily serve economically distressed urban areas. The NBA provides services and information to its members and advocates on legislative and regulatory matters that concern MWOBs and their communities.

Regulatory Agencies

Federal Deposit Insurance Corporation

Division of Compliance and Consumer Affairs
550 17th Street NW, Room F139

Washington, DC 20429

202-942-3437; fax: 202-942-3098

www.fdic.gov

The FDIC is the agency that provides insurance to all federally insured financial institutions. It also supervises state-chartered banks that are not members of the Federal Reserve System. It has eight regional offices: in Atlanta, Boston, Chicago, Dallas, Kansas City, Memphis, New York, and San Francisco.

In 1990, the FDIC created a Community Affairs program to promote compliance with the Community Reinvestment Act, the Home Mortgage Disclosure Act, and other fair lending laws. There are Community Affairs Officers in each of the agency's regional offices, as well as a Community Affairs Division at its headquarters. The community affairs officer, backed by a staff of three to five, is responsible for promoting fair lending activities through educational outreach to financial institutions, community groups, and government agencies; for fostering and maintaining relationships with community development initiatives; and for providing technical assistance to the examiners who examine the FDIC-regulated financial institutions.

Federal Housing Finance Board

1777 F Street NW

Washington, DC 20006-5210

202-408-2500

www.fhfb.gov

www.fhlbanks.com (links to regional banks)

The Federal Housing Finance Board regulates the twelve FHL Banks and their over 7,000 members that constitute the FHL Bank system. The FHL Bank system was created in 1932 to make credit accessible to home-buying Americans. The System is made up of twelve district banks (in Atlanta, Boston, Chicago, Cincinnati, Dallas, Des Moines, Indianapolis, New York, Pittsburgh, San Francisco, Seattle, and Topeka) and a central Office of Finance, all of which are supervised and regulated by the Federal Housing Finance Board.

Each Federal Home Loan Bank is government chartered but privately financed and managed. Its stockholders are other financial institutions, which, as "members," are eligible to receive advances from the Bank. All federally chartered thrifts are members of one of the twelve Banks.

In 1989, following the savings and loan crisis and bailout, Congress created the Affordable Housing Program and the Community Investment Program, operated through the Federal Home Loan Banks, to provide low-cost loans to community projects producing low-income housing and community economic revitalization.

Each Bank has a Community Investment Officer (CIO) who is responsible for these programs as well as overseeing its members' community investment activities and providing technical assistance and outreach to ventures benefiting disadvantaged populations and communities.

Federal Reserve System

Constitution Avenue and 20th Street, NW

Washington, DC 20551

202-452-3378

www.bog.frb.fed.us

The Federal Reserve System was created in 1913 by Congress to provide the nation with a safer, more flexible, and more stable monetary system. The Federal Reserve System consists of the Federal Reserve Board and a network of twelve Federal Reserve banks in major cities across the country: Atlanta, Boston, Chicago, Cleveland, Dallas, Kansas City, Minneapolis, New York, Philadelphia, Richmond, St. Louis, and San Francisco. The Federal Reserve seeks to ensure the stability of the American economy by influencing the flow of money and credit in the nation's economy and by supervising the activities of certain state-chartered banks and bank holding companies. It is also charged with writing regulations for the major consumer credit laws.

Since 1981, the Federal Reserve System has devoted significant resources to community affairs efforts to help financial institutions better understand and respond to community development and reinvestment needs in their communities, and to related fair lending issues. Altogether the Board and reserve banks have a community affairs staff of about ninety people. Community affairs officers and staff design and sponsor conferences, seminars, and training workshops and produce publications on community development and reinvestment topics for bankers and government officials and representatives from community development and small business organizations. Each Federal Reserve district publishes its own newsletter. Staff also convene outreach meetings and provide hands-on technical assistance to bankers and community development groups to help them structure effective community development and reinvestment programs and entities, such as multibank loan consortia and bank-owned community development corporations.

Office of the Comptroller of the Currency

Community Development Division

250 E Street SW

Washington, DC 20219

202-874-4930; fax: 202-874-5566

email: bud.kanitz@occ.treas.gov

www.occ.treas.gov

The Office of the Comptroller of the Currency (OCC) charters, regulates, and supervises national commercial banks. To encourage national bank involvement in community and economic development activities, the OCC offers policy guidance, training, and technical assistance to financial institutions; manages several community development projects; and maintains an electronic database of bankers and others who have an interest in community and economic development.

Each of the six regional district offices is staffed with two community reinvestment and development specialists, whose responsibility is to encourage national banks to form partnerships with community developers and to fulfill their obligations under the Community Reinvestment Act. At headquarters, the director of community relations and his staff are responsible for the Comptroller's outreach activities to promote community development, consumer protection, and fair lending.

Office of Thrift Supervision

U.S. Department of the Treasury
1700 G Street NW, 6th floor
Washington, DC 20552
202-906-7087; fax: 202-906-5735
email: community@ots.treas.gov
www.ots.treas.gov

OTS, a bureau of the U.S. Department of the Treasury, charters and regulates the federal thrift institutions (which are members of the Federal Home Loan Bank system); it regulates many state-chartered thrifts as well.

Established in 1993, the OTS Community Affairs program seeks to build partnerships to stimulate thrift industry participation in community development activities. The staff includes three professionals in headquarters and two in each of the five regional offices (Atlanta, Chicago, Dallas–Ft. Worth, Jersey City, and San Francisco). Their work involves outreach, including holding conferences, training, research, developing educational materials on specific topics such as lending on tribal lands, and providing direct assistance to financial institutions and support of OTS examiners. The staff also oversees an OTS program to support and promote minority-owned thrifts.

The OTS sponsors a yearly Financial Regulatory Fellows Program; in some years, Fellows are selected to work in the Community Affairs program.

Secondary Market Institutions

Secondary-market institutions help replenish lenders' supply of money so that the lenders can continue to make loans. The secondary-market institution accomplishes this by buying loans (including mortgages) from the original lender, pooling the loans, then selling pieces of the loan pools or bonds backed by the loan pools to private investors.

AFL-CIO Investment Trust

1717 K Street NW, Suite 707
Washington, DC 20006
202-331-8055; fax: 202-331-8190
www.aflcio-hit.com

The AFL-CIO Housing Investment Trust is an investment company that pools investments from nearly 350 union pension funds to finance, through secondary-market purchases, the construction or rehabilitation of union-built residential, commercial, and industrial properties.

Community Reinvestment Fund

821 Marquette Avenue, Suite 2400
Minneapolis, MN 55402-2903
800-475-3050 or 612-338-3050; fax: 612-338-3236
email: info@crf.com
www.crfusa.com

The Fund, founded in 1988, is a nonprofit organization that serves a secondary-market function for community economic development loans. CRF purchases development loans from community-based development organizations and government agencies. It pools the loans and issues and sells bonds backed by the loan pools to private investors. The loans it purchases include economic development loans, affordable housing loans, and other loans originally funded under a number of public programs (including the Community Development Block Grant program), as well as loans originally funded by private philanthropy.

Fannie Mae

Office of Low- and Moderate-Income Housing
3900 Wisconsin Avenue NW
Washington, DC 20016
202-752-4320
www.fanniemae.com

Fannie Mae is a private, publicly traded corporation chartered by Congress as a secondary-market institution to help ensure a steady flow of mortgage funds in America's housing market; it is known as a government-sponsored enterprise. In 1987, Fannie Mae created its Office of Low- and Moderate-Income Housing to promote and coordinate its several initiatives designed to improve access of low- and moderate-income people to decent, affordable housing. Fannie Mae works with local governments, nonprofits, mortgage lenders, and mortgage insurers. The Fannie Mae Foundation supports affordable housing initiatives and sponsors a yearly competitive award program for effective nonprofit projects producing low-income housing.

Federal Home Loan Mortgage Corporation

Community Development Lending Department
8250 Jones Branch Drive
Mail Stop A67
McLean, VA 22102
800-424-5401 or 703-918-5151

www.freddiemac.com

Freddie Mac is a private, publicly traded corporation chartered in 1970 by Congress as a secondary-market institution; it is known as a government-sponsored enterprise. It facilitates the flow of funds to mortgage lenders in support of homeownership and rental housing by buying first and second mortgages from private lenders.

Syndicators

The Enterprise Social Investment Corporation (ESIC)

10227 Wincopin Circle
Columbia, MD 21044
410)-964-9552
www.esic.org

A leading investor in nonprofit housing, ESIC finances, develops and acquires affordable housing and other community development projects in underserved neighborhoods nationwide. ESIC organizes limited partnerships with corporate investors in order to make tax credit equity investments in affordable housing projects. The company also offers long-term mortgage and financing to developers of affordable multifamily housing and develops affordable rental and for-sale housing. ESIC is a subsidiary of The Enterprise Foundation.

National Equity Fund

547 West Jackson Boulevard
Chicago, IL 60661-5701
312-697-6000
www.nefinc.org

National Equity Fund, Inc. is the nation's largest nonprofit syndicator of low-income housing tax credits. NEF has raised $3 billion since 1987 to help build more than 47,000 homes across the country. An affiliate of Local Initiatives Support Corporation (LISC), NEF provides technical assistance and equity for new construction or rehab of multifamily, single family, assisted living, supportive housing, public housing and historic projects in both urban and rural areas

Sources of Further Information

Interfaith Center for Corporate Responsibility

Clearinghouse on Alternative Investments
475 Riverside Drive, Room 566
New York, NY 10115
212-870-2926

The ICCR represents a national ecumenical coalition concerned with issues of economic justice. The Clearinghouse acts as an advocacy and educational group for churches involved in investing in projects that promote economic growth and stability in low-income communities. The Clearinghouse is a source of information about church and community partnerships.

Social Compact

5225 Wisconsin Avenue NW, Suite 204
Washington, DC 20015
202-686-5161; fax: 202-686-5593
email: email@socialcompact.org
website: www.socialcompact.org

The Compact's mission is to mobilize private-sector leadership and capital to strengthen America's neighborhoods. As part of its program, it makes substantial cash awards each year to alliances between businesses and community organizations that are models of successful community investment. The published materials about the awards offer a gold mine of information about financial institutions and strategies with a community focus.

Woodstock Institute

407 South Dearborn, Suite 550
Chicago, IL 60605
312-427-8070; fax: 312-427-4007
email: woodstock@wwa.com
www.nonprofit.net/woodstock

Founded in 1973, the Woodstock Institute is a research and technical assistance organization that specializes in reinvestment and economic development strategies for low-income urban communities. The Institute engages in applied research, policy analysis, technical assistance, public education, and program design and evaluation. Its areas of expertise include CRA and fair lending policies, financial and insurance services, small business lending, community development financial institutions, and economic development strategies including local employment. The Institute also has an extensive publications list.

This appendix offers information about the networks representing city and state public-sector agencies whose work is important to community development. You can also use this list to contact national organizations to request the names and addresses of their member agencies in the city or state in which you may want to find a job. The second part of the appendix describes the federal agencies most directly active in community development.

As is true for all the appendices in this guide, this listing does not represent the entire universe, but it is a useful base from which you can do further networking and information gathering. Entries are listed alphabetically.

Local and State Agencies

National Association of Local Housing Finance Agencies
1200 19th Street NW, Suite 300
Washington, DC 20036
202-857-1197; fax: 202-857-1111
www.alhfa.org
Founded in 1982, ALHFA is a national nonprofit association of housing finance agency professionals, from city and county agencies that finance affordable single-family, multifamily, and special needs housing for low-income people. Other members include nonprofits and private firms, such as underwriters, bond counsels, and financial advisors. ALHFA provides technical assistance, education, and training; lobbies at the federal level; and advises members of public policy changes.

Council of State Community Development Agencies
Hall of the States
444 North Capitol Street, Suite 224
Washington, DC 20001
202-624-3630; fax: 202-624-3639
email: jsidor@sso.org
www.coscda.org
COSCDA advocates for, assists, and serves as a clearinghouse for executive agencies in state government working on holistic community development, administering, for example, the HOME, Community Development Block Grant, and Community Services Block Grant programs, as well as the Low Income Housing Tax Credit program.

International City Management Association
777 North Capitol Street NE, Suite 500
Washington, DC 20002-4201
202-289-4262; fax: 202-962-3500
www.icma.org
ICMA is the professional and educational organization for more than eight thousand appointed administrators serving cities, counties, other local governments, and regional entities throughout the world.

National Association of Counties
440 First Street NW
Washington, DC 20001
202-393-6226; fax: 202-393-2630
email: sbullard@naco.org
www.naco.org
NACo is the only national organization representing county governments. Its membership (of about 1,800 of the 3,104 U.S. counties) come from urban, suburban, and rural counties across the country. It has forty-eight state associations, which provide training, consultation, and technical assistance to county governments.

National Association for County Community and Economic Development
1200 19th Street NW, Suite 300
Washington, DC 20036-2422
202-429-5118; fax: 202-857-1111
email: leslie-suarez@dc.sba.com
www.nacced.org
NACCED is a nonprofit national organization composed of nearly 120 county governments that administer community development, economic development, and housing programs, including the Community Development Block Grant and HOME programs. It assists counties in implementing programs through publications, educational conferences, structuring projects, and providing advice on agency structure and program administration.

National Association of Development Organizations
444 North Capitol Street NW, Suite 630
Washington, DC 20001-1512
202-624-7806; fax: 202-624-7806
email: nado@sso.org

www.nado.org

NADO promotes economic development in America's small cities and rural areas and provides information and training for development professionals and local officials. Its members are regional and state organizations that provide community development assistance to local governments and the private sector. NADO publishes a weekly newsletter on legislation and a monthly report on rural economic development.

National Association of Housing and Redevelopment Officials

630 Eye Street NW
Washington, DC 20001-3736
202-289-3500; fax: 202-289-8181
email: nahro@nahro.org
www.nahro.org

NAHRO is a professional membership organization representing local housing authorities, community development agencies, and individual professionals in the housing, community development, and redevelopment fields. It provides information on federal policy, legislation, regulations, and funding, as well as professional development and training programs.

National Association of Latino Elected & Appointed Officials

5800 South Eastern Avenue, Suite 365
Los Angeles, CA 90040
323-720-1932; fax: 323-720-5919
www.naleo.org

NALEO, founded in 1976, is a nonprofit membership organization with more than 5,400 Latino members who hold elected or appointed positions. It holds annual and regional training seminars to help current and future Latino political leaders attain public office and govern effectively; has a research and educational arm; and also sponsors a summer legislative intern program in which young Latinos can gain experience in public policy. NALEO publications include newsletters and a directory of the nation's elected Latinos.

National Association of Towns and Townships

444 North Capitol Street NW, Suite 294
Washington, DC 20001-1512
202-624-3550; fax: 202-624-3554
www.natat.org

NATaT is a nonprofit membership organization that offers public policy support to local government officials from more than thirteen thousand small communities (with populations of fewer than ten thousand residents) across the country, by educating lawmakers and public policy officials about how small-town governments operate and by advocating policies on their behalf in Washington, D.C. It has several state associations.

The National Center for Small Communities, formerly an affiliate of NATaT and now a separate organization, serves small-town leaders by producing educational publications and training modules.

National Community Development Association

522 21st Street NW, #120
Washington, DC 20006
202-293-7587; fax: 202-887-5546
email: ncda@ncdaonline.org
www.ncdaonline.org

NCDA is a national nonprofit organization with 560 members; most members are public agencies from entitlement cities and counties (that is, areas that are large enough and have a substantial enough population of lower-income households to receive funds under certain federal programs). The member agencies administer the Community Development Block Grant program; some also administer the HOME program, as well as several homeless assistance programs. The national association provides information and technical support.

National Conference of Black Mayors, Inc.

1430 West Peachtree Street NW, Suite 700
Atlanta, GA 30309
404-892-0127; fax: 404-876-4597
email: ncbmmay@aol.com
www.votenet.com/blackmayors

NCBM is a national nonprofit membership and service organization for black mayors. It provides technical assistance, conducts seminars and workshops, and sponsors executive management training for mlack mayors at its annual Leadership Institute for Mayors.

National Conference of State Legislators

1560 Broadway, Suite 700
Denver, CO 80202
303-830-2200; fax: 303-863-8003
email: info@ncsl.org
www.NCSL.org

NCSL is a bipartisan organization serving the lawmakers and staffs of the nation's fifty states, as well as its commonwealths and territories. The Conference is a source for research, publications, consulting services, meetings, and seminars.

National Council of State Housing Agencies
444 North Capitol Street NW, Suite 438
Washington, DC 20001
202-624-7710; fax: 202-624-5899
email: ncsha1@sso.org
www.ncsha.org

Each state has established a quasi-public but self-supporting Housing Finance Agency that provides funds, usually at below-market interest rates, to help finance the development of affordable housing for low-income households. The agencies raise capital through the issuance of bonds and through appropriations from the state legislature. They administer, underwrite, and monitor the Low Income Housing Tax Credit program; many serve as the state recipient of HOME funds, and some administer the state allocation of Community Development Block Grant funds. At the national level, the Council (NCSHA) takes positions on federal legislation and policies, serves as a clearinghouse of information, and sponsors meetings and conferences.

National Governors' Association
444 North Capitol Street
Washington, DC 20001
202-624-5300
email: nga@nga.org
www.nga.org

NGA is the only bipartisan national organization of, by, and for the nation's governors. Its members are the governors of the fifty states, the commonwealths of the Northern Mariana Islands and Puerto Rico, and the territories of American Samoa, Guam, and the Virgin Islands. Through NGA, the governors identify priority issues and deal collectively with issues of public policy and governance at both national and state levels.

National League of Cities
1301 Pennsylvania Avenue NW
Washington, DC 20004
202-626-3020; fax: 202-626-3043
email: pa@nlc.org
www.nlc.org

NLC is the oldest, largest, and most representative organization providing advocacy, education, and training to city officials throughout the United States. Its primary constituency is local elected officials (mayors and council members), as well as their staffs. Through a network of state leagues, NLC serves more than seventeen thousand communities.

Public Housing Authorities Directors Association
511 Capital Court NE
Washington, DC 20002
202-546-5445; fax: 202-546-2280
email: information@phada.org
www.phada.org/house

PHADA is the trade association, founded in 1979, representing local housing authorities across America; currently, its membership totals nearly seventeen hundred. It lobbies on national policy issues and legislation, produces publications that track major developments in the housing industry, sponsors training and networking conferences, and offers workshops on issues such as resident initiatives. Its newsletter and Web site announce job opportunities in housing authorities nationwide.

U.S. Conference of Mayors
1620 I Street NW, 4th floor
Washington, DC 20006
202-293-7330; fax: 202-293-2352
email: info@usmayors.org
www.usmayors.org/USCM/

The United States Conference of Mayors is the official nonpartisan organization of cities with populations of thirty thousand or more. Its members are mayors of those cities. The principal roles of the USCM are to aid the development of effective national urban policy, strengthen federal-city relationships, ensure that federal policy meets urban needs, and provide mayors with leadership and management tools of value in their cities.

Federal Agencies

The following are just a few, though the most important, of the federal agencies with a community development mandate. Many other federal agencies, such as the Small Business Administration and the U.S. Department of Justice, have programs that affect communities; and there are many regional agencies, such as the Appalachian Regional Commission, with community development programs. See also appendix F.

U.S. Department of Agriculture
Rural Development
14th Street and Independence Avenue SW
Washington, DC 20250
www.usda.gov
Rural Development (formerly known as Farmers Home Administration)

is the USDA agency that provides loans and grants for low-income and farm worker housing and home ownership. It also provides business loans and grants to help spur rural economic development.

U.S. Department of Commerce

Economic Development Administration
14th Street and Constitution Avenue NW
Washington, DC 20230
202-482-5112
www.doc.gov

The Economic Development Administration (EDA), a Department of Commerce agency, provides financing for economic development facilities and initiatives in distressed communities.

U.S. Department of Health and Human Services

Hubert H. Humphrey Building
200 Independence Avenue SW
Washington, DC 20201
202-401-3333
www.hhs.gov

Among the three hundred programs administered by HHS are a number of direct importance to community development. The Office of Community Services (see telephone number above) administers programs that serve low-income people and communities including the Community Service Block Grant program, the Social Services Block Grant program, the Low-Income Home Energy Assistance program, and programs to test new approaches to reduce welfare dependency. This office also runs a competitive award program for community development organizations. The Office of Family Assistance administers, among others, the Job Opportunities for Low Income Individuals program (telephone 202-401-9275).

U.S. Department of Housing and Urban Development

451 Seventh Street SW
Washington, DC 20410
202-401-0388
www.hud.gov

HUD is the federal agency whose programs are most focused on community development. These programs include: the Community Development Block Grant Program; the HOME program; a Brownfields Economic Development Initiative, and other programs to support economic development; special needs housing programs; the Housing for Persons with AIDS program; Empowerment Zones and Enterprise Communities; fair housing; the YouthBuild initiative; housing counseling programs; and more. The Federal Housing Administration (FHA) within HUD operates programs that provide insurance for housing developments. HUD has recently set up a special office to encourage nonprofit and faith-based community development initiatives.

Another source of information on HUD programs and resource materials is an information center established by HUD: Community Connections, 800-998-9999, www.commcon.org.

U.S. Department of Transportation

400 7th Street SW
Washington, DC 20590
202-366-4000
www.dot.gov

The Transportation Efficiency Act for the 21st Century (TEA-21) is a DOT program that, in addition to financing highways, provides funding for mass transit (including an Access for Jobs program) and neighborhood-scale transportation approaches that benefit communities. DOT also runs a Livable Communities Initiative.

Environmental Protection Agency

401 M Street SW
Washington, DC 20460
202-260-2090
www.epa.gov

Among the initiatives administered by the EPA that directly affect communities are the Environmental Equity Grants program and the Sustainable Development Challenge Grants program.

This appendix describes some Internet sites with broad information relevant to the community development world. In addition, other appendices describe national community development organizations and agencies, including their Web addresses, and appendix I lists Web sites with job postings relevant to community development.

Action Without Borders (formerly Contact Center Network)
www.idealist.org
email: info@idealist.org
Idealist, a project of Action Without Borders, is an electronic resource for nonprofits, with direct links to over fourteen thousand organizations in hundreds of countries.

Benton Foundation
www.benton.org
email: benton@benton.org
The Benton Foundation, devoted to promoting the use of communications technology on behalf of the public interest, maintains a wide variety of useful links.

Centre for Community Enterprise
www.cedworks.com
This Web site of a Canadian nonprofit community development center sells a selected number of books on community development—it's trying to serve as the amazon.com of the field.

Change Communications
www.change.org
Operated by a company that sells Internet consulting services to nonprofits, this Web site offers a prodigious hyperlink to just about every online community development resource. The links are well organized, continuously updated, and extremely useful.

Community Connections
www.comcon.org
This is the Web site of the U.S. Department of Housing and Urban Development's Office of Community Planning and Development. Its information covers HUD program alerts, HUD publications, and a calendar of training and conferences sponsored by HUD and national organizations.

Community Development Online
www.nw.org
The Community Information Exchange maintains twelve online databases, covering best-practice models gathered over the past fifteen years, and financial, technical, and written resources for community development projects, as well as news and opportunities for those working in community development. The service is maintained by the Neighborhood Reinvestment Corporation and is available through its Web site (see NeighborWorks Network below).

Community Development Society
www.comm-dev.org
email: alboss@scn.org
A few documents on this site are restricted to CDS members, but the vast majority, including pages of links, list-servs, and job opportunities, are open to all. To join their discussion group, follow directions on their Web site.

Comm-Org
www.comm-org.utoledo.edu
This online information resource, sponsored by the Urban Affairs Center and Department of Sociology of the University of Toledo, focuses mainly on community organizing.

Enterprise Foundation Web Page
www.enterprisefoundation.org
The Enterprise Foundation's Resource Center's online information includes a database of model documents, how-to manuals, program descriptions, and recommended practical information resources in low-income housing and community development. The Enterprise Foundation is developing a new forum of Internet resources and links on housing and community development for World Wide Web users. Also see the Enterprise Foundation's online news magazine: www.horizonmag.com

Handsnet on the Web
www.handsnet.org/
email: hninfo@handsnet.org
Handsnet is a network of five thousand public interest and human services organizations across the country; its Web site is a collection of forums on human services issues ranging from children, youth, and families, to housing and community development. Each forum is managed by staff of leading organizations in each field.

Housing Information Gateway
www.colorado.edu/plan/housing-info
email: willem@spot.colorado.edu
Those interested specifically in housing concerns will find a wealth of electronic resources catalogued here, including links to organizations, individuals, and list-servs.

Internet Non-profit Center
www.nonprofits.org
email: mgilbert@nonprofits.org
The Internet Non-profit Center is an extensive source of information about nonprofits for donors and volunteers.

Neighborhoods Online
www.libertynet.org/nol/natl.html
Philadelphia's Libertynet and Institute for the Study of Civic Values created this national online resource, which covers organizing, cities, community development, education, crime, jobs, environment, health, and human services. Check out the host organizations as well.

NeighborWorks Network
www.nw.org
email: webmaster@nw.org
Maintained by the Neighborhood Reinvestment Corporation on behalf of a national network of community development organizations, this site has an outstanding series of hypertext links to other online organizations and resources, as well as information about training opportunities, publications, job postings, news releases, and even a glossary.

NonProfit Gateway
www.nonprofit.gov
email: gateway@comcat.org
The NonProfit Gateway is a directory of federal government services and information pertaining to nonprofits.

Philip Walker's Nonprofit Resources Catalogue
www.clark.net/pub/pwalker
email: pwalker@clark.net
This site, with information compiled by one dedicated individual, is a comprehensive general guide to online nonprofit activity and resources.

PRAXIS
www.ssw.upenn.edu/~restes/praxis.html
email: restes@caster.ssw.upenn.edu
This is one professor's compilation of online resources for all aspects of social and economic development.

This appendix describes the organizations that offer job postings, as well as other employment services; the newsletters that routinely carry job announcements; and the Web sites that are the best sources for community development job announcements.

Employment Resources Organizations

Access: Networking in the Public Interest
1001 Connecticut Avenue NW, Suite 838
Washington, DC 20036
202-785-4233
email: commjobs@aol.net
www.communityjobs.org
ACCESS is a national nonprofit organization that serves as a national clearinghouse on nonprofit-sector jobs. It produces a national paper called *Community Jobs*, which includes job postings from all regions. It also offers one-on-one career counseling. Its Web site links to the *Wall Street Journal's* online postings, which include nonprofit job openings.

Peace Corps Career Resources Center
Office of Domestic Programs
1111 20th Street NW, 1st floor
Washington, DC 20526
email: dpinfo@peacecorps.gov
www.peacecorps.gov/rpcv/careertrack/index.html
The Peace Corps has a wealth of job resources available to returned Peace Corps volunteers (RPCVs). The Returned Volunteer Services (RVS) office in Washington, D.C., coordinates and provides career information through print materials, videos, workshops, and the Internet. The Washington RVS office collaborates with regional Peace Corps offices to provide career information to RPCVs all over the nation. Only RPCVs have access to career information at the national and regional offices.

A particularly valuable resource available only to RPCVs is *Hotline*, a six-page bulletin. Published twice a month, *Hotline* contains private- and public-sector job announcements from various employers who are specifically interested in hiring RPCVs, and from educational institutions that are looking to enroll RPCVs in graduate programs.

Career TRACK (Transitions, Readjustment, Assistance, Connections, and Know-How), the Web page maintained by RVS, is a useful career resource for anyone with access to the Internet. On Career TRACK you will find everything from sections about the preliminary job search to Internet career resources to outlines of different career paths. As well as being an excellent general career resource, Career TRACK has many sections devoted to nontraditional careers in the federal and nonprofit sector, law, and community services.

Newsletters

Grassroots Economic Organizing Newsletter
RR1, Box 124A
Stillwater, PA 17878-9989
800-240-9721
email: wadew@epix.net
Predominantly focuses on organizing and economic justice kinds of jobs.

Left Hand/Right Hand
National Community Capital Association
924 Cherry Street, 2nd floor
Philadelphia, PA 19107
215-923-4754
email: NCCA@communitycapital.org
Lists jobs in community development financial institutions nationwide.

LISC Links
Local Initiatives Support Corporation
733 Third Avenue
New York, NY 10017
202-455-9800
Contains job listings from the CDCs and other organizations that are located in the geographical areas where LISC concentrates its activities.

NNC Voice
National Neighborhood Coalition
1875 Connecticut Avenue NW, Suite 710
Washington, DC 20009
202-986-2096
email: nncnnc@erols.com
Usually offers the greatest number and variety of job announcements from all the related sectors of the community development world.

Planners Network

c/o Pratt GCPE
200 Willoughby Avenue
Brooklyn, NY 11205
718-636-3486
email: wintonp@ix.netcom.com
Lists job announcements for planners and planning professors, including positions in CDCs and intermediaries, and more.

Public Sector Job Bulletin

P.O. Box 1222
Newton, IA 50208
515-791-9019
As the name states, the jobs listed are all in the public sector; it includes positions such as planner, redevelopment agency director, and community development director.

Resources

National Congress for Community Economic Development
1030 15th Street NW, Suite 325
Washington, DC 20005
877-44NCCED or 202-289-9020
Carries announcements about CDC and related jobs. Six issues per year.

Roundup

Low Income Housing Coalition Information Service
1012 14th Street NW, Suite 1200
Washington, DC 20005
202-662-1530
email: info@nlihc.org
Specializes in jobs relating to low-income housing, from organizing to development to policy.

SCANPH

3345 Wilshire Boulevard, Suite 1000
Los Angeles, CA 90010
213-480-1249
email; scanph@earthlink.net
The bimonthly newsletter of the Southern California Association of Non-Profit Housing; lists job announcements from housing and service organizations in Southern California.

Shelterforce

P.O. Box 3000
Denville, NJ 07834
973-678-9060
email: nhi@nhi.org
The bimonthly newsletter of the National Housing Institute; lists job announcements from housing and community and economic development organizations, mostly in metropolitan areas in the eastern United States.

Washington Update

National Association of Affordable Housing Lenders
2121 K Street NW, Suite 700
Washington, DC 20036
202-293-9850
email: naahl@naahl.org
Lists jobs in the housing and community economic development finance world, from banks and loan funds to foundations and regulatory agencies. Monthly, or more often when warranted.

Web Sites

Note that many of the organizations listed in other appendices also run job announcements on their Web sites.

www.careerpath.com

The Careerpath site has the largest number of current, online job postings on the Internet. This site is more difficult to use than other sites listed in this appendix because it does not have a "nonprofit jobs" search category. However, some categories on this site that may lead to jobs in community and economic development are: activism, fundraising, environment, and social services.

www.dbm.com/jobguide

Described as "one of the best internet sites of its kind," the Riley Guide (written by career expert Mary Riley) provides excellent links to informational and job listings sites of all categories, including those in the nonprofit sector. In addition to providing general information about the job search process, the Riley Guide has links to summer opportunities, internships, and fellowships.

www.eco.org

Maintained by the Environmental Careers Organization, this site lists jobs and paid internships in the green sector.

www.essential.org/goodworks/jobs

This online version of Good Works, an authoritative national directory of social change organizations and more than one thousand jobs in the nonprofit sector, provides listings of nonprofit jobs, searchable by state.

www.idealist.org

Maintained by Action Without Borders, the Idealist Web site provides a large directory of jobs, internships, and volunteer opportunities in the nonprofit sector, searchable by location, required skills, and job description. This site also provides an email notification service for those who want to be contacted when new jobs are added to the directory, as well as links to other Web sites listing nonprofit job opportunities.

www.nonprofitcareer.com

The Nonprofit Career Network is an online job search and posting service with an annual usage fee of $40 for individuals. The Network provides an extensive database of nationwide nonprofit job listings, searchable by state. It also provides organizational profiles of nonprofits, answers to frequently asked questions, and information regarding job fairs. An online résumé posting service is available for those who want to make their résumés available to potential employers via the Internet.

www.nonprofitjobs.org

This Web site calls itself the Community Career Center. It is the premier site for job and résumé postings in the community development world. It can be searched by keywords such as community development, neighborhoods, and housing, as well as by region, skills, and many other characteristics.

www.nptimes.com

The NonProfit Times Online is a monthly online periodical. Its classifieds section lists jobs in the nonprofit sector.

www.nw.org

This home page of the National Reinvestment Corporation (see appendix D) also posts job opportunities.

www.opnocs.org

The New England Nonprofit Organizations Classifieds, a Web site maintained by the Executive Service Corps of New England, a consulting organization for nonprofits, has listings of jobs, internships, and volunteer opportunities available in the New England nonprofit sector.

www.opportunitynocs.org

Described as one of the "best sources nationally for nonprofit jobs," the National Opportunity Nonprofit Organization Classifieds (NPOC) has hundreds of job listings searchable by state, date of posting, and keywords. The NPOC site is maintained by the Management Center of San Francisco with regional affiliates in Los Angeles, Dallas, Philadelphia, Atlanta, and Boston.

www.peacecorps.gov/rpcv/careertrack/index.html

The Peace Corps job Web site contains many listings relevant to community development.

www.philanthropy.com

The online version of the Chronicle of Philanthropy, updated every other week, lists nonprofit jobs by category with a focus on fundraising.

www.pj.org

The Philanthropy News Network (formerly *Philanthropy Journal*) online provides job listings and a résumé posting site. Most of the jobs relate to fundraising.

www.sanetwork.org

Maintained by Sustainable America, this site carries job listings relating to the environment, economic justice, and community development. It also has links to other related job sites.

www.servenet.org

Maintained by Youth Service America, Servenet is an online resource for volunteers and the service community that lists job opportunities by state.

In addition to the national and local programs listed here, the organizations listed in appendix D, "National and Regional Organizations," often sponsor informal or unpaid internships without having a set program.

Many universities provide internship or fellowship opportunities as part of their graduate programs. An example is the MIT Department of Urban Studies and Planning Internship. Schools with special centers or institutes that engage in field study and community service are likely to include experiential and often paid opportunities for students. Examples include the internship offered through Cleveland State University's Center for Neighborhood Development; the University of Delaware's Center for Community Development & Family Policy's paid placements of graduate students in community development projects; and a program in San Diego in which several local colleges and universities collaborate in offering community development internships.

There are numerous community service programs, such as Campus Compact, that you might be able to shape to allow you to work in a community development initiative.

Use the entries below to start your search. The world of internships changes frequently, and it seems more internships are offered each year.

National Programs

Listed alphabetically by sponsoring organization

AmeriCorps
1201 New York Avenue
Washington, DC 20525
800-942-2677
www.americorps.org

AmeriCorps is the national service program, passed by congressional legislation in 1993, that allows people of all ages and backgrounds to earn money for education in exchange for a year of service. AmeriCorps members meet community needs with services that range from housing renovation and neighborhood policing to child immunization. There are three AmeriCorps programs: AmeriCorps VISTA, AmeriCorpsState, and AmeriCorps National Civilian Community Corps (NCCC).

AmeriCorps VISTA members or interns serve in sponsoring agencies in low-income communities. This is the category most relevant to community development—and the program through which national and other community development organizations create their internships, described below.

VISTA members are involved in all facets of nonprofit work: helping to develop new programs, recruiting and training volunteers, writing grants, and organizing fundraisers and media campaigns. (The AmeriCorps VISTA program is based on the original Volunteers In Service To America program launched as part of the War on Poverty in the 1960s.)

AmeriCorpsState volunteers are recruited directly by national, state, and local nonprofit organizations. The recruiting nonprofit organization provides training and guidance for volunteers who serve in one of four program areas: education, public safety, human services, and the environment.

AmeriCorpsNCCC, the National Civilian Community Corps, is a ten-month residential, national service program for youth between the ages of eighteen and twenty-four. AmeriCorpsNCCC members focus their service on the environment, education, public safety, unmet human needs, and disaster relief. Members work in teams and live together in housing complexes on AmeriCorpsNCCC campuses, located in Perry Point, MD; Washington, DC; Charleston, SC; Denver, CO; and San Diego, CA. After an initial three-week training, members serve on a variety of team-based projects lasting from one day to six weeks. Members may be temporarily relocated from their campuses to projects in communities throughout their region of service.

AmeriCorps members receive a modest living allowance (from $600 to $800 a month) and health coverage. After completing a year of service, they receive an education award of $4,725, which can be used to pay off student loans or to finance college, graduate school, or vocational training. Interns can serve for two years and earn two years' worth of education awards.

Further, a number of national, state, and local organizations, including those from the community development world, operate AmeriCorps programs. The current national programs are described throughout this appendix. Since these programs change from year to year, you can also refer to the AmeriCorps Web site for the latest information. Go to the Web site, then to Joining AmeriCorps, and then to the AmeriCorps VISTA link, which includes a list of state offices, if you want to find out whether a community development-related AmeriCorps program is located in your state or community.

Association of Community Organizations for Reform Now
117 West Harrison, 2nd floor
Chicago, IL 60605
312-939-7488; fax: 312-939-8256
email: fielddirect@acorn.org
www.igc.org/community

The Association of Community Organizations for Reform Now (ACORN), the largest grassroots community organization in the United States, has three-month and year-long internships in community organizing. Interns learn about the field of community-based organizing by working closely with experienced organizers and by participating in ACORN's trainings. ACORN member organizations are running cutting-edge campaigns in housing, redlining, school reform, living wages, welfare, toxic chemicals, sexual assault, and other local community concerns.

BRIDGE Housing Corporation

Donald Terner Residency Program
1 Hawthorne Street, Suite 400
San Francisco, CA 94105-3901
415-989-1111; fax: 415-495-4898
www.bridgehousing.com

This fellowship honors the founder and former CEO of BRIDGE Housing Corporation, who died in a plane crash in 1996 along with U.S. Secretary of Commerce Ron Brown and others on an international development mission to Eastern Europe. The program provides a paid full-time, two-year residency with the organization, one of the leading nonprofit housing developers in the nation. Candidates must be outstanding individuals, with a minimum of a college degree and demonstrated commitment to the betterment of communities.

Christmas in April USA

1536 16th Street NW
Washington, DC 20036
202-483-9083; fax: 202-483-9081
email: SINAUSA@erols.com
www.christmasinapril.org

This organization encourages and assists communities to launch volunteer programs to help repair the homes of low-income, elderly, and disabled homeowners in the area.

City Year

285 Columbus Avenue
Boston, MA 02116
617-927-2500
www.city-year.org

Founded in 1988, City Year is a national organization that places youth, aged seventeen to twenty-three, in public and nonprofit organizations dedicated to community service. The ten-month program runs from September through June. Participants receive a weekly stipend of $125 and upon graduation are eligible for up to $4,725 in post-service awards for higher education. In addition to stipends, City Year provides volunteers with a comprehensive education program, including GED classes, college and career counseling, and overnight training retreats. City Year currently operates in Boston; Chicago; Cleveland; Columbia, SC; Columbus, OH; Detroit; Philadelphia; Providence, RI; San Antonio; San Jose/Silicon Valley, CA; and Seattle/King County, WA. Call the Boston office for more information.

Coro Foundation

Fellowship in Public Affairs
1706 Marin Avenue
Berkeley, CA 94707
888-GO-4-CORO, fax: 510-528-6875
email: national@coro.org
www.coro.org

The Coro Foundation Fellowship in Public Affairs is a full-time (at least fifty hours a week), nine-month, graduate-level program conducted each year at centers in Los Angeles, New York, Pittsburgh, St. Louis, and San Francisco. Forty-eight fellows (twelve at each center) receive experience-based leadership training in public affairs through a series of internships, interviews, public service projects, seminars, and retreats. Coro seeks applicants with at least a bachelor's degree or equivalent who have demonstrated leadership ability, integrity, and a commitment to public service. Tuition for the Fellows Program is $3,500, but assistance is available for tuition and living expenses based on financial need. Coro also operates a summer internship program in Kansas City as well as Exploring Leadership, Leadership New York, and Women in Leadership programs.

Peter A. Drucker Foundation for Nonprofit Management

Hesselbein Fellows Program
320 Park Avenue, 3rd Floor
New York, NY 10022-6839
212-224-1174; fax: 212-224-2508
email: info@pfdf.org
www.pfdf.org

Each year, the Drucker Foundation's Frances Hesselbein Community Innovation Fellows Program recognizes the accomplishments and supports the professional development of three to six social-sector leaders who have a demonstrated record of leadership and entrepreneurial performance, and who are engaged in projects or programs that demonstrate community innovation. Program activities are designed to foster Fellows' professional and leadership development by providing the following opportunities: leadership workshops and consultations; educational conference participation (particularly the annual Drucker Foundation Leadership Conference); access to Drucker Foundation personnel and educational resources; complimentary receipt of Drucker publications; and post-fellowship opportunities. The Foundation pays for all travel expenses to conferences.

Echoing Green Foundation

Public Service Fellowship
198 Madison Avenue, 8th floor
New York, NY 10016
212-689-1165; fax 212-689-9010
www.echoinggreen.org

The Echoing Green Foundation applies a venture capital approach to philanthropy by supplying seed money and technical support to social entrepreneurs starting innovative projects in the areas of education, the arts, health, and human and civil rights. The Echoing Green Public Service Fellowship offers one-year, $20,000 stipends to fellows applying through participating undergraduate university and college programs and community-based organizations, and one-year, $30,000 stipends to individuals applying though participating graduate school programs. In addition to stipends (which can be renewed for a second year) and technical support, fellows receive training and support at conferences and workshops. Eligible applicants must be in a current graduating class or have completed a degree program within the last ten years from an Echoing Green participating undergraduate university or college, graduate school, or community-based organization.

Environmental Careers Organization

286 Congress Street, 3rd floor
Boston, MA 02210
617-426-4375
www.eco.org

ECO's mission is to protect the environment through the promotion of careers and the development of professionals. Through one of its programs, it helps place more than three hundred college students, recent graduates, and entry-level job seekers in short-term, paid "associate" positions of three to eighteen months' duration with federal, state, and local agencies, corporations, consulting firms, and nonprofit groups. Salaries range from $300 to $700 per week or more. The placements may include community development work such as brownfields redevelopment, historic preservation, and community economic development. ECO maintains offices in the Great Lakes, California, the Pacific Northwest, and Florida, as well as Boston.

Fannie Mae Foundation

Fellowship Program
North Tower, Suite One
4000 Wisconsin Avenue NW
Washington, DC 20016-2804
202-274-8000; fax: 202-274-8100
www.fanniemaefoundation.org

The Fannie Mae Foundation Fellowship Program is intended to enhance the management and decision-making skills of experienced senior public and nonprofit housing officials who are committed to increasing affordable housing opportunities, developing vital communities, and improving housing conditions in the United States. Fannie Mae fellows attend the Program for Senior Executives in State and Local Government, a three-week summer program at the Kennedy School of Government at Harvard University, and attend special housing sessions coordinated by the Kennedy School's Joint Center for Housing Studies and the Fannie Mae Foundation. Fellowship funds cover the costs of the summer program, including deposit, tuition, and room and board.

Fannie Mae Foundation

James A. Johnson Fellows Program
North Tower, Suite One
4000 Wisconsin Avenue NW
Washington, DC 20016-2804
202-274-8000; fax: 202-274-8100
www.fanniemaefoundation.org

This program awards substantial fellowship grants to six outstanding leaders each year, in recognition of their dedication and contribution to the affordable housing or community development fields. In addition to the grants, the Fellows are given stipends for educational travel and study, and their organizations may receive additional grants for costs related to the temporary loss of the Fellow. The Fellow's year is to be spent researching issues, traveling, studying, and pursuing other activities to enhance skills and learning.

Habitat for Humanity, International

121 Habitat Street
Americus, GA 31709
912-924-6935, x118
www.habitat.org

Habitat for Humanity is a Christian housing organization that works with volunteers to build and renovate homes that are sold to those in need for no profit. Thousands of people participate in Habitat projects every year. Interested volunteers should call the national office for contact information on local affiliates.

Jesuit Volunteer Corps

P.O. Box 25478
Washington, DC 20007
202-687-1132
www.jesuitvolunteers.org

The Jesuit Volunteer Corps (JVC) places volunteers in nonprofits dedicated to serving the poor in sixty cities across the United States. Applicants must be at least twenty-one years of age and open to Christian ideals. During the year-long commitment, volunteers receive room and board and a personal expense allowance of $80 to $100 per month. Applications are accepted on a rolling basis through July, with priority given to applications postmarked by March 1. Call the Washington, D.C., office for more information.

W.K. Kellogg Foundation

National Leadership Program
P.O. Box 550
Battle Creek, MI 49016-0550
www.wkkf.org

The Kellogg Foundation National Leadership Program is engaged in evaluating its leadership programs and is not currently accepting applications.

Local Initiatives Support Corporation

733 Third Avenue, 8th floor
New York, NY 10017
212-455-9324; fax: 212-986-1857
www.liscnet.org

Internships with community development corporations combined with training and mentoring are available through the AmeriCorps program run by the Local Initiatives Support Corporation (LISC). Interns participate in significant development work, ranging from counseling first-time home buyers to assisting staff members in creating a neighborhood day care center. The internship runs for one year (1,700 hours); interns may apply for a second year as well. A stipend of $15,890 and health benefits are provided; interns may also apply for a $4,725 education award after completing their year of service. In addition to the hands-on learning from their day-to-day assignments, interns attend a week-long national training course, two four-day regional trainings, and twelve monthly meetings.

An average of ninety internships are open each year. Apply either by contacting LISC in New York or by getting a list from the New York office of the participating CDCs in your area; the CDC makes the intern selection.

The Lutheran Volunteer Corps

1226 Vermont Avenue NW
Washington, DC 20005
202-387-3222
www.lvchome.org

The Lutheran Volunteer Corps (LVC), founded in 1979, places volunteers in nonprofits dedicated to addressing issues of social injustice. Applicants must be at least twenty-one years old to be eligible for the year-long service assignments. Volunteers receive room and board and a personal expense allowance of $85 per month. Currently, LVC places volunteers in the following cities: Baltimore, MD; Chicago, IL; Milwaukee, WI; Minneapolis–St. Paul, MN; Seattle, WA; Tacoma, WA; Washington, DC; and Wilmington, DE. Applications are accepted at the Washington, D.C., office for all sites. Applications are available every October and are processed beginning March 1. Positions are available on a first-come basis.

National Congress for Community Economic Development (NCCED)

Emerging Leaders and Summer Internships Programs
1030 15th Street NW, Suite 325
Washington, DC 20005
877-44NCCED or 202-289-9020; fax: 202-289-7051
www.ncced.org

The National Congress for Community Economic Development (NCCED), the trade association for community-based development organizations, offers two programs for students interested in community development: summer internships for graduate students and the Emerging Leaders Program.

In partnership with the Association for Public Policy Analysis and Management, NCCED offers a summer Community Development Internship Program that matches graduate students studying public policy or urban planning with community development corporations for ten-week terms of service. Students work on activities ranging from small-business lending and the development of employment-generating businesses, to affordable housing development, homelessness prevention, and community development public policy. NCCED places students and provides a national three-day training seminar as well as mentorships. Students receive a stipend of $4,500. Applications are due by mid-February.

NCCED's Emerging Leaders Program is a five-day conference, held in conjunction with the NCCED's Washington Policy Conference on Community Economic Development, for undergraduate and graduate students interested in learning more about the field of community economic development. Through workshops and meetings with top executives and renowned leaders in the field, student leaders learn about careers in community economic development.

National Federation of Community Development Credit Unions

120 Wall Street, 10th floor
New York, NY 10005
800-437-8711; fax: 212-809-3274
email: ddelrio@natfed.org
www.natfed.org

The NFCDCU operates two programs that place AmeriCorps VISTA interns with community development credit unions. One program is national in scope. In that program, interns help a local community development credit union to create either an Individual Development Account program or a micro-enterprise program. IDAs are restricted savings accounts (similar to Individual Retirement Accounts) that provide incentives and support services to help low-income people save and build assets that they can use to buy or repair a home, start a business, or obtain an education; the program raises matching funds for the IDAs. Micro-enterprise lending refers to relatively small loans made to self-employed people or very small businesses.

The second NFCDCU AmeriCorps VISTA program is for New York State. In that program, interns work on a wide variety of projects, such as setting up youth savings programs in local schools and recruiting new credit union members.

At the start, all NFCDCU AmeriCorps VISTAs and their supervisors go through a three-day training with the Corporation for National Service (as all AmeriCorps interns do). In addition, NFCDCU provides national VISTAs with one or two days of specialized training in IDAs and micro-enterprise lending. VISTAs also come together for a second training meeting when they are about three to six months into their service. NFCDCU also sponsors monthly meetings for VISTAs in the New York City area, where NFCDCU is located.

The internship runs for one year (1,700 hours) and may be renewed for up to three years of service. A biweekly stipend ranging from $300 to $360, depending upon the city, and health benefits are provided; interns also choose between a $4,725 education award and a $1,200 stipend after completing their year of service. A number of AmeriCorps VISTAs have been hired as staff by their community development credit unions. An average of forty-five positions are available each year, although the numbers vary. Apply by contacting NFCDCU.

National Urban Fellows, Inc.

Management Fellowship
55 West 44th Street, Suite 600
New York, NY 10036
212-921-9400; fax: 212-921-9572
www.nuf.org

The National Urban/Rural Fellows, Inc., Fellowship is a full-time, fourteen-month graduate degree and internship program for post-baccalaureate, mid-career professionals who have worked in public management or administration for three or more years, and who are committed to addressing urban and rural economic and community development issues. The Fellowship combines course work at Bernard M. Baruch College, City University of New York, with nine-month internship assignments with senior executives at public and nonprofit agencies nationwide. A master of Public Administration from Baruch College's School of Public Affairs is awarded to Fellows upon successful completion of academic and program requirements. (See the "Business, Nonprofit Management and Public Administration" section in appendix B for information about the required curriculum delivered through Baruch College.) A mid-year conference offers further networking and career exploration opportunities. The Fellowship provides a monthly stipend, full tuition assistance, book allowance, and reimbursement for program-related expenses.

Net Impact

609 Mission Street
San Francisco, CA 94105
415-778-8366; fax: 415-778-8367
email: mail@Net-Impact.org
www.Net-Impact.org

Net Impact (founded under the name Students for Responsible Business) is a network with fifty local chapters connecting emerging business leaders committed to using the power of business to create a better world. The Internship Program places about seventy-five graduate business students in paid positions in a variety of organizations, including community development corporations and intermediaries and new social ventures. Net Impact solicits internship positions nationwide and posts the descriptions of these positions on their Web site and through career development offices at over one hundred business schools. They solicit applications from students, who select the projects they would like to work on, and Net Impact matches students to projects. Internship pay ranges from $5,000 for a nonprofit placement and $7,500 for a for-profit organization for a ten-week project assignment.

Points of Light Foundation

Youth Engages in Service Ambassadors Program
1737 H Street NW
Washington, DC 20006
202-729-8160
www.pointsoflight.org

The Points of Light Foundation Youth Engages in Service (YES) Ambassadors Program is an internship program for youth (ages eighteen to twenty-five) who have a variety of volunteer experiences; are creative, articulate, and organized; and have a passion for community change and youth leadership. YES Ambassadors help develop and run programs that promote youth service, leadership, and service learning with state-level partners of the Points of Light Foundation. YES Ambassadors receive an annual stipend of $20,000 with medical and dental benefits, a travel stipend, and year-round training. Contact the national office of the Points of Light Foundation for a list of state-level contacts.

Pro Bono Students America

110 West 3rd Street, 2nd floor
New York, NY 10012
212-998-6222
www.pbsa.org

PBSA is a national network of law schools that fosters law student community service. Its database allows students at each of the more than seventy member law schools to search for service opportunities ranging from short-term uncompensated volunteer positions during school semesters to full-time summer internships, post-graduate jobs, and fellowships.

Public Allies

1511 K Street NW, Suite 330
Washington, DC 20005
202-293-3969
www.publicallies.org

Public Allies, founded in 1991, is currently operating in Chicago; Cincinnati; Durham, NC; Los Angeles; Milwaukee, WI; New York City; the Silicon Valley area of California;Washington, DC; and Wilmington, DE. At each site, groups of young adults, ages eighteen to thirty, spend four days per week over ten months in paid professional internships with public and nonprofit agencies. The fifth day of the week is spent in training and team-based community improvement projects. A living allowance, health and child care benefits, and an education award are provided. The Web site contains local contact information.

The Rockefeller Foundation

Next Generation Leadership Fellowship
420 Fifth Avenue
New York, NY 10018
212-852-8407
www.nglnet.org

The Rockefeller Foundation Next Generation Leadership Fellowship is intended to develop a corps of leaders with shared values for the twenty-first century. This Foundation provides two-year fellowships for people of color between the ages of twenty-five and forty who have already achieved a high level of accomplishment in fields ranging from community development to the arts, education, and information technology. Those selected participate in intensive experiential and discussion sessions that focus on issues of American democracy, race relations, and community building during the first year. During the second year, fellows develop and work on projects in their own communities that put learned theories of democracy into action. Candidates are nominated by a group of civic leaders; individuals cannot apply or nominate themselves.

YouthBuild U.S.A

58 Day Street
Somerville, MA 02144
617-623-9900
www.youthbuild.org

Founded in 1988 out of the Youth Action Program in East Harlem, YouthBuild U.S.A. is a comprehensive youth and community development program and alternative school. Designed to run on a twelve-month cycle, YouthBuild offers job training, education, counseling, and leadership development opportunities to unemployed and out-of-school young adults, ages sixteen to twenty-four, through the construction and rehabilitation of affordable housing in their own communities. Many graduates go on to construction-related jobs or college. The buildings that are rehabilitated or constructed during the program are usually owned and managed by community-based organizations as permanent low-income housing. Currently, there are over one hundred YouthBuild programs operating in thirty-four states around the United States. Call the Massachusetts office to get a list of participating communities. Students receive compensation, as determined by their program site.

Local and Regional Programs

Northeast

Massachusetts Association of Community Development Corporations

99 Chauncy Street, 5th floor
Boston, MA 02111
617-426-0303; fax: 617-426-0344

The Massachusetts Association of Community Development Corporations (MACDC), the trade association and network for community development groups throughout the state, has three programs to encourage individuals to enter and stay in the field of community development: Community Development Apprenticeship Program; Minority Fellows Program; and Minority Community & Regional Economic Development Internship and Training (CREDIT) Program.

The Community Development Apprenticeship Program places VISTA members in jobs with community development corporations throughout Massachusetts. First, Massachusetts CDCs apply to MACDC for a VISTA position. Then CDCs that are selected for VISTA funding recruit individuals from the communities they serve to fill their VISTA position.

The Minority Fellows Program is designed to expand the pool of people of color working at CDCs in mid- to senior-level positions. MACDC provides partial funding for resources necessary to maintain Fellows as permanent staff members. Individuals of color in adult programs at Massachusetts-area universities and those with experience in fields related to community and economic development (e.g., real estate, banking) are encouraged to apply.

The Minority Community & Regional Economic Development Internship and Training (CREDIT) is a joint program of the Massachusetts Institute of Technology's Department of Urban Studies and Planning (DUSP) and MACDC. MACDC, member CDCs, and MIT recruit talented college seniors and graduates of color to apply to the master of Community Planning program at DUSP. When students are admitted, their admission is deferred for two years while they work in a paid full-time position at a CDC in Massachusetts. After two years of work, the interns matriculate at the master of Community Planning program, with full tuition coverage by MIT. Graduates are required to work in the community development field for at least two years after receiving their degree.

State of New Jersey Department of Community Affairs/First Union Bank

Housing Scholars Program
William Ashby Community Affairs Building, CN 800
Trenton, NJ 08625-0800
609-292-6831; fax: 609-292-6831
www.njit.edu/CDS/hsp

The New Jersey Department of Community Affairs, with First Union Bank as the cosponsor, offers a paid, ten-week, full-time placement with nonprofit organizations engaged in housing and community development for selected students during the summer. The program supports about twenty-five graduate and undergraduate students enrolled in the development-related programs of collaborating New Jersey colleges and universities. A pre-employment training seminar is provided to the scholars, covering the development process and the emerging role of nonprofits in affordable housing and community and economic development. (The New Jersey Institute of Technology is the program's administrative agency.)

South

Atlanta Neighborhood Development Partnership, Inc.

34 Peachtree Street NW, Suite 1700
Atlanta, GA 30303-2210
404-522-2637
www.andpi.org

Atlanta Neighborhood Development Partnership (ANDP) is a nonprofit that provides technical assistance to community development organizations in the Atlanta, Georgia, metro area and promotes community development in the city's public and private sectors. ANDP has fellowship and internship programs. The purpose of both programs is to increase the number of people—especially people of color, women, and neighborhood residents—working in community development in Atlanta's neighborhoods.

The Fellowship Program is intended for recent graduates, master's degree candidates, and people with long-term experience in other fields of work or professional experience in a field related to community economic development. Up to three Fellows are selected each year and placed at participating organizations as full-time, year-long employees. Fellows receive stipends of $28,000 plus benefits.

The Internship Program is intended for undergraduate and graduate students and residents of communities with existing community development groups. Interns are placed at host agencies for a period of three to nine months. Interns receive a stipend of $10 per hour.

In addition to on-the-job experience, ANDP provides fellows and interns with training at its Community Development Institute, mentorship opportunities with experienced community development practitioners, peer support, and networking opportunities.

Lower Mississippi Delta Service Corps

1321 Highway 8 West, Suite 12
Cleveland, MS 38732
662-846-1113
email: lmdsc@techinfo.com

The Lower Mississippi Delta Service Corps is a tri-state partnership of Arkansas, Louisiana, and Mississippi designed to meet the critical needs of communities by providing volunteers to nonprofits involved in the areas of education, the environment, health care, human services, and public safety. Volunteers must be over eighteen and must perform at least 1,700 hours of service over the year (from September through August). Volunteers can draw up to $8,340 for living allowances and, at year's end, are eligible for educational awards of approximately $4,300.

Midwest

Metropolitan State University, Center for Community-Based Learning

c/o Local Initiatives Support Corporation
Hamline Park Plaza
570 Asbury Street, Suite 101
St. Paul, MN 55104
651-649-1109; fax: 651-649-1112

Metropolitan State University's Center for Community-Based Learning, with the support of the Twin Cities office of the Local Initiatives Support Corporation, has two programs for people of color interested in community development work: the Career-Ship Program and the Internship Program.

The primary goal of the Career-Ship Program is to attract peoples of color with some work and college experience to community development work. Career-Ship apprentices work one day a week for eighteen months at a community development corporation, attend seminars on community development, and fulfill college and technical course work as necessary for future work in the field. Apprentices receive a $12,000 stipend for the eighteen-month program and some assistance for college and technical course work.

The primary goal of the Internship Program is to attract people of color in college or graduate studies to community development work. Interns are placed in community development corporations. The program provides compensation.

Minnesota State University Student Association

Penney Fellowship
108 Como Avenue
St. Paul, MN 55103
651-224-1518; fax: 651-224-9753

Sponsored by the Minnesota State University Student Association (MSUSA), the Penney Fellowship is a need- and merit-based scholarship program that encourages students to serve Minnesota communities by financially supporting their involvement in community and public service internships. Students enrolled full- or part-time in any Minnesota state college or university institution (state universities and community and technical colleges) are eligible. Eligible students must also have a cumulative GPA of 2.5 or above and have secured a low-paying or voluntary internship in the public sector.

Neighborhood Progress, Inc.

1956 West 25th Street, Suite 200
Cleveland, OH 44113
216-830-2770; fax: 216-830-2767
email: Nprogressi@aol.com

Neighborhood Progress, Inc., a citywide intermediary for Cleveland's many CDCs, offers summer and one-year internships to graduate students from Cleveland State University, Case Western Reserve, and other universities and institutions. The interns are placed with host CDCs, where they work on neighborhood revitalization projects. Stipends are available.

West

California Community Economic Development Association

1611 Telegraph Ave., Suite 402
Oakland, CA 94612
510-251-8065; fax: 510-251-8068
and
1541 Wilshire Blvd.
Los Angeles, CA 90017
213-353-1676, fax: 213-207-2780
email: cceda213@aol.com
www.cceda.org

CCEDA, the trade association and network for California's community development organizations, offers several internships each year in its office. The nature of the assignment will vary for each intern.

Los Angeles Trade-Technical College

Community Development Technologies Center
2433 South Grand Avenue

Los Angeles, CA 90007
213-763-2520, fax: 213-763-2729
www.cdtech.org

The Center, in partnership with the Los Angeles office of the Local Initiatives Support Corporation (and with funding from the National Congress for Community Economic Development's Human Resources Development Initiative), provides a year's paid fellowship for four students of the certificate and associate degree programs in community planning and economic development. While studying, the fellows get work experience in community-based organizations, foundations, intermediaries, and government agencies, giving them a broad experience in all aspects of community-building activities.

The Center also provides fellowships under HUD's work-study program for Hispanic-serving institutions.

Non-Profit Housing Association of Northern California

369 Pine Street, Suite 350
San Francisco, CA 94104
415-989-8160; fax: 415-989-8166
email: dianne@nonprofithousing.org
www.nonprofithousing.org

An internship program for area students is under development.

Portland State University, Center for Academic Excellence

Community Service Internships
P.O. Box 751
Portland, OR 97207-0751
503-725-8316; fax: 503-725-5262

The purpose of the Community Service Internship Project at Portland State University (PSU) is to provide technical assistance to community-based organizations by placing students as volunteers.

More Information:

Directory of Special Programs for Minority Group Members, published by Garrett Park Press (P.O. Box 190-B, Garrett Park, MD 20896, 301-946-2553) contains leads to more than 2,800 sources of job training opportunities, scholarships, fellowships, and internships.

Internships: The Guide to On-the-Job Training Opportunities for Students and Adults is published annually by Peterson's (800-338-3282).

The National Directory of Internships, published by the National Society for Internships and Experiential Education (3509 Haworth Drive, Suite 207, Raleigh, NC 27609-7229, 919-878-3263) has thousands of listings of internships covering seventy-five fields, from the arts to the environment, women's issues, and international opportunities.

This appendix describes national training and professional development programs, faith-based training programs, specialized training programs, organizing and advocacy training programs, and training programs that are specific to an area of the country.

A good number of universities offer certificate programs that provide professional development and training, so refer to appendix B, too. Many national organizations hold conferences that are in essence training opportunities (see appendices D, F, and G.) Further, training is usually provided as part of an internship or fellowship program (see appendix J).

National Training

Community Development Training Institute

50 Washington Square
Newport, RI 02840
401-849-7053; fax: 401-849-5193
www.ncdaonline.org/cdti

The Institute is the training and technical assistance affiliate of the National Community Development Association (see appendix D). The Institute offers training in community economic development, management of nonprofit neighborhood groups, and housing. Training is offered to groups around the country through the state or local government agency. Cost depends on number of participants, resources required, etc.; a three- to seven-day training program costs from $6,000 up.

Community Revitalization Training Center

P.O. Box 82267
Columbus, OH 43202
1-800-282-2782 or 614-262-6662; fax: 614-262-3282
email: crtc@crtc.com
www.netwalk.com/~crtc

CRTC offers training and technical assistance with a focus on affordable housing, including rehabilitation, construction, and housing finance. It is funded by groups involved in affordable housing funding, including units of state government, the U.S. Department of Agriculture, and the U.S. Department of Housing and Urban Development. Training is aimed at a broad range of those involved in affordable housing, including individuals, community development groups, and professionals from the public or for-profit sector. It is a highly interactive adult education model of training. Training is offered all over the United States; costs run $75–$200 per day (usually between two and five days). A certificate of course completion is conferred.

Development Training Institute

2510 St. Paul Street
Baltimore, MD 21218
410-338-2512; fax: 410-338-2751
email: info@dtinational.org
www.dtinational.org

DTI is a national nonprofit training organization specializing in community development. Its national Leadership and Management Program (LAMP) trains management-level staff of community-based organizations to expand their skills in project development and comprehensive community planning and development. DTI's Project Development Program provides training and technical assistance on-site in local communities, under contract; the PDP training, tailored to fit the needs of each community, enhances the skills of local community groups to implement housing and economic development projects. DTI also provides training for financial institutions; see "Training in Community Development Lending," below.

Enterprise Foundation

10227 Wincopin Circle, Suite 500
Columbia, MD 21044
410-964-1230; fax: 410-964-1918
email: mail@enterprisefoundation.org
www.enterprisefoundation.org

The Enterprise Foundation, a large national community development intermediary, holds numerous training workshops covering technical development areas; topics range from introduction to nonprofit housing development, construction management, and hazardous materials treatment to comprehensive neighborhood-based community planning, fundraising, and business planning for nonprofit organizations. In addition, the organization's annual national conference includes a broad series of workshops on housing and community development topics (for example, linking jobs and child care to communities) for which training certificates are offered.

Harvard Institute of Affordable Housing

Harvard University Graduate School of Design
48 Quincy Street
Cambridge MA 02138
617-495-1680; fax: 617-496-0154
email: pd@gsd.harvard.edu
www.gsd.harvard.edu

Harvard's annual Institute of Affordable Housing covers the development process, financing, design, and management of affordable housing. Sessions in past years have included The State of the Nation's Housing, Reclaiming the Inner City, and Managing Affordable Housing. The Institute is held in mid-July; it runs for a week, Monday through Friday. Cost for tuition and materials is $1,225 plus a nonrefundable $30 application fee.

Heartland Center for Leadership Development

941 O Street, Suite 920
Lincoln, NE 68508
800-927-1115 or 402-474-7667; fax: 402-474-7672
email: mw4137@aol.com
www.4w.com/heartland

This research and training center, which focuses on rural community survival, offers a Community Economic Development Training Program. Two training sessions are offered annually, in the fall and in the spring. The fall program, titled Helping Small Towns Succeed, is a five-day intensive program designed for those who are relatively new to the community economic development profession. It is held in October at the Snow King Resort in Jackson, Wyoming; cost is $750, which includes materials and food. The spring program, titled Skill Building for Stronger Communities, is a 3 1/2-day program designed for more experienced CED professionals. It is also held at the Snow King Resort during May or June; cost is $475.

Housing Assistance Council

1025 Vermont Avenue NW, Suite 606
Washington, DC 20005
202-842-8600; fax: 202-347-3441
email: hac@ruralhome.org
www.ruralhome.org

HAC offers two types of training in rural housing development: a national conference approximately every other year, and regional training five to six times a year, except years with a national conference. Training is open to both groups and individual practitioners. Topics cover rural housing and some economic development issues. The National Rural Housing Conference is held every two or three years in December in Washington, D.C., and includes thirty to forty workshops on such topics as how to do a housing needs assessment, water and wastewater issues, government rental programs, how to do home buyer counseling and education, economic issues, and some special interest workshops in housing. Cost ranges between $235 and $340 depending on nonprofit or for-profit status and registration before or after the mid-October deadline. Limited scholarship funds are available.

Management & Community Development Institute

Lincoln Filene Center for Citizenship & Public Affairs
Tufts University
Medford, MA 02155
617-627-3453; fax: 617-627-3401
www.ase.tufts.edu/lfc

A training institute is held yearly during the early summer in the Boston area, offering about thirty-eight one- to four-day courses. Areas include: community economic development (e.g., Business Planning and New Enterprise Development; Reclaiming Brownfields); affordable housing production and management (e.g., Developing Resident-Controlled Housing); organizational capacity building (e.g., Introduction to Media Strategies); individual skills development (e.g., Public Speaking); and board leadership training. An exhibit space includes career resource materials. A Young Professionals Career Development Certificate is conferred for attending a course for young professionals on developing their careers plus three additional days of courses. Cost is $150/day; scholarships are available.

Ms. Foundation for Women

120 Wall Street, 33rd floor
New York, NY 10005
212-742-2300; fax: 212-742-1653
email: program@ms.foundation.org
www.ms.foundation.org

A multi-issue women's foundation with a special Economic Security program that provides grants, training, and technical assistance to grassroots and national organizations. The Foundation sponsors an annual National Institute on Women and Economic Development that provides training in practical skills, including, for example, organizational development, community organizing, and mentoring.

National Affordable Housing Training Institute

1200 19th Street NW, Suite 300
Washington, DC 20036
202-857-1113; fax: 202-223-4579
www.nahti.org

This nonprofit organization provides training to help communities use the resources of the HOME Investment Partnership program of the U.S. Department of Housing and Urban Development. The organization is composed of eight member groups representing cities, counties, and states.

National Council for Urban Economic Development

1730 K Street NW, Suite 915
Washington, DC 20006
202-223-4735; fax: 202-223-4745
email: mail@urbandevelopment.com
www.cued.org/cued/conference

CUED's courses are designed to provide economic development practitioners with the training and accreditation they need for their professional and personal growth. The CUED training program focuses on the building blocks of economic development—business retention and expansion, job creation, strategic planning, finance, management and marketing, and up-to-date tools and techniques to meet the challenges of the future. Course examples are Economic Development, Finance, and Business Retention and Expansion; a professional certificate is also offered: Economic Developer Certificate Program.

National Development Council

211 East 4th Street
Covington, KY 41011
606-291-0220; fax: 606-291-3774
email: ndccvg@aol.com
www.ndc-online.org

NDC, which focuses on the development of small towns and small businesses, provides a four-week economic development certification program and a three-week housing development finance program for people from the private sector and city government. Cost is $875 for one week.

Neighborhood Reinvestment Corporation

Training Institute
1325 G Street NW, Suite 800
Washington, DC 20005-3100
800-438-5547 or 202-220-2300; fax: 202-376-2168
email: nrti@nw.org
www.nw.org

The Institute offers more than sixty courses four to five times each year, in different locations across the country, providing intensive, professional training. Courses cover leadership and management (e.g., Developing a Winning Business Plan); affordable housing (e.g., Using the Low Income Housing Tax Credit program); community building (e.g., Addressing Youth Concerns); construction and production management; economic development (e.g., Small-Business Development); community planning; community development lending; and home ownership (e.g., Reaching Immigrant Populations). Tuition is $140 per day; some scholarships are available. Individuals including newcomers to the field will find these courses extremely informative.

New Hampshire College

Community Economic Development Program
2500 North. River Road
Manchester, NH 03104
603-644-3103; fax: 603-644-3130
http://merlin.nhc.edu

The CED Program, which offers master's and PhD (see appendix B), launched its training program in 1996. The faculty provides training in community economic development, offering a year-long track of short courses, each focused on a different technique. The three modules offered are: financing community economic development; nonprofit management; and nonprofit real estate. A certificate is conferred upon completion of these areas. This training is useful in several ways: as an introduction to the field, for professionals with an adequate educational background who want to learn more about a specific approach, or for people who cannot commit to a degree program. Individual two-day courses cost $300. Each module costs from $2,160 to $3,105.

Northwestern University

Institute for Policy Research
Asset-Based Community Development Institute
2040 Sheridan Road
Evanston, IL 60208-4100
847-491-8711; fax: 847-467-4140
email: earlee@nwu.edu
www.nwu.edu/IPR/abcd.html

The Asset-Based Community Development (ABCD) Institute, a project of Northwestern University's Institute for Policy Research, offers resources to redirect community revitalization from looking at "needs and deficiencies" to more effective methods of community building using neighborhood assets. The approach is based on research by professors John McKnight and John Kretzmann. The institute offers trainings, workshops, and consultations by the ABCD faculty and publishes technical workbooks on such topics as how to develop a neighborhood's economy, design and implement capacity inventories, do evaluations, and create neighborhood information exchanges. In addition, ABCD offers an electronic discussion group on capacity building issues and distributes John McKnight and John Kretzmann's resource book, *Building Communities from the Inside Out: A Path Toward Finding and Mobilizing a Community's Assets*. A training video, based on that book, is also available.

Rocky Mountain Institute

Economic Renewal Program
1739 Snowmass Creek Road
Snowmass, CO 81654
970-927-3807; fax: 970-927-4510
email: kinsley@rmi.org
www.rmi.org
This nonprofit research and educational organization provides training sessions on sustainable community economic development.

Urban Land Institute

1025 Thomas Jefferson Street, Suite 500 West
Washington, DC 20007-5201
202-624-7000; fax: 202-624-7140
email: webmaster@uli.org
www.uli.org
ULI provides training in commercial and for-profit housing development, including a two-day workshop offered twice a year (cost: $800), one large conference a year, and a four-day real estate program once a year (cost: $1,600).

Woodstock Institute

407 South Dearborn, Suite 550
Chicago, IL 60605
312-427-8070; fax: 312-427-4007
email: woodstock@woodstockinst.org
www.woodstockinst.org
Woodstock, a nonprofit research institute, also provides training for community groups and banks on the Community Reinvestment Act (the federal legislation requiring lending institutions to invest in their communities).

Faith-Based Training Programs

The Faith-Based Community Development program of the National Congress for Community Economic Development maintains and updates a listing of most faith-based training opportunities. See its program description below.

P.F. Bresee Foundation

Institute for Urban Training
Urban Partners
3401 West Third Street
Los Angeles, CA 90020
213-387-2822; fax: 213-385-8482
email: dhojsack@bresee.com
www.bresee.com
The Institute for Urban Training provides urban ministry training for undergraduate and graduate students who relocate to Los Angeles for a semester or a full year of courses and a twenty-five-hour-a-week internship. Tuition is paid to the student's home school. Urban Partners offers training for churches and nonprofits (both groups and individuals) that are working on a community development project. Training is offered twice a year for six Saturdays; cost is $250.

Center for Urban Resources

Inner City Impact Institute
990 Buttonwood Street, 6th floor
Philadelphia, PA 19123
215-236-4100
email: cur@libertynet.org
www.liberty.net/cur
The Inner City Impact Institute provides training as well as technical and practical assistance to churches on how to access resources (both human and financial) from beyond their congregations for community outreach programs. It is open to urban pastors and leaders who desire to implement or expand needed community service programs. Three-hour workshops are held weekly from October to January (fourteen weeks) at the CUR training facility.

Christian Community Development Association

3827 West Ogden Avenue.
Chicago, IL 60623
773-762-0994; fax: 773-762-5772
email: ccdaexdir@aol.com

www.ccda.org

CCDA holds a training conference yearly over a five-day period in late October or early November, offering a variety of workshop tracks. Areas can include: fundraising, youth, church ministries, economic development, welfare reform, racial reconciliation, education, leadership, health care, and housing, as well as how-to workshops such as how to start a health center, learning center, or substance abuse center. The location of the conference varies from year to year; it is always held in a U.S. city with an organization that is a member of CCDA. The cost of the conference is around $70; CCDA sometimes covers the cost for participants.

Community Training and Assistance Center

30 Winter Street, 7th floor
Boston, MA 02108
617-423-1444; fax: 617-423-4748
email: ctac@ctacusa.com
Web site under development

CTAC provides technical assistance to a range of organizations, including church-based groups that plan community development programs. The cost is sometimes covered by grant funds; otherwise, a contractual agreement is entered into with a sliding scale fee.

Congress of National Black Churches

Economic Development Program
1225 I Street NW, Suite 750
Washington, DC 20005
202-371-1091; fax: 202-371-0908
email: webmaster@cnbc.org
www.cnbc.org

The mission of the Economic Development Program is to increase economic opportunities for African-American churches and their members. Working with churches, the EDP has three main focuses for its training: to increase homeownership among African-Americans through intensive training programs and educational curricula (this program is made possible through a partnership with the U.S. Department of Housing and Urban Development); to develop employment and job skills training for churches in response to welfare reform legislation; and to increase church access to information, resources, and technical assistance related to economic development.

Direct Action Research Training Center

137 Northeast 19th Street
Miami, FL 33137
305-576-8020; fax: 305-576-0789
email: dartcenter@aol.com
www.fiu.edu/~dart

DART's mission is to support the building and long-term maintenance of large, institutionally based, direct action organizations that arise out of and are supported by local community congregations. DART offers two types of workshops to congregations on community organizing: Orientation workshops are offered twice a year (one in Florida and one in the Midwest), and once a year an advanced workshop is offered to groups in DART's network to provide ongoing localized training. Workshop topics include fundraising, leadership, and organizing. Cost for the orientation workshop is $400; the advanced workshop is $500/day for groups not in the network.

Gamaliel Foundation

203 North Wabash, Suite 808
Chicago, IL 60601
312-357-2639; fax: 312-357-6735
email: gamaliel@frontiernet.net
www.gamaliel.org

Gamaliel provides staff and leadership training to clergy and membership of church-based community organizations to organize and address social and economic injustice and to bring about community development solutions.

Harvard Divinity School

Summer Leadership Institute
Center for the Study of Values and Public Life
45 Francis Avenue
Cambridge, MA 02138
617-495-5766; fax: 617-495-9489
email: csvpl@div.harvard.edu
http://divweb.harvard.edu/csvpl

The Summer Leadership Institute is an intensive two-week session beginning in mid-June that is designed for clergy and lay leaders involved in local church-based community and economic development, as well as church-sponsored development initiatives. The Institute draws on the faculty and resources of Harvard University, the Massachusetts Institute of Technology, and other relevant institutions. Courses are divided into four modules: Theology, Ethics and Public Policy; Organizational Development and Management; Housing and Community Development; and Finance and Economic

Development. Applications are due May 1; the program fee is $3,300, which includes room and board; a limited number of need-based grants are available.

Howard University

International Faith Community Information & Training Center
1400 Shepard Street NE, Suite 295
Washington, DC 20017
202-806-0750; fax: 202-635-4904
email: kleathers@howard.edu or ischus@aol.com
www.iscfccn.net

The Lawrence N. Jones Training Institute offers an intensive one-day training program for churches on topics that vary according to the interests of participating churches—from the use of technology to church outreach ministry to substance abuse prevention. The Training Institute provides consultants and experts in the field. The program takes place at the church location; cost varies based on number of people, travel, and other factors.

Industrial Areas Foundation

220 West Kinzie Street, 5th floor
Chicago, IL 60610
312-245-9211; fax: 312-245-9744

IAF, the oldest and largest institution for community organizing in the United States, organizes nationwide through churches and also trains in general organizing, working with churches, community leaders, and democratic organizations. The focus is on helping mediating institutions, particularly religious congregations, to build broad-based community organizations to exercise the power to redress issues of importance to each community.

The Jewish Organizing Initiative

37 Temple Place, 5th floor
Boston, MA 02111
617-350-9994
email: Mbrown7387@aol.com
www.Jewishorganizing.org

This initiative provides training for Jews twenty-two to twenty-nine years of age, secular or observant, in social justice organizing skills and Jewish tradition. It then places them in a Boston-area grassroots social justice organization for a year-long paid internship.

McCormick Theological Seminary

African American Leadership Partnership Program
1313 East 60th Street
University of Chicago Campuses
Chapin Hall
Chicago, IL 60637
773-753-2470; fax: 773-753-2480
email: aalpinfo@aalpinc.org
www.aalpinc.org

The African American Leadership Partnership program offers a Community Leadership Development Program: an eight-module process for a church to develop community development leadership principles, practices, and skills to effect change in its community. The curriculum follows a hands-on approach, often involving individual projects, which the AALPP helps the church to develop and carry out. The motto is: "Transforming communities from the inside out." This training is offered according to the schedules of the individual churches, in Chicago and Indiana. The fee is nominal.

The program also offers an MA in theological studies with an emphasis on faith-based community development topics. (See appendix B.)

National Catholic Council for Hispanic Ministry

2929 North Central Avenue, Suite 1500
Phoenix, AZ 85012
602-266-8623; fax: 602-266-8670
email: ncchm1@aol.com
www.ncchm.org

The Council has developed and piloted a Hispanic Leadership Development Initiative, which provides leadership training for community development, including both practical skills and theological aspects of leadership. It is a fifteen- to eighteen-day training program offered in both English and Spanish. The NCCHM plans to have the materials from this pilot program published for the use of other organizations wishing to conduct similar training.

National Congress for Community Economic Development

1030 15th Street NW, Suite 325
Washington, DC 20005
877-44NCCED or 202-289-9020; fax: 202-289-7051
email: winstead@ncced.org
www.ncced.org

NCCED operates a special initiative, the Faith-Based Community Development Program, to encourage and assist faith-based institutions to work with their communities to engage in community development. As part of that program, it provides training, both at conferences across the country and in specific localities on request.

New Hampshire College Community Economic Development Program

2500 North River Road
Manchester, NJ 03106-9967
603-644-3103; fax: 603-644-3130
http://merlin.nhc.edu

The Community Economic Development Program of New Hampshire College (see appendix B) provides a specialized education in community economic development for religious and lay leaders who are working with religious institutions. In addition to the program's standard curriculum, required courses for the faith-based specialization are Faith-Based Economic Development and Strategic Management of Faith-Based Community Economic Development. Students in this specialized program carry out an actual activity in their community that will strengthen their institution's community economic development capacity.

ORGANIZE Training Center

442-A Vicksburg
San Francisco, CA 94114
415-821-6180; fax: 415-821-1631

OTC conducts up to three four-day sessions per year in various locations across the country, mostly in California, Colorado, and Nebraska, providing training for congregation-based groups in community organizing and the values underlying organizing.

Pacific Institute for Community Organization

171 Santa Rosa Avenue
Oakland, CA 94610
510-655-2801; fax: 510-655-4816
email: baumannpico@aol.com
www.pico.rutgers.edu

PICO serves a national network of congregation-based community organizations, providing two six-day national leadership training sessions a year on leadership development and organizing training.

Pittsburgh Theological Seminary

Metro-Urban Institute
616 North Highland Avenue
Pittsburgh, PA 15206
412-362-5610, x2163; fax: 412-363-3260

The Seminary offers intensive weekend-long seminars three times a year around three broad topics: family, education, and economics. Tuition for weekend seminars is $30. Seminars are open to those from out of state as well as those from Pittsburgh.

Unitarian Universalist Association

Faith In Action Department
25 Beacon Street
Boston, MA 02108
617-742-2100, x450; fax: 617-742-7025
email: wgardiner@uua.org
www.uua.org

The Association offers a Social Justice Empowerment Program that educates congregations on developing and putting into action programs on issues of social justice. It is offered on a Friday night and all day Saturday. The UUA conducts about fifteen such programs each year, once or twice for each congregation. The program is free; expenses, when applicable, are no more than $200.

United Methodist Church

Board of Global Ministries
Community Ministries Department
475 Riverside Drive
New York, NY 10115
212-870-3600; fax: 212-870-3948
email: websysop@gbgm-umc.org
www.gbgm-umc.org

The Department is a resource for training, funding, and technical assistance for United Methodist churches around the United States that wish to undertake community development projects.

Urban Leadership Institute

620 West Olympic Boulevard, Suite 220
Los Angeles, CA 90015
213-746-2211; fax: 213-746-5308
email: mmata@usc.edu
www.cst.edu/

The Urban Leadership Institute, associated with the Claremont School of

Technology, offers a number of workshops, training events, conferences, consultations, and action-research projects to religious leaders and institutions. The training is tailored to the organizations or coalition of organizations and ranges from an asset-based community analysis to strategic planning to program development and implementation. The projects can range from services to jobs creation and housing development. The ULI orients the organizations to community development approaches as part of a requested consultation. Most of the time training is referred to Urban Partners (offered by the Bresee Institute; see above) and is offered when a critical mass of organizations (ten or more) request the training. Cost is about $250 per individual for a six-session program, usually held on Saturdays. The sites of the training vary but usually are in inner-city churches or at the offices of the respective programs.

Specialized Training

Training in Housing Management

Consortium for Housing and Asset Management
c/o Enterprise Foundation
10227 Wincopin Circle, Suite 500
Columbia, MD 21044
410-964-1230; fax: 410-964-1918
email: dfromm@cham.org
www.enterprisefoundation.org
CHAM offers training and technical assistance to aid community-based nonprofits in managing affordable housing. It offers a certificate program as well as workshops in five regions across the United States. CHAM is sponsored by the Enterprise Foundation, the Local Initiatives Support Corporation, and the Neighborhood Reinvestment Corporation.

The National Center for Housing Management
1010 Massachusetts Avenue NW
Washington, DC 20001
202-872-1717; fax: 202-789-1179
www.nchm.org
The Center offers classes on all aspects of housing management, including a core set of certifications covering dealing with tenants, finances, regulations, and maintenance, culminating in a Registered Housing Manager (RHM) program. Classes are offered at many locations across the country. Most cost between $500 and $600.

Training in Community Development Lending

Chicago Community Loan Fund
111 West Washington Street, Suite 1221
Chicago, IL 60602
312-345-1770; fax: 312-345-1737
email: cclf@aol.com
Chicago Community Loan Fund, whose work focuses primarily on financing community-based housing and economic development, runs a three-pronged Gateway to Community Development Program that focuses on financing as well as technical assistance program. The program includes Project Readiness Workshops for community development organizations four times a year.

Development Training Institute
2510 St. Paul Street
Baltimore, MD 21218
410-338-2512; fax: 410-338-2751
email: info@dtinational.org
www.dtinational.org
In partnership with local universities and financial institutions, DTI offers training for bankers to help them master the art of underwriting loans that benefit their communities and generate a healthy profit.

Fannie Mae Foundation Mortgage Finance Program
4000 Wisconsin Avenue NW, North Tower, Suite One
Washington, DC 20016-2804
202-274-8096; fax: 202-274-8111
www.fanniemaefoundation.org
The Foundation has developed a Mortgage Finance Program, currently available in three community colleges: Cuyahoga Community College in Cleveland; Miami-Dade Community College in Miami; and LA Trade-Technical College in Los Angeles. The mission is to prepare minorities for a career in mortgage finance. The program is a one-year certificate degree program; course credits can be applied toward a degree. In each city, a local industry collaborative will link the community college program to the local mortgage lending institutions for job placement opportunities. Financial incentive packages are available for qualified students.

National Community Capital Association
924 Cherry Street, 2nd floor
Philadelphia, PA 10107
215-923-4754; fax: 215-923-4755

email: nacdlf@aol.com
www.communitycapital.org

This national nonprofit-membership organization provides training to community development financial institutions. Individuals may attend training sessions. Sessions are offered once a year at an annual training conference held in October in a different city each year. There are also targeted training sessions during the year, which are one-day events held in different cities on specific topics, such as creating a CDFI business plan.

Training for Youth Construction Programs

YouthBuild USA

58 Day Street
Somerville, MA 02144
617-623-9900; fax: 617-623-4331
www.youthbuild.org

This national nonprofit organization trains CDCs and other community-based groups to set up YouthBuild programs in their communities. These programs integrate construction job training and education with other social services for at-risk youth, preparing them for careers in construction.

Training in Organizing and Advocacy

Association of Community Organizations for Reform Now

739 8th Street SE
Washington, DC 20003
202-547-2500; fax: 202-546-2483
email: fielddirect@acorn.org
www.acorn.org

ACORN, the nation's largest community organization, hires and trains new organizers to work in low- and moderate-income neighborhoods across the country. ACORN seeks to build grassroots, participatory organizations, developing leadership from within the communities.

Center for Third World Organizing

1218 East 21st Street
Oakland, CA 94606
510-533-7583; fax: 510-533-0923
email: ctwo@igc.org
www.ctwo.org

CTWO promotes the development of grassroots organizing in communi-

ties of color by providing training in community organizing, technical assistance, campaign planning, issue analysis, and networking. It also sponsors a Minority Activist Apprenticeship Program, which involves a seven-week field internship with community organizations around the country and includes a stipend.

Community Training and Assistance Center

30 Winter Street, 7th floor
Boston, MA 02108
617-423-1444; fax: 617-423-4748
email: ctac@ctacusa.com

CTAC's national training and technical assistance delivers customized training programs to organizations, enabling groups such as public housing tenant councils, neighborhood task forces, and civic associations to strengthen their citizen participation in education, housing, economic development, and human services.

Consensus Organizing Institute

130 Seventh Street, 8th floor
Pittsburgh, PA 15222
412-201-2036; fax: 412-201-2039
email: coi1@aol.com
www.censusorganizing.org

Active in over two dozen cities, the Institute works with selected community organizations and trains staff to become consensus organizers. This method of organizing builds consensual relationships among relevant public and private institutions to work toward community improvement, in contrast with confrontational organizing, which uses an "us vs. them" approach.

Gamaliel Foundation

See "Faith-Based Training."

Massachusetts Association of CDCs

99 Chauncy Street, 5th floor
Boston, MA 02111
617-426-0303; fax: 617-426-0344
email: macdc@gbls.org

MACDC competitively awards grants, training, and technical assistance to CDCs throughout the state to increase their emphasis on community organizing as a central part of community development. See also appendix J.

Midwest Academy

225 West Ohio Street, #250
Chicago, IL 60610
312-645-6010; fax: 312-645-6018
email: mwacademy1@aol.com
www.mindspring.com/~midwestacademy

The Midwest Academy provides training sessions on how to organize to build coalitions at the city, state, and national levels; work with other groups and institutions; and develop sound administrative and financial systems. Its training also probes issues relating to community economic development. It offers specific training programs for minorities and students. It has also published the manual, *Organizing for Social Change*.

National Training and Information Center

810 North Milwaukee Avenue
Chicago, IL 60622
312-243-3035; fax: 312-243-7044
email: ntic@ntic-us.org

NTIC provides training on neighborhood organizing and on confrontation, negotiation, and public relations strategies. Its focus is on redlining by financial institutions, insurance companies, and government agencies. It is affiliated with National People's Action (see appendix D, Part I).

Pacific Institute for Community Organization

See "Faith-Based Training."

Regional and Statewide Training

In addition to the training opportunities listed below at the regional and state level, there are many citywide courses in larger cities. In Washington, D.C., for example, a coalition of nonprofit housing groups sponsors periodic training institutes. To find out about citywide courses, contact some of the national organizations listed in appendix B and ask for their local field office or member group.

Northeast

New York CED Network

c/o Senses
275 State Street
Albany, NY 12210
518-463-5576; fax: 518-432-9073

email: SENSESMJ@aol.com

The Network, the trade association and network for community development groups across the state, holds periodic training workshops, focusing on loan funds, micro-enterprise, and public policy issues, especially those relating to welfare reform.

Philadelphia Neighborhood Development Collaborative

1234 Market Street, Suite 1800
Philadelphia, PA 19107
215-563-6818; fax: 215-575-0436
email: jtaylor@pndc.net
www.pndc.net

The Community Development Management Certificate Course is a year-long, comprehensive management training program designed to help build the capacity of CDC senior and middle managers to deal with increasingly complex organizational and management issues as their organizations grow. The program, sponsored by the citywide Philadelphia Neighborhood Development Collaborative, is delivered by the Development Training Institute and housed at Eastern College's urban campus. The training, open to selected Philadelphia-area CDC senior and middle managers, covers affordable housing and economic development, finance, asset management, organizational management, community building, and professional leadership development. The course spans six 3 1/2-day workshops. It feeds into Eastern College's graduate Economic Development Program; upon completion, participants receive a certificate and may apply course credits toward the College's master's degree. Scholarships may be available.

Rhode Island Association of Nonprofit Housing Developers

8 Aurora Street
Providence, RI 02908
401-521-1461; fax: 401-521-1478

The Association, the trade association and network for community development groups across the state, offers periodic training in partnership with the Local Initiatives Support Corporation, the housing finance agency, and others, covering development techniques and topics.

University of Delaware

Center for Community Development and Family Policy
Graduate School of Urban Affairs and Public Policy
297 Graham Hall
Newark, DE 19716

302-831-6780; fax: 302-831-4225
email: CCDFP@diamond.net.udel.edu
www.udel.edu/chep/

The Center offers training in three areas of professional development. Community-based development training is offered through one course, taught by University faculty, in the fall semester; a certificate is awarded upon successful completion. (Students matriculating in the college's graduate program may take this training for credit.) In the spring, four two-day Community Development Institute workshops, one each month, provide additional training. The Center also offers training in nonprofit management and in housing development. Training is open to novices and people with intermediate skills; there are also advanced courses intended for those with a higher level of experience. (See also appendix B.)

South

Atlanta Neighborhood Development Partnership

Community Development Institute
34 Peachtree Street, Suite 1700
Atlanta, GA 30303
404-522-2637, x26; fax: 404-523-4357
www.andpi.org

The Institute, a joint project of the Atlanta Neighborhood Development Partnership and Clark Atlanta University, pays special attention to leadership development and community development skills particular to the South. It offers quarterly training workshops of three days' duration, in which interactive training techniques are stressed. Examples of topics covered include: nuts and bolts of community economic development; financing economic development; low-income housing tax credits; developing a business plan; grant and proposal writing; community organizing; and social services that support community development. The Institute is open to everyone. Cost is $325.

Highlander Research and Education Center

1959 Highlander Way
New Market, TN 37820
423-933-3443; fax: 423-933-3424
email: hrec@igc.org
www.highlandercenter.org

The Center works mainly in southern states and Appalachian regions, providing leadership training and technical assistance. One of its priorities is the restoration of the ecology of those areas.

Institute for Community and Economic Development

North Carolina Community Development Initiative
316 West Edenton
Raleigh, NC 27603
919-834-8480; fax: 919-834-8018
email: ncinitiative@ncinitiative.org
www.ncinitiative.org

The Institute, a collaboration of Saint Augustine's College, the North Carolina Community Development Initiative, and the North Carolina State Association of CDCs, offers two training opportunities. The Community Economic Development Studies program is an eleven-week survey course, which awards a certificate and covers topics ranging from the history of community economic development to structuring a commercial real estate deal. The Life Long Learning System enables people to custom-tailor training to their needs; a certificate may be awarded. The Initiative also offers an undergraduate degree program (see appendix B).

Maryland Center for Community Development

1118 Light Street
Baltimore, MD 21230
410-752-6223; fax: 410-752-1158
email: mccd@mccd.org
www.mccd.org

The Center, the trade association and network for community development organizations across the state, provides ongoing professional development training, including certification courses and regular workshops and forums.

Tennessee Network for CED

P. O. Box 23353
Nashville, TN 37202-3353
615-395-4341; fax: 615-256-9836

The Network, the trade association and network for community development organizations across the state, provides ongoing training to board members, volunteers, and staff, on a range of topics relevant to community economic development.

Midwest

Chicago Rehabilitation Network

53 West Jackson
Chicago, IL 60604
312-663-3936; fax: 312-663-3562

email: crn@uic.edu

The Network offers a community development and empowerment training series touching on topics such as planning, organization, and project analysis. It is a certificate program.

Community Development Academy

University of Missouri Extension Teaching
103 Whitten Hall
Columbia, MO 65211
800-545-2604; fax: 573-884-5371
www.mucourses.missouri.edu

The Community Development Academy is a series of three courses, each five days long, for professionals and volunteers in community work. The courses are held in a retreat setting, which allows the student to build support networks of faculty and participants. They are sequential, each course building on the content of the prior courses. Undergraduate or graduate credits are available upon successful completion. Courses range from Community-Based Development Concepts and Impact Assessment to Change, Ethics, and Sustainability.

Michigan State University

Center for Urban Affairs
1801 West Main Street
Lansing, MI 48915-1097
517-353-9555; fax: 517-484-0068
email: cua@msu.edu
http://uap.msu.edu

The Center has established a Community and Economic Development Program to offer training and technical assistance to local communities throughout the state.

Ohio CDC Association

35 East Gay Street, Suite 400
Columbus, OH 43215
614-461-6392; fax: 614-461-1011
email: ohiocdc@ohiocdc.org

The Association, the trade association and network for community development organizations across the state, each year offers entry-level training in affordable housing development for anyone new to the field. The training spans twenty days; it is possible to attend one week at a time or all at once. A certificate is awarded to trainees who pass a test at the end of the training.

Training at the entry level is also offered in how to design and administer a micro-enterprise program; this course is not offered every year. The Asso-

ciation also provides ten-day courses in community economic development strategy. Training costs vary, from $300 to $1,000; scholarships are available.

Southwest

Texas Development Institute

824 West 105th Street, Suite 107
Austin, TX 78701-2039
512-478-6067; fax: 512-478-1601

This nonprofit organization's mission is to promote and assist community economic development in low-income communities across Texas. As part of its work, it provides basic training workshops, usually without cost, for community development groups and their partners, such as financial institutions and public agencies, as well as individuals. It helps groups organize, get started, and increase their skills in economic and housing development work. It also provides research and policy analysis.

Texas Southern University and Third Ward Community Development Corporation

Community Development Leadership Program
See appendix B.

West

Northwest Community Development Academy

c/o Portland State University
School of Extended Studies
P.O. Box 1491
Portland, OR 97207-1491
503-725-4864 or 800-547-8887, x4864 or 4849
email: pat@ses.pdx.edu
www.pdx.edu

The School of Extended Studies at Portland State University, in partnership with the College of Urban and Public Affairs and the Communitas Group (a Washington State–based nonprofit agency with an interest in professional and community development), sponsors the Northwest Community Development Academy, which provides training that serves the needs of professionals in both Oregon and Washington.

Several workshop sessions are held each year, focusing on community organizing, organizational management, and those specific skills necessary to strengthen the business of CDCs in housing, economic development projects, and community involvement in solving complex and interrelated problems. The NWCD Academy training includes two key weekend sessions and six optional one-day events. In addition, some of the one-day events are available via distance delivery over educational television networks. The Association of Oregon Community Development Organizations, the statewide CDC association, consults with the Academy on its training curriculum.

Oregon Community Development Training Consortium
The Neighborhood Partnership Fund
1020 Southwest Taylor Street, Suite 585
Portland, OR 97205
503-226-3001; fax: 503-226-3027
email: npf@teleport.com
Web site under development

The Neighborhood Partnership Fund coordinates a training consortium, which includes other funders as well as trainers in the region. Trainings offered by the consortium include one- and two-day workshops, informal monthly discussion groups, and the extensive training program offered through Portland State University's Northwest Community Development Academy.

Rural Community Assistance Corporation (RCAC)
2125 19th Street, Suite 203
Sacramento, CA 95818
916-447-2854

Oregon Office:
921 Southwest Morrison Street, Suite 529
Salem, OR 97205
503-279-1477; fax: 503-279-1472
email: cmarko@rcac.org, lshelby@rcac.org
www.rcac.org

RCAC, a nonprofit organization dedicated to improving the quality of life for rural communities and disadvantaged people, provides training and technical assistance in the areas of affordable housing, organizational capacity building, community economic development, and environmental services. It works with rural agencies and organizations and native tribes in Hawaii, Alaska, and the twelve states that make up the Southwest and West.

San Diego State University, College of Extended Studies
Certificate Program in Community Economic Development
5250 Campanile Drive
San Diego, CA 92182-1920
619-594-5821; fax: 619-594-8566
email: extended.std@sdsu.edu
www.ces.sdsu.edu

San Diego State University's College of Extended Studies offers a Certificate in Community Economic Development. This program seeks to teach students the skills and knowledge necessary to develop efficient, productive, and profitable community-based ventures and enterprises that unite the talents of residents with public needs. The certificate program consists of seven courses taken September through May, two Saturdays each month. The seven courses for the recent academic program were: Introduction to Community Economic Development; Marketing Strategies; Accounting; Organizational Management; Financing Community Economic Development; Legal Structures for Community Economic Development; and Small Business Development. Students also completed hands-on community projects.

Twelve additional seminars on special topics were also offered. The total cost for the program is $1,150. The certificate program is also offered in an intensive one-week institute format every year. All courses are the same, but institute students do not complete community projects. The institute fee is $895, which includes all courses and printed materials.

Appendix L: Recommended Readings

The following books will give you a fine grounding in the ideas, history, and practice of community development. They could be considered a basic library. Once you are ready to read and learn more, there are hundreds of other books from which to choose, from the theoretical to the practical and technical, depending on which path you take in the field.

Basics of Organizing, Shel Trapp (National Training and Information Center, Chicago, 1986).

Written by a master of community organizing, this book describes organizing's basic themes: interest-generating tactics, leadership development, consciousness raising, collective action, and information development.

Beyond the Market and the State: New Directions in Community Development, Severyn T. Bruyn and James Meehan, editors (Temple University Press, Philadelphia, 1987).

A political economics text whose theme is that both the capitalist market system and state socialism are seriously flawed, but another alternative is available: free enterprise, people-oriented, community-based or -focused strategies. Two introductory chapters provide an analysis of the flaws of traditional and statist economics. The following chapters, each written by an expert practitioner, illustrate an alternative strategy, from community land trusts to community development corporations to worker cooperatives and more.

Coming of Age: Trends and Achievements of Community-Based Development Organizations, Carol Steinbach (National Congress for Community Economic Development, Washington, DC, 1999).

This short booklet summarizes the findings of the fourth national census of community-based nonprofit development organizations, sponsored by the National Congress for Community Economic Development. Highly readable, it reports on the rapid growth of the community development industry, its aggregate accomplishments, and its current activities. It also illustrates the whole field with profiles of eleven very different local CDCs.

Communities on the Way: Rebuilding Local Economies in the United States and Canada, Stewart Perry (State University of New York Press, Albany, NY, 1987).

The most in-depth history of rural and urban community development corporations; this book provides an analytic framework of different approaches used to address persistent poverty and shows how CDCs fit into the community development landscape. Using case studies, the book also identifies common forces, financing sources, and problems, such as the CDC tendency to forsake holistic approaches.

Community-Building: Coming of Age, G. Thomas Kingsley, Joseph B. McNeely, and James O. Gibson (Urban Institute Press, Washington, DC, 1997).

This monograph is a good discussion of the "community-building" approach to community revitalization—which is holistic (rather than focused on one strategy such as developing housing), emphasizes building social capital, is driven by neighborhood residents, emphasizes and draws on the community's assets (rather than its needs), and seeks to connect the isolated neighborhood to the broader society.

Corrective Capitalism, Neal Peirce and Carol F. Steinbach (Ford Foundation, New York, 1987).

An outstanding history, using case studies, of the growth of community development corporations from "diamonds in the rough" in the 1960s to competent achievers in the economic marketplace of the 1980s. The book shows how CDCs are vehicles of corrective capitalism, capable of redressing market failure in low-income neighborhoods.

The Death and Life of Great American Cities, Jane Jacobs (Random House, New York, 1961).

This great book explains the ingredients for a city neighborhood full of social and economic vitality (in contrast to the "marvels of dullness" of the suburbs and the "hopelessness" of low-income planned projects). The author also explains the differences between a small town and a city neighborhood, cautioning that what works in one will not work in the other.

Meaning and Action: Community Planning and Conceptions of Change, Peter Marris (Routledge & Kegan Paul, London, 1987, 2nd edition).

The theme of this seminal book is that community economic development plans and initiatives do not achieve their goals because they "reflect the disintegration of a dominant paradigm of social policy ('liberal capitalism'), which had defined the relationship between meaning and action for a generation." The author discusses the two new paradigms facing society today: on the one hand, the internationally integrated economy of corporate giants; on the other, the decentralized, sustainable development of mutual accommodation, which "recognizes the attachments which bind people to each other and to places."

Re-Building the Inner City: A History of Neighborhood-Based Initiatives to Address Poverty in the United States, Robert Halpern (Columbia University Press, New York, 1995).

This book examines neighborhood initiatives from settlement houses and community action agencies up through today's CDCs and Empowerment Zones. This history and analysis show how each generation of initiatives has been constrained (if not undermined) by the contexts of scarcity and exclusion that mark impoverished neighborhoods, and it calls for better national policies rather than placing the responsibility for poverty and discrimination solely on the suffering areas.

Reveille for Radicals, Saul Alinsky (Vintage Books, New York, 1949).
Rules for Radicals, Saul Alinsky (Vintage Books, New York, 1972).

The organizer's bibles. These are must-reads for inspiration and education. The first book is a recounting of Alinksy's experiences in Chicago. The latter volume, written at the end of his life, is designed as a guide book with a rhetorical flair.

The Truly Disadvantaged: The Inner City, the Underclass, and Public Policy, William Julius Wilson (University of Chicago Press, Chicago, 1987).

Looking at the "underclass" of impoverished minority individuals isolated in inner cities, this book discusses the social and economic forces that brought about this situation and the outcome of the destruction of the community's job networks and social institutions.

Appendix M: Glossary

Words and phrases used in this guide that may be unfamiliar to some readers are defined in the following paragraphs. For a complete glossary, consult the Neighborhood Reinvestment Corporation's Web site, www.nw.org, and click on the word "glossary." We have used many of the Web site's definitions here.

Affordable housing is generally defined as housing for which the occupant pays no more than 30 percent of gross income for gross housing costs, including utilities; the term is used interchangeably with *low-income housing*.

Blockbusting was a racist practice in which real estate agents panicked homeowners on a block into selling their homes by alleging that the neighborhood was undergoing drastic racial change.

Community development, sometimes called *community economic development*, refers to the range of activities carried on by nonprofit organizations and their partners whose purpose is to improve the social, physical, and economic conditions of low-income urban and rural communities.

Community development corporation (CDC) is a private, nonprofit, incorporated organization serving a low-income community or constituency, governed by a community-based board, and producing affordable housing and commercial and/or industrial business-enterprise development and related services to improve the community.

Community development finance institution (CDFI) is a nongovernmental financing entity with the primary mission of promoting community development by investing in and providing financial services to low-income communities or other underserved markets.

Community land trust is a means of restricting use of land and housing through not-for-profit ownership of land with leases to the land users. It is often used to protect low-income housing from speculation.

Community Reinvestment Act (CRA), legislated by the U.S. Congress in 1977, requires federally regulated banks and savings and loans to offer credit and deposit services to all who reside or do business within their geographic service area.

Development bank is a commercial bank organized to use the tools of a private financial institution to stimulate market forces and create a systematic cycle of economic development in its neighborhood.

Disinvestment is the withdrawal of financial resources, by the public and/or private sectors, from a neighborhood.

Equity is the permanent capital invested in a project. A buyer's initial ownership interest (downpayment) in a house or other property is one form of equity.

Gentrification is the migration of middle-class residents into a deteriorating area. This migration may help to revitalize an area, but it also tends to "squeeze out" lower-income families by boosting property values.

Intermediary is an organization that mediates between grassroots groups and larger-scale sources of capital. Intermediaries function at the city, regional, and national levels, aggregating capital (from sources such as foundations, corporations, and government agencies), then disbursing that capital to grassroots groups along with technical assistance.

Low-income refers to a definition based on family income as a percentage of an area's median income. Different programs may set different percentages. Section 8 of the U.S. Housing Act of 1937 defines a low-income household as one whose annual income adjusted for family size is at or below 80 percent of the median income in a particular metropolitan area, as determined by HUD.

Low-Income Housing Tax Credit is a credit against taxes owed by investors in the development and preservation of multifamily rental housing affordable to low- and very-low-income households. The tax credit provides an incentive for those investments.

Micro-enterprise is a business that is "smaller than small" and operated by a person on a full- or part-time basis, usually out of a home, e.g., carpenters, day care providers, craftspeople, caterers.

Mortgage banker is an entity who originates, sells, and services mortgages.

Portfolio—in the context of this book—is a collection of loans held for servicing or investment.

Redlining is the arbitrary denial of real estate loan or insurance applications in certain geographical areas without considering individual applicant qualifications. In former years, before it was outlawed, some banks literally drew red lines around neighborhoods that they considered too risky for investment.

Secondary market refers to a financial strategy that helps replenish lenders' supply of money so that they can continue to make loans. The secondary market accomplishes this through a set of institutions, such as Fannie Mae and Freddie Mac, that buy loans (including mortgages) from the original lender, thus replenishing the lender's supply of money; the secondary-market institutions then pool the loans and sell pieces of the loan pools or bonds backed by the loan pools to private investors.

SRO (Single-Room Occupancy) housing is one of the country's oldest forms of affordable housing for single and elderly low-income people. Typically, an SRO room will have a sink and a closet. Bathroom, shower, kitchen, and other rooms are usually shared. Today many SRO units are equipped like an efficiency apartment, with their own bath and kitchenette.

Sweat equity is the monetary value given to the labor homeowners contribute to the building of their home instead of cash. Lenders generally want the owner to have some of his or her own cash invested but will occasionally recognize the value of the work done.

Syndication is a method used to sell equity interests (shares) in real estate projects to investors other than the original developers. The concept extends generally to any group of investors who have contributed funds for the common purpose of carrying out a real estate project requiring concentration of capital. It can take several business forms, but the most common is the limited partnership.

Technical assistance is the advice and consultation provided to organizations to help them achieve a particular objective, carry out a particular program, or operate more effectively.

Transitional housing provides shelter for homeless individuals and families for six months to two years in an environment of security and support designed to help residents progress toward self-sufficiency. It is a middle point between emergency shelter and permanent housing.

Underwrite is the process used to determine in what amounts and on what terms an insurance company or bank will accept the risks, such as defaults, when investing its money.

Index